Dedicated
TO
My Parents and Wife

AN INTRODUCTION TO PHYSICAL GEOGRAPHY

Nizamuddin Khan
Department of Geography
Aligarh Muslim University,
Aligarh (India)

CONCEPT PUBLISHING COMPANY, NEW DELHI-110059

ISBN 81-7022-899-9(PB)

First Published 2001

Published and Printed by
Ashok Kumar Mittal
Concept Publishing Company
A/15-16. Commercial Block, Mohan Garden,
NEW DELHI-110059

Phone : 25351460, 25351794
Fax : 091-11-253557103
Email : publishing@conceptpub.com

CONTENTS

Page

Preface vii

List of Figures xiv

1. **Introduction** 1

Definition of Geography, Classification of Geography, Physical Geography, Human Geography, Approach of Study in Geography, Systematic Geography, Regional Geography, Geomorphology, Historical Evolution of Geomorphic Ideas in Geography, Prospects of Geomorphology in Future, Geomorphology and Man, Geomorphological Features and Environment, Geomorphology and Transportation, Geomorphology and Agriculture, Geomorphology and Settlement, Geomorphology and Planning.

2. **Origin of the Earth** 17

Solar System, Theories or Hypothesis Regarding the Origin of the Earth; Nabular Hypothesis, The Planetesimal Hypothesis, Tidal Hypothesis of Jeans and Jaffreys, Otto Schemidst's Inter-Steller Dust Cloud Hypothesis, Comparative Study of Various Hypothesis of Origin of the Earth between Nabular and Planetesimal Hypothesis, Tidal and Plantation Hypothesis and Tidal and Dust Cloud Hypothesis.

3. **Interior of the Earth** 29

Sources of the Information about the Interior of the Earth, Indirect Sources, Structure and Composition of the Earth.

4. **Origin of Continents and Ocean Basins** 35

Features of the Earth Surface, Tetrahedral Theory, Wegner's Continental Drift Theory, Plate Tectonic Theory.

5. **Materials of the Earth Crust** 49

Classification of the Rocks — Igneous Rock, Sedimentary Rock, Metamorphic Rocks, Rock Cycle.

6. Earth Movement 61
Causes of Earth Movements, Type of the Earth-Movement, Epirogenic Earth Movement, Orogenic Earth Movement, Sudden Earth Movements, Classification of Volcanoes, Plate Tectonic Movement, Earthquakes

7. Mountain Building or Orogeny 80
Geographical Classification, Classification of Mountains on the Basis of Process of Origin, Theories of Orogen of Fold Mountains — Geosynclin Orogen Theory of Kober, Thermal Contraction Theory of Mountain Building by Jeffreys, Continental Drift Theory of Mountain Building by Alfred Wegner, Mountain Buildings by Thermal Convection Currents by Arthur Holmes, Radio-active Theory of Mountain Building of Joly, Continent Sliding Hypothesis of Mountain Building of Daly, Plate Tectonic Theory.

8. Process of Gradation 97
Works of Denudation, Weathering, Process of Denudation, Running Water (River), Process of Erosion by Rivers, Transportation by Rivers, Erosional and Depositional Work Done by Rivers, Water Fall, River Meanders and Oxbow Lakes, Floor Plain and Levees, Delta, Cycle of Erosion, Landforms Formed by Rejuvenation of the River, Development of River Valley, Drainage Basin, Type and Pattern, River Capture.

9. Glaciers 120
Definition and Meaning, Kinds of the Glaciers, Denudational Works of Glaciers, Work of Erosion, Transportation, Deposition, Fluvio-glacial Deposition, Landforms Formed by Glacier Erosion.

10. Winds 130
Process of Erosion, Erosional Works Done by Winds, Depositional Works Done by Winds.

11. Underground Water 137
Conditions or Elements of Underground Water, Hydrological Cycle, Sub-surface Water Zones, Well, Springs, Groundwater

PREFACE

This book is an introduction to Physical Geography. I have been engaged in teaching physical geography to the students preparing for +2 and B.A./B.Sc. degree in geography. During the period of my teaching and interaction with students, I felt the need to write a book explaining the fundamentals of physical geography, so that the students may understand the subject easily. Many students take admission to B.A. (Geography) without having any fundamental geographical knowledge. This may also be useful for the students who offer Geography as a subject in the competitive examination.

I take the opportunity to thank my colleagues and research scholars who have contributed in various ways to the making of this book. I am highly grateful to Prof. K. Z. Amani who always inspired me to read and write. He always tought me about academic values and helped me at different level of my academic development. Thanks are also due to Prof. Ali Mohammad, Chairman, Department of Geography, A.M.U. Aligarh, who has been constant source of inspiration and has helped me in the preparation of this book. Mr. Mohammed Javed and Mr. Munne Khan, Cartographers in the department of Geography, A.M.U. Aligarh, who prepared the maps and diagrams also deserve my appreciation.

I owe a deep debt of gratitutde to the various well known writers on the subject whose works I have freely consulted and used. Suggestions for the further improvement of this book will be highly appreciated and acknowledged.

Nizamuddin Khan

of Major Climatic Types, Evaluation of Koppen's Classification of Climate.

24. Configuration of Bottom of Ocean and Seas 303
Continental Shelf, Continental Slope, Abyssal Plain or Deep Sea Plain, Ocean Deeps, Submarine Canyons, Submarine Ridges, Relief Features on Floor of Individual Oceans, Pacific Ocean, Atlantic Ocean, Indian Ocean.

25. Characteristics of Ocean Water 315
Physical Characteristics, Temperature of Ocean Water, Distribution of Temperature in the Ocean, Vertical Distribution of Temperature, Salinity — Composition of Salinity, Sources of Salinity, Determinants of Salinity, Distribution of Salinity, Salinity in Individual Ocean.

26. Circulation of Ocean Water 327
Ocean Currents — Determinants of Ocean Currents, General Pattern and Characterstices of Ocean Currents, Distribution of Ocean Currents; Ocean Currents in Atlantic Ocean — Ocean Currents in Pacific Ocean, Ocean Currents in Indian Ocean, Effects of Ocean Currents, Tides, Genesis of Ocean Tides, Types of Tides, Duration of Tides, Tidal Currents, Tidal Waves and Their Movement.

27. Ocean Deposits 344
Significance of Ocean Deposits, Classification of Ocean Deposits — Terrigenous Deposits, Pelagic Deposits, Organic Deposits, Inorganic Deposits, Chemical Precipitate, Coral Reef — Ideal Conditions for Coral Polyps and Coral Reefs, Kinds of Coral Reefs, Origin of Coral Reefs, Subsidence Theory of Darwin, Glacial Control Theory of Daly, Theory of Murray.

Glossary 355

References 378

Bibliography 382

Index 385

LIST OF FIGURES

2.1	Nabular Hypothesis	18
2.2	Planetesimal Hypothesis of Chamberlin and Moulten	19
2.3	Tidal Hypothesis of Jeans and Jaffreys	23
2.4	Formation of Disc of Dark Matter and Embryos of the Planets According to Schemidst.	24
2.5	Final Stage of the Formation of Planets According to Schemidst.	25
3.1	A Typical Seismogram	30
3.2	A Cross-section of Earth with the Help of Seismic Waves	32
4.1	Distribution of Land and Water on Tetrahedron	37
4.2	Phases of Continental Drift by A.Wegner	40
4.3	Reassembled Continents, Prior to Start of Continental Drift	42
4.4	Jig Saw fitting Jaxtaposition of South America and Africa	42
4.5	Six Principal Plates were Proposed by Le Richan, 1984	45
4.6	Types of Plate Boundary	46
5.1	Various Types of Igneous Rocks	51
5.2	Lopolith	52
5.3	Phacolith	53
5.4	Illustration of Contract or Thermal Metamorphism	53
5.5	Example of Dynamic Regional Metamorphism	59
5.6	Rock Cycle	60
6.1	Groups of Folds	64
6.2	A Normal Fault and its Components	66
6.3	Kinds of Faults	67
6.4	Horst or Block Mountain	68
6.5	Rift-valley or Graben	69
6.6	Cinder Type Volcano	70
6.7	Dome Shaped Volcano	71
6.8	Sheild Volcano	71
6.9	Composite Volcano	71
6.10	Elastic Rebound Theory	74

as Agent of Denudation, Landforms Formed by Solution Action or Erosion of Underground Water, Depositional Features of Underground Water.

12. Wave Action 149

Formation of Waves, Movement of Water Particles in Sea Wave Motion, Kind of Waves, Process of Erosion of Sea Coast by the Sea Waves, Erosional Features Formed by Sea Waves, Depositional Features Formed by Sea Waves, Coastlines.

13. Cycle of Erosion 163

W. M. Davis' Concept of Cycle of Erosion, Evaluation of Davis' Cycle of Erosion, Penck's Cycle of Erosion, Evaluation of Penck's Model, Distinguish between Davis and Penck's Model of Cycle of Erosion.

14. Soils 172

Components or Elements of Soils, Characteristics or Properties of Soils, Soil Profile, Soil Forming Process and Factors, Landforms or Topography, Classification of Soils.

15. Climatology 186

Elements of Climate and Weather, Weather and Man, Weather and Agriculture, Weather and Culture, Weather and Transportation, Weather and Health, Weather and Hydrology, Weather and other Human Activities.

16. Atmosphere 198

Significance of the Atmosphere; Composition of the Atmosphere, Structures of the Atmosphere.

17. Insolation and Temperature 208

Factors Affecting the Distribution of Insolation on the Earch Surface, Process of Heating and Cooling of the Atmosphere, Distribution of Insolation, Heat Budget, Latitudinal Heat Balance, Temperature, Factors Determine the Temperature, Distribution of Temperature, Thermal Anomaly.

18. Atmospheric Pressure 225
Significance of Atmospheric Pressure, Measurement of Pressure, Variation of Atmospheric Pressure, Determinants of Atmospheric Pressure, Distribution of the Pressure Winds, Determinants of Winds, Kinds of Winds, Monsoon.

19. Airmass and Fronts 245
Characteristics of Air Masses, Significance of Air Masses, Factors which Determine the Characteristics of Air Masses, Modification of Air Masses, Classification of Air Masses, Fronts — Classification of Fronts.

20. Atmospheric Moisture 255
Source of Water Vapours in the Atmosphere, Measurement of Humidity, Some Important Features Associated with Atmospheric Humidity, Adiabatic Change of Temperature, Fogs, Clouds, Precipitation, Types of Precipitation, Distribution of Precipitation, Seasonal Distribution of Precipitation.

21. Atmospheric Equilibrium 271
Significance of the Study of Atmospheric Equilibrium, Determinants of Atmospheric Equalibrium, Conditional Instability.

22. Atmospheric Disturbances 279
Classification of Atmospheric Disturbances, Extra-Tropical or Mid-Latitude Atmospheric Disturbances, Extra-Tropical Cyclone, Origin of Extra-Tropical Cyclones, Size and Structure of Extra-Tropical Cyclone, Movement of Extra-Tropical Cyclone, Weather Conditions Associated with Track of Extra-Tropical Cyclone, Area of Occurrence of Extra-Tropical Cyclones, Extra-Tropical Anticyclones, Tropical Disturbances, Major Tropical Disturbances, Minor Tropical Disturbances, Thunderstorms, Types of Thunderstorms, Distribution of Thunderstorms, Tornado.

23. Classification of Climates 295
Koppen's System of Climatic Classification, Sub-groups

6.11	World Distribution of Earthquakes and Volcanoes	75
7.1	Geosyncline Model of Mountain Building of L.Kober	83
7.2	Covection Current Hypothesis of A.Holmer	89
7.3 (a)	Enlarged earth during Molten State of Sima	92
7.3 (b)	Shortened Global Radius due to Resolidification of Sima (Substratum) and Contraction of Volume	92
7.4	Formation of Mountain by Crumpling of Geo-synclinical Rock Layers and Sliding of Broken Block of Earth's Crust (1926)	94
8.1	Water Fall	104
8.2	Formation of Rapids	104
8.3	River Potholes and Rock Gorge	104
8.4	Formation of Pot Hole	106
8.5	Alluvial Fan	106
8.6	Alluvial Cone	106
8.7	Meander of River	106
8.8	Formation of Oxbow Lake	107
8.9	Flood Plain and Levees	107
8.10	Braided River	108
8.11	Formation of Deltas	108
8.12	Example of Paired River Terraces	110
8.13	Development of River Valley	111
8.14	Dendritic Drainage Pattern	115
8.15(a)	Trellis Drainage Pattern	115
8.15(b)	Radial Pattern	115
8.16	Centripetal Pattern	116
8.17	Annular or Ring Like Pattern	116
8.18(a)	River A and B Flow on either side of Water Shed	118
8.18(b)	River B is Captured by A	118
9.1	Glacial Deposits	124
9.2	Fluvio Glacial Deposits	124
9.3	U-Shaped Valley	126
9.4	Hanging Valley	126
9.5	Formation of Cirque	127
9.6(a)	Cirque matter Horn A.R. etc.	127
9.6(b)	Horn, Aretes and Cirques	128
9.7	Roche Moutonne	128
9.8	Crag and Tail	129
10.1	Mushroom Rock	133

10.2	Zeugen	133
10.3	Yardang	133
10.4	Burchan	135
10.5	Transformation of Burchans into Seif Dunes	135
11.1	The Hydrologic Cycle	139
11.2	Distribution of Underground or Sub-surface Water	139
11.3	Arterian Well	141
11.4	Scarp Foot Spring	143
11.5	Fissure Spring	143
11.6	Formation of Fault Spring	144
11.7	Hot Spring and Geyser	145
11.8	Sink Hole	147
11.9	Lapies	147
11.10	Cavern	147
11.11	Interior View of Limestone Cave	148
12.1	Formation of Sea Waves	150
12.2	Wave	150
12.3	Wave Motion	151
12.4(a)	Development of Breakers	152
12.4(b)	Swash and Backwash Current	153
12.5	Notch	155
12.6	Wave Cut Platform	155
12.7	Sea Cliff	156
12.8	Formation of Sea Arch	157
12.9	Cave and Bar	157
12.10	Stack	157
12.11	Bar and Spit	159
12.12	Trombolo	159
12.13	Rai Coast	161
12.14	Fiord Coast	161
12.15	Dalmation Coast Line	161
12.16	Divisions of Sea Coast and Sea Shore	161
13.1	The Geographical Cycle of Erosion by Davis	165
13.2	The Represemtation of Penck's Cycle of Erosion over a Long Period	168
14.1	Soil Texture	174
14.2(a)	Soil Horizons and Bedrock	176
14.2(b)	A Soil Profile	176
14.3	Soils of the World	181

16.1	Structure of Atmosphere on Temperature Basis	203
17.1	Inclination of Sun's Rays and Efficiency of Insolation	209
17.2(a)	Incidence of Sun's Rays at various Latitudes at the Winter Solastice	211
17.2(b)	Incidence of Sun's Rays at various Latitudes	211
17.3	Heat Budget	216
17.4	Distribution of Temperature (January) in °C	220
17.5	Distribution of Temperature (July) in °C	222
17.6	Laps Rate and Inversion of Temperature	223
18.1	Zonal Distribution of Atmospheric Pressure	229
18.2	Distribution of Pressure (July) in Millibars	232
18.3	Distribution of Pressure (January) in Millibars	232
18.4	Planetary Wind System	235
18.5	Summer and Winter Monsoon in India	240
18.6	Sea and Land Breezes	242
18.7	Valley and Mountain Breezes	243
18.8	Foehn or Shinook	243
19.1	Representation of Fronts by Symbols	251
19.2	Warm Front	251
19.3	Cold Front	252
19.4	A View of Typical Occluded Fronts	253
20.1	Convectional Precipitation	267
20.2	Orographic Precipitation	267
20.3	Frontal Precipitation	268
21.1	Stability of Air	275
21.2	Instability of Air	275
21.3	Absolute Instability	276
21.4	Instability of Air (Conditional)	277
22.1	Stages in the Life Cycle of a Middle Latitude Cyclone as proposed by J.B. Jerknes	282
22.2	A Model of Cyclone	283
22.3	Structure of Tropical Cyclone	285
22.4	Tropical Disturbances in the Form of Easterly Wave	289
22.5	Thunderstorm	291
24.1(a)	Hypsographic Curve	304
24.1(b)	The Ocean Floor	304
24.2	Distribution of Sea Trenches in Pacific Basin	311
24.3	The Configuration of the Atlantic Ocean	312
24.4	The Configuration of the Indian Ocean	313

25.1	Surface Salinity of the Atlantic Ocean	324
25.2	Surface Salinity of the Indian Ocean	324
25.3	Surface Salinity of the Pacific Ocean	326
26.1	Currents of the Atlantic Ocean	331
26.2	Currents of the Pacific Ocean	333
26.3(a)	Currents of the Indian Ocean (Summer)	336
26.3(b)	Indian Ocean: Distribution of Ocean Currents (Winter)	336
26.4	Origin of Tide	338
26.5	Spring Tide	340
26.6	Neap Tide	340
26.7	To show Why the Time Intervals of Consecutive Tide is 12h 26m.	342
27.1	Centralized Profile of the Ocean Floor showing the General Distribution of Marine Deposits	345
27.2	The World Distribution of Marine Deposits based on a map by Sir John Murray, The Ocean (London)	349
27.3	Fringing Reef	351
27.4	Barrier Reef	351
27.5	Atoll	352
27.6	Theories of Atoll Formation	353

1

INTRODUCTION

Geography is a very interesting and fascinating subject which acknowledges the human beings about the distinct natural and cultural features of the world. It appears very essential to know and understand the origin, definition, evolution and the scope of the subject especially for a common geographers. The geneology and the evolution of the subject have been passed through various phases and many new ideas and concepts were added from time to time. Consequently it becomes very complex and tidious for a common man to understand the discipline of geography.

The term geography is derived from Latin words Geo and Graphie. Geo means the earth and Graphie refers to the discourse. Thus Geography is defined as the subject which deals with the description of the earth.

During the early period, before Christian era, the scholars recorded the geographical knowledge about natural facts on earth's surface through the observation and speculation. Different old civilizations contributed to the development of geographical concepts. Geography was considered simply as knowledge of topographical features, like mountains, rivers, places of one's own country and its boundaries. The scientific approach was added to the subject during late Greek period especially Erastosthene and Hipparchus who made their scientific observation about size, shape and the circumference of the earth. Hecataeus gave first systematic description of the world by writing Ges Periods (description of the earth) due to which he is known as the father of Geography. Thus the Greek scholars developed mainly physical and mathematical geography. The Roman scholars contributed mainly to historical and regional geography. Strabo is the first man who conceived the idea of complete geographical treatise comprising four branches of disciplines mainly mathematical, physical, political and historical geography. While the Ptolemy stressed on the need of cartography in geography. He viewed that geography is the science which deals with the art of making map. Same idea was also given by Erastothenes and Hipparchus before Ptolemy. The contributions of Arab geographers cannot be underestimated. Their achievements in the field of climatology, geomorphology and mathematical geography are very much appreciable.

During the era of exploration and discoveries (15th to 18th century) many expeditions were made by Arabs and Europeans through navigation, new geographical areas were explored. The impact of exploration strengthened the need of map making and the world boundaries were extended. The descipline was diversified and different new approaches of geographical studies of an area are added by different scholars during the same period. Varnius, a German scholar, developed sound scientific base for different theories of his days. He divided geography into general and special section which led to the development of systematic and regional geography. The foundation of dichotomy of systematic vs regional geography was laid down by him. Branching of geography into physical and human took place also during the period of discoveries. The period of, especially, Varnius and Kant, is also known as classical period of modern geography.

Eighteenth century and onwards witnessed a dramatic and remarkable change in the nature and approach of the geography discipline. Alexander van Humboldt and Carl Ritter, known as the father of modern geography emphasised on the interrelationship of different phenomena in a given space or geographical area. Humboldt opined that the geography is the description of the earth which deals with interrelationship of phenomena that exist together in an area. While the Carl Ritter defined that Geography is that branch of science which deals with globe in all its features, phenomena and relations as an independent unit and shows the connections of this unified whole with man and mans' creator. He looked geography is to describe the earth as the home of man. Importance of man was visualized even by Ratzel during late 19th century. He thought that man is an important ingredient of discipline of geography. In diverse physical conditions man adjusts himself, so in order to have complete understanding of geography, synthesis between diverse physical and cultural phenomena is inevitable. Alfred Hettner viewed geography as chorological science or study of regions. He thought that in geography different kinds of physical and cultural phenomena must be described in an areal context as history in time.

During 20th century, before First World War, a tremendous contributions were made to geography by German, French, British and American geographers. Various models had been developed in urban, agricultural, political and social geography. Walter Christaller put forward the Central Place Theory with the objective of functional organisation of space. Von Thunen developed Agricultural Land Use Location Theory. Meckinder gave Heartland theory. Man-environmental relationship dominated during early 20th and late 19th century. Miss Semple, Vidalde La Blach, Brunch, Huntington etc. advocated the theory of Determinism, Possiblism and Neo Determinism in human geography. Economic geography, agricultural geography, historical geography and social geography were well developed during inter-wars period. There occurred a shift from physical side to the social aspects. Post second war period witnessed new concept like human ecology, chorology and functional organisation of the space, quantification, as well as social relevance which became very dominant. The period of 1960s witnessed an encouraging trend in geographical research to use the language of mathematics with a view to postulate the theory in more scientific and precision ways. Thus the use of statistical techniques in geography, both in physical and human aspects, continuously increased in modern period. Behavioural and social relevance approaches were further added to geography to make the research more acceptable and relevance to human society.

Definition of Geography

Geography is defined, simply, as the description of the earth. Because the descipline geography is derived by two Latin words, Geo and Graphie which mean the earth and discourse about respectively. The concept and definition of the subject changed with the passage of time in response to the changing philosophy and approaches used by geographers since Greek to modern period of geographical historical evolution. Many geographers defined the subject into following ways:

1. Geography is described as mother of all sciences. — *Sir Patrick Gaddes*.
2. Geography is the science which deals with the art of map making. — *Ptolemy, Erastosthenes* and *Hipparchus*.
3. Geography is the description of the earth which deals with the inter- relationship of phenomena that exist together in an area. — *Alexander von Humboldt*.
4. Geography is the branch of science which deals with globe in all its features, phenomena and relations as an independent unit and shows the connection of this unified whole with man and man's creator. — *Carl Ritter*.

5. Geography is a chorological science or it is the study of regions. It is a field in which things have to be described in their areal context on the face of earth, such as history is the field in which things are to be considered in their time context. — *Alfred Hettner*.
6. Modern geography has been described as a science, an art and a philosophy. It is concerned with the interrelationship of man and his environment, with the earth as home of man and with the relationship of these features of the earth's surface which owe their origin due to work of man. — *L.D. Stamp*.
7. Geography is concerned to provide accurate, orderly and rational description and interpretation of variable characters of the earth's surface. — *Hertshorn*.

Above definition reveals that the man is the pivot of the subject matter of the descipline. He is surrounded and abound by various realms of natural and cultural environments. The interaction and inter-relations of man with biotic and abiotic environment vary from region to region and various regions may be differentiated on the basis of similarity or dissimilarity of different characters. Hence, geography does not only deals with the description of earth's features but it studies the spatial variations of various kinds of physical and cultural phenomena in a very scientific and rational way.

Classification of Geography

Geography was considered as a bundle of knowledges and informations collected by different scholars during pre-historic periods. It was in crude form consisting of philosophical, political, sociological and natural scientific ideas. During the Greek and Roman period geography was mainly concerned with the illustration of natural environment like landforms, climates, biotic life, oceans and different natural catastrophic phenomena on the earth surface. Later on man was emphasised very much and homocentric concept was given especially by Arabs, German and French scholars. Earth was considered as the home of man. Man-environment relation became the focus of geographical themes. Thus on the basis of content of the subject, the geography may be broadly divided into two branches (i) Physical Geography, and (ii) Human geography.

PHYSICAL GEOGRAPHY

Physical geography may be defined as the branch of Geography which deals with the distribution and origin of natural phenomena on the earth surface. The effect of these phenomena on the human beings may be also included in this descipline. Physical geography was developed during Greek period because the Greece was surrounded by land of various types of topographical and physical features. In fact Greece itself is a hilly region. Greek people were able to make tremendous advancement in the knowledge of geomorphology, climatology and oceanography. Hecataeus gave more weightage to physical geography. Physical geography can be divided into different sub-divisions as given below —

(i) Geomorphology	(iv) Climatology
(ii) Meteorology	(v) Hydrology
(iii) Oceanography	(vi) Biogeography

HUMAN GEOGRAPHY

Human geography deals with the cultural phenomena which came into being after the interaction of man with the different realms of physical environment in which he lives. Branching of Geography into human geography may be traced back to the period (17th and 18th century) of Varenius and Kant. In the later part of 19th century F. Ratzel, known as father of human geography, recognised that man is an integral part of the discipline of Geography. He stressed on unity in diversity, stating that man is adjusting himself in various physical conditions. So in order to have a complete geographical view of the earth surface, a synthesis between physical and cultural phenomena is essential. Thus many schools developed in Germany, France and America to discuss man and environment relations. Deteminism, possibilism, neodeterminism and social Darwinism etc. are excellent example of man-environment relation. Miss Ellen Churchil Semple, Huntington, Taylor, Vidal de La Blache, Jean Brunches and so on were leading human geographers. Human geography is defined into various ways. Ratzel defined human geography as the "synthetic study of relationship between human societies and the earth's surface". Miss Semple advocated that human geography deals with changing relationship between unresting man and unstable earth. Vidal de La Blache stated that human geography offers a new conception of inter-relationship between earth and man. While Huntington defined human geography as the study of nature and distribution of the relationship between geographical environment and human activities and qualities. H.H. Burrow recently described human geography as human ecology.

However the human geography includes the contents of social, economic, political, cultural, transportation, and many other tangible and intangible subjects related to human societies. It studies mainly three subject matters.

(i) Population-distribution, density and composition in time and soace,
(ii) The relationship between man and environment, and
(iii) The regional analysis of the inter-relation of man and environment in the world.

Moreover, the population and the inter-relationships of man with environment (Natural) vary from place to place. Man adjusts and adopts with natural conditions which also show spatial variations. Thus the human geography deals with spatial distribution of human activities on earth surface. Human geography may be sub-divided as (i) Population geography, (ii) Social geography, (iii) Urban geography, (iv) Marketing geography, (v) Industrial geography, (vi) Political geography, (vii) Behavioural Geography.

Approach of Study in Geography

There are broadly two kinds of approaches by which a geographical fact can be studied. They are (i) regional, and (ii) systematic approaches. So on the basis of approaches used for study the geography may be divided as regional and systematic geography. The concept of these approaches was given as early as back to Greek period. But Varenius divided geography into general and special sections which led to the development of systematic and regional geography. He is considered as the founder of dichotomy of regional vs systematic geography.

Systematic Geography

Systematic geography is a type of the discipline in which an individual element of geographical attributes are dealt for whole world. The terms systematic, general and topical are applied to the systematic geography. Systematic is the British usage which have been borrowed from botany which used it to describe the classification of species of plants. General geography is the oldest established of the three terms. This term was used more than three hundred years ago by German writer Bernhard Varenius (1650). While the topical terms used by American geographers for the study of specific topics. After the understanding of the system and process of functions of an element, an universal law is formulated which can be indiscriminately applied in every part of the world. Such approach is mainly applicable to natural phenomena where generalization can be made like natural sciences. It is a deductive approach. Climatology, meteorology, oceanography etc. are the systematic geography. Systematic geography is also known as General Geography. It drew inspiration from the adjacent systematic sciences with a search for general or universal concept. Important propounders of this approach are Varenius and Humboldt who are known as founder of systematic appraoch in geography.

There has been a strong tendency in recent decades to increase the range of systematic geography. Human geography has been subdivided into various special branches following the systematic approach selecting any individual element of cultural phenomena. Economic geography is an excellent example of systematic geography. Thus one systematic branch, economic geography again has been splitted into several parts, agricultural geography, manufacturing geography, transportation geography, marketing geography, geography of resource etc. Systematic geography have some measure of importance as applied sciences, contributing towards the understanding and solution of current problems or planned developments in a variety of fields like economic, social and political.

Regional Geography

Regional geography is the branch of geography in which the earth is divided into different units or regions and all geographical elements, physical and cultural, are studied together to understand holistic view of the area. It is also known as particular or specific geography because a particular area is studied with reference to general geographical outlook. Man-environment relation and their combined effects are taken together in regional analysis of the any country or region.

Regional geography has long been accepted as one of the principal form of geographical study. It is the means of attaining the geographer's main objective of understanding the spatial differences of various kinds of natural and cultural phenomena from one place to another on the earth surface. It is the complementary of the systematic geography which provides the base and help in the formation of formulation of hypothesis and structured ideas. Regional geography is concerned to show how these elements (biotic, abiotic and cultural) interact together within individual areas on the earth's surface.

The genesis and development of concept of regional geography can be traced as back as to Greek period but the real beginning took place after the work of Varenius, a German scholar during 17th century. He made clear distinction between systematic and regional geography. Later on Humboldt, Ratzel, Carl Ritter, Hettner in Germany, Vidal de la Blache, Bouch in France, Meckinder and A.J. Herbertson in Great Britain were important contributors of regional geography.

Regional geography has great utility for government, administrators, trade, commerce and planning. The importance of this branch has increased in the modern era of regional economic development. Besides, it is also greatly concerned with the economists, sociologists, geological scientists besides geographers because it involves the wide field of spatial organisations. It reflects nevertheless the tendency for convergence between geography and other social sciences because the problem investigation is best approached by inter-disciplinary subjects.

Geomorphology

Geomorphology is an important branch of physical and systematic geography. It is derived from three Greek terms, i.e. Geo, Morphe and Logos which means "a discourse on land forms". Thus geomorphology may be defined as the science which deals the distribution of different kinds of landforms on the earth surface. Some geomorphologists viewed that origin and evolution of topographic features caused by physical and chemical process operating at or near the earth's surface.

P.G. Worcester defined geomorphology as the subject of interpretative description of the relief features of the earth's surface A.L. Bloom viewed that geomorphology is the discipline which describes systematic analysis of land scapes and the process involved for their evolution. Further stated by another geomorphologist that the scientific discipline, which is concerned with the explanation of varied morphology of land surface, is geomorphology. According to him, it has not been principally concerned with whole of the earth's surface but also with a comparatively small part of it. Continents and their shallow submarine extension, are also included in geomorphological study. Geomorphology is not only concerned with simply description of land features but it deals with their spatial variation too.

Some scholars thought that the geomorpholgy has come about as a result of dissatisfaction with term physiography which was formerly applied to this subject. Some one thinks that geomorphology is geology (A.K. Lobeck, 1939). However, the geomorphology studies about spatial distribution of various land forms on land surface and the cause of origin and process of evolution of these features. The effects of geomorphological features on human aspects is also subject matter of it.

Historical Evolution of Geomorphic Ideas in Geography

The subject geomorphology and the geomorphic ideas grew up through the span of long history. This idea has been rooted in ancient thinking of Greek and Roman scholars. After a long gap of dark age, the geomorphic thoughts were contributed to subject by Arabs, Europeans and American scholars from time to time. The beginning of the geomorphology could not take place as descipline before Hutton (1726-1797), though the word Geomorphology was first used by Albrecht Penck (1910).

During Ancient period, before the beginning of Christian era, both Greek and Roman scholars like Herodotus, Aristotle, Strabo and so on have given many simple geomorphic observations especially in terms of fluvial action. Herodotus (485-425 B.C.) was well aware of physical process that modified the surface of the earth. He insisted on the formation of delta by the deposition of sediments brought by Nile river from Ethopia. Aristotle (384-322 BC) observed the phenomena of land expansion in shallow seas by filling up of the sediments and formation of delta at the mouth of rivers. He rightly pointed out the process of alluvium deposition by which in so many places land is gaining on the sea and especially in the sea of Azov which, he thought, was continually being shallower and would be one day entirely dry. Strabo (54 BC - 25 AD) pointed out that different kinds of changes have taken place on the earth surface due to transgression and regression of the sea. Moreover, the concept of abrasion work by river was given by Seneca. He viewed that rivers abrade their own valley in the rocks where they flow.

Besides, Arab geographers after the end of dark age in Europe, offered important geomorphological observations regarding the process of transformation and changes which took place on the face of the earth surface. Ibn-Sina (980-1037) geve the idea of differential rate of erosion by rivers due to variation in hardness of the rocks in which they flow. He sought the formation of mountains as standing resistent rock amongst the softer surrounding fast eroded rocks. Kitab-al-Hind

(1030 AD) of Al-Biruni recorded the excellent geomorphological speculation in India. In this book he wrote that the rounded stones of alluvium along Himalayas were formed by rolling up along the torrential mountain streams. Concepts of sorting of deposited materials by river is also given by him. He sought that coarser materials are deposited along foothills and finer in textures away from the mountains. Al-Masudi gave the idea of comparative study in geomorphology. He made comparative study of landforms and made an analytical study of processes concerned in their formation. In his views landforms are visualized to pass through a cycle of development from youth stage to maturity and finally old stage — the stage of pene-planation. The role played by cycle of erosion and the adjustment of streams with the structure in evolution of landforms have been observed by him.

The contributions of Arabs in geomorphology were not known in Europe. Many European scholars, showed great interest in the analysis of land forms and the process of their formations. They saw rivers as the main agents to sculpture the surface of the earth into various forms. Origin, flow, velocity, work of erosion, transportation, and deposition, etc. of the rivers were highlighted by different scholars in Europe. During 15th century and onward various geomorphological ideas were given on which modern geomorphological concepts developed. Leonardo da Vinci (1452-1519) expressed his views that the rivers cut their valleys and transport the eroded materials, from one place and deposit to other place. Frenchman Buffon (1725-1815) and De-Saussure (1740-1799) and so on are other European scholars who presented the process especially about fluvial erosion. Buffon recognised the powerful erosive ability of river to destroy the highland with view reduce it at sea level. Tozetti explored the formation of irregular streams or channels of river due to differential erosion as related to varying geologic structure and materials. Desmarest was first apparently to give idea of development of landforms through the successive stages of evolution.

Thus, on the basis of preceding literature and experience James Hutton (1726-1797) established some laws and principles regarding the development of landforms on the face of earth surface. He evolved a concept that "present is key to the past" and he founded the doctorine of uniformitarianism in opposite to catostrophism. This doctorine reveals that the same physical processes and laws that operate today, operated throughout geological time but not necessarily always with the same intensity as now". John Play Fair and Charls Lyell were the great followers of Hutton's Uniformitarianism. Hutton is known as the father of modern geomorphology. The contemporary scholars in Europe showed great interest in the study of glaciation of geological past when the northern Europe was covered by ice sheet. Concept of marine erosion was given by Ramsay (1814-1891) and Richothofen (1833-1903) in 19th century.

The era of 20th century began with the systematic and scientific explanation in geomorphology. In America J. W. Powell (1834-1902) G. K. Gilbert (1843-1913) and G. E. Dulton (1841-1912) were pioneer thinkers in geomorphology. They provided foundation on which W.M. Davis (1899) put his foundation of concept of geomorphic cycle. His cycle became so prominent that the American school of geomorphology was known as Davisian school of geomorphology. Powell is known as the father of American geomorphology. His contributions were related to the base level erosion classification of rivers on the basis of rock structure and their origin. Gilbert introduced quantitative approach to study the relationship between stream loads, velocity and lateral erosion made by a river in any area. During 1920s 1930s Walther Penck and his followers staged a minor revolt against some of the Davis ideas and found some adherence. Towards, the middle of 20th century, climatic geomorphology especially in Germany and France, arose on premise that distinctive land farms and processes are associated with the certain climatic regions, Geomorphology, after the Second World War became diversified. Many new sub-fields Karst geomorphology, coastal geomorphology, glacial and fluvial geomorphology sprung up.

Prospects of Geomorphology in Future

The geomorphic knowledge and investigation seem to be an essential element for understanding the problems of man in a given eco-system. Nature of topography, drainage pattern, soil characters, hydrological behaviour etc. are very important natural milieus to determine the socio-economic development of the area. Man-environment relation has been gradually very complex due to increasing man's interference with the nature. Soil erosion deforestation underground water depletion, filling of river channels for construction of building in urban areas, embankment of river channels for controlling the floods etc. are crucial intervention, of man. Due to lack of geomorphic informations about the distribution of various geomorphic features and the process of their evolution, sometimes, would result in very serious natural and socio-economic hazards like floods, land-slides, earthquakes, subsidence of land etc.

The important of geomorphological studies and informations would appear very useful for evaluation of natural resources and for the mitigation of diverse problems of human societies in a particular geomorphological environment. The scope of this descipline is very bright in future. Management of natural hazards, planning of rural and urban settlements cannot be made without appropriate scientific knowledge of geomorphology of the given area. It is also greatly needed in construction industries. Slope of terrain direction of river flow, nature of Karst topography, structure of rocks and availability of water etc. are very essential knowledge needed for civil engineers and architectures before planning the construction of houses, road or railways in any region. Planning of regional development is also preferred on the basis of geomorphic region like river basins of watersheds. Landuse plan in agriculture can be optimally made with the help of geomorphic informations. Type of land forms, degree of slops, areas of floods, nature of shifting of sand dune and coastal geomorphology may determine the type of agriculture and agricultural practices in an area. So the Geomorphology is probably looked in applied forms. Environmental geomorphology, urban geomorphology, agricultural geomorphology, anthropo-geomorphology etc. are the excellent example of well prospect of the descipline in future. Quantitative revolution has also enriched the subject in order to make very rational and scientific study of relationship between man and different geomorphological features. This scientific explanation would enable the planners to construct the strategies for socio-economic development of different natural regions on the basis of local ecological condition in a given geomorpholgical region.

GEOMORPHOLOGY AND MAN

As being a subject of dealing with distribution of geomorphological features and their relation with the man, the geomorphological features and the geomorphology have very intrinsic relations with different aspects of human civilization or cultural environment. The contribution of geomorphological study may be understood in the following ways.

Geomorphological Features and Environment

Natural environment and the cultural environment are very much affected by distribution of landforms, mountain building movement, various kinds of erosional and denudation process and consequent features on the different parts of land surface. Mountainous topography of young origin presents very rugged, inaccessible topography with variable founa and flora from foothill to top. Climatic variations, immature and thin soils and slopy terrain are also distinct characters of mountainous topography. Soil erosion, landslides, occurrence of earthquakes etc. may be also effect of mountain building process in the weak zones.

Contrary to this plain regions provides a human occupance environment. Accessibility, availability of water, good soil, and montonous surface present the good agroclimatic environment. About 70 per cent world population reside in either river basins or coastal regions.

Plateau region is usually storehouse of minerals and the distribution of denuded hills and peneplains are very common.

Besides, different varied geomorphological process like flooding of rivers, erosion of watersheds by river; shifting of sand dunes in aeolian region and the distibution of different type of Karst topographic features present district environment in which a particular type of human life can be survived. Surface and underground water, lithology and the vegetation characters may also be very distinct owing to geomorphological variation on the earth surface. Distribution and orientation of landforms like mountains also reflect an influence on the amount of temperature and precipitation. Usually windward slope of mountains has higher amount of precipitation rather than the area of *leeward* (rain shadow zone). Altitudinal variation of temperature is very common. Difference in temperature of India and Central Asia during winter is only due to presence of Himalayas.

Geomorphology and Transportation

Network of roads, railways and airstrips are determined by distribution of various kinds of geomorphological features on the land surface. Construction of roads and railway tracks require very suitable plain and hard rock region. Transportational networks especially surface transport have very high density in plain and coastal lowland areas as compared to mountain and plateau regions where construction of roads and railway tracks are very difficult due to rugged topography, hardrocks, and presence of a large number of unpassable transverse hillocks. Construction of roads are very difficult and expansive as a large number of tunnels have to be made to connect the accessible places in rough topography region. There are many places in Himalayan region which are still not connected either by roads or railways. Very few roads are constructed in North Eastern Hill areas. While the North Indian states like U.P., Bihar, M.P., West Bengal show high concentration of roads and railways. Valley region is more preferred for construction of road in hilly areas to avoid unnecessary breaking of rocks which present obstacles in the way. Desert area exhibits low density of roads and railways due to loose soils and high temperature which expands the railway tracks and melts the charcol of metalled roads.

Moreover, Karst topography, glacial topography and fluvial geomorphic region also affect the nature of transportation system. The presence of permafrost in preglaciated region especially in Arctic & Taiga climate and the subterranean cave, cavern and rivers, poses serious danger for road construction. A knowledge about nature, process and distribution of permafrost is very essential for engineers before construction of highways or railways. Deforestation and the construction may cause thawing of permafrost, consequently, collapsing of ground and constructed structure may take place. In the initial stage, Trans-Siberian Railway suffered from damage due to thawing of permafrost. Similarly, Karst region of limestone also dangerous for road construction. Roofs of caves and blind valley may collaps down and the built up structure would be damaged. So such regions, are usually devoid of construction.

Flooded region also discourages the construction of roads and railways. Repeated floods, weaken and break up the structure (roads). A huge amount of capital is also requires for construction of bridges across the rivers which are in abundance in lower basins.

Geomorphology and Agriculture

Geomorphological features have also very paramount effects on the agriculture practised in different geomorphological regions. Agriculture is not well developed in mountainous or hilly tract area. Steep slope, rugged topography, thin and immature soil and inaccessibility have discouraged agricultural practices. Longitudinal valleys, intermountain valleys and gentler slop with terracing are only areas where cultivation of some cereals or cash crops are grown. Horticulture, animal rearing, sericulture, plantation and agriculture are mainly practised in mountainous region. Apple gardening in J & K and Himachal Pradesh, rubber and oil palm cultivation in Malaysia, Tea in China, Sri Lanka and India are some examples of agriculture in rugged topographic region.

River basins, coastal lowland, alluvial plain areas and the peneplain in plateau areas are generally following very intensive agriculture system. Every bit of arable lands are brought under cultivation. Lowlands or delta region is alloted to paddy cultivation whereas uplands, where water logging may persist, is considered suitable for wheat if the climatic conditions are favourable. Availability of both surface and underground water for irrigation and good alluvium fertile soil, black or chernozeom soils etc. are encouraging features for cultivation of cereals, pulses, oilseeds and various other commercial crops. Coastal geomorphological region witnesses the cultivation of paddy and coconut. Aquaculture and fishing are the main type of diversification of agriculture here. Karst topographic region associated with sink holes, swallow hole, polje, doline collapsed caverns, flooding of rivers especially in middle and lower courses, area of meanders; shifting of sand dunes in aeolian topographic region and the glaciated geomorphic area are usually associated low concentration of agriculture. In desert area where soil is very loose and consists of hard pane of calcium carbonate in subsoil zone discourage the deep ploughing and intensive agriculture. Animal rearing and nomodic harding are important economic activities in the tropical desert of world. Hence, type of crops to be grown, landuse pattern and intensity, type of agricultural practices and the planning for management of soils, water, and crops can be made at optimum level with the help of geomorphic informations from applied geomorphology.

Geomorphology and Settlement

Geographological features have very unforgettable impact on the nature, type and pattern of both urban and rural settlement. All the ancient civilization did not flourish in hilly or rugged topographic region but in river basins. Settlement in mountainous region is very scattered in the valleys. Along the gentler slope of hills, some houses are also being constructed but with a great risk. Dense forests, presence of wild animals, unfavourable climatic conditions breathing problems at higher altitudes due to rareness of air, and inaccessibility to transportation facilities and potable water have discouraged the concentration of human inhabitation. River valley, coastal low land, alluvial plains, less plain and delta region attracted the population and consequently dense settlement flourished since ancient periods. Nile basin, Euphrate Tigiris basin, Indo-Gangetic plain, Hwang Ho plain etc. are important areas of compact settlement. Oasis in desert also is the area of compact settlement. Again, Karst region, flooded regions, weak zone of occurrence of earthquakes, volcanism, subsidence, coastal region at the time of rising of sea level, area of migratory sand dunes are usually avoided for inhabitation due to fear of damage of life and properties at the time of natural hazards. Moreover, type of houses, used materials for construction and the pattern of settlements are the effects of geomorphological features. Small, light, wooden and temporary houses are constructed especially in earthquake zones. Stones are used for construction in plateau and mountainous regions, bricks in plain areas and bamboo or woods in forest hilly tracts. Drainage lines must be planned after studying the slope of terrain. Thus geomorphic informations are very much essential for town planning and rural settlement planning in any country or region.

Geomorphology and Planning

Geomorphic knowledge is very useful for planning of sustainable development of any part of the earth's surface. A geomorphological region or ecosystem consists of various kinds of abiotic and biotic elements and these elements have established very inseparable interrelation among themselves. Man, himself, has adjusted well following the natural process of the given area. The strategy of socio-economic development of different areal unit, would be more effective if the geomorphological conditions and their inherent processes are taken into consideration. The use of geomorphic informations, i.e. nature of landforms, slope, lithological or soil condition, hydrological characteristic, landslides, soil erosion, fauna and flora, erosional and depositional features of various agents of denudation etc. was, first started during 1930s in U.S.A. River basin was selected as basis of regional development. Tennesse Valley Authority (TVA) was commissioned in 1933 for total watershed management to control soil erosion and floods for optimum utilization of lands, reforestation, water conservation, employment generation and the resource exploitation in discussed basin.

Such practices now became very popular in many countries. Damodar Valley Authority in India is the example of watershed management like TVA. Many other river basins like Chamble basin, Godavari basin, Ganga basin etc. can be used as unit of regions rather than administration unit for better rational and sustainable development. Karst, glacial and coastal geomorphological regions may be also used for planning of regional development on the distinct geomorphological features and system prevailed in particular region. Problem oriented planning like flood area development programme, drought prone area development programme, sand dune area development programme would be very important example of planning in response to geomorphological feature and processes. Such area cannot well developed without a scientific and systematic study of geomorphological characters and system of these regions.

2

ORIGIN OF THE EARTH

The Earth is an important planet of solar system. It is the only planet which has life and atmosphere. It is also called as watery planet. It is one of the inner or terrestrial planets with a diameter of 12740 km and lies at an average distance of 149 million km from the Sun. The age of the earth (solar system) is believed to be about 4.7 billion years. It revolves around the Sun in 365 ¼ days. The origin of the earth is matter of great speculation and discussion. Exact process of her origin has not been known till now. As a member of solar system the process of its origin can be understood in the genesis of the solar system. Different scientists have put forward various theories to explain the origin of the earth (solar system). Hence, a brief knowledge about some facts of solar system is also essential.

Solar System

The solar system consists of nine planets, and together with satellites, asteroids and comets revolves around the Sun. Following are some of the important characteristics associated with solar system :

(i) All the planets revolve around the Sun in the same direction in ecliptical or almost circular orbits. Most of their satellites also revolve in the same direction;

(ii) The planets, except venus and uranus, rotate in the same direction as their revolution around the Sun;

(iii) The distance of each planet from the Sun is roughly twice that of next planet closer to the Sun;

(iv) The Sun contributes 99.9 per cent of total mass of solar system but it holds only 1 per cent angular momentum; and

(v) The planets consist of two groups, i.e. (i) terrestrial or inner planets, and (ii) outer or giant planets. Mercury, Venus, Earth and Mars are inner planets and have relatively higher density and smaller sizes while the giant planets include Jupiter, Saturn, Uranus, Naptune and Pluto which are generally of bigger sizes and low density.

Theories or Hypotheses Regarding the Origin of the Earth

Gaseous Mass Theory

Many theories regarding the origin of solar system have been given by the scientists over the centuries. The modern approach to the problem began in 1755 when the German philosopher Immanuel Kant, developed a theory to explain the origin of the earth on the basis of assumed primeval slowly rotating cloud of gases (called as Nabula) and Newton's law of gravitation and rotation. (Figs. 2.1 and 2.2) Kant's hypothesis contains following assumption :

(i) Primordial cold and motionless matter were scattered in the space;
(ii) Due to gravitational force of the matters and mutual attraction, sudden motion took place among the matters;
(iii) The collision of matters caused an increase in temperature and with the passage of time, these matters became hot and gaseous rotating mass, now called as Nabula;
(iv) Increase in temperature resulted in expansion of volume and angular velocity of Nabula. Due to angular velocity, the matter from the Equator of Nabula bulged so much that rings began to separate from the gaseous mass due to centrifugal force. After the cooling of mass, rings of planets were formed; and
(v) Remaining part of gaseous Nabula was known as proto-sun. By the repetition of same process many sub-rings were separated from newly formed planets and the materials of each ring condensed to form satellites.

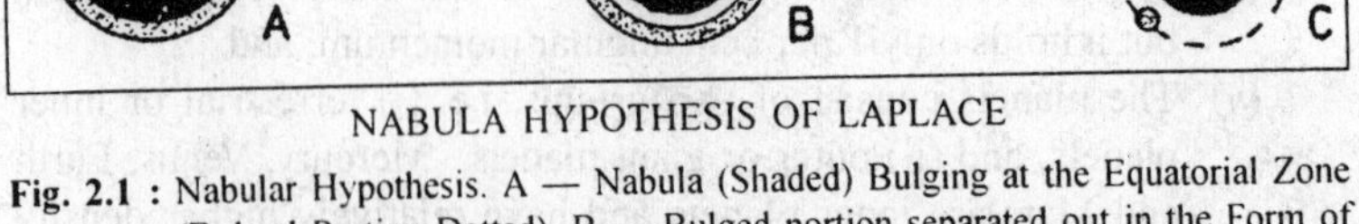

NABULA HYPOTHESIS OF LAPLACE

Fig. 2.1 : Nabular Hypothesis. A — Nabula (Shaded) Bulging at the Equatorial Zone (Bulged Portion Dotted). B — Bulged portion separated out in the Form of a Ring. C — Ring coalsced to form a Globe or Planet which rotate around Nabula in its orbit (dotted line).

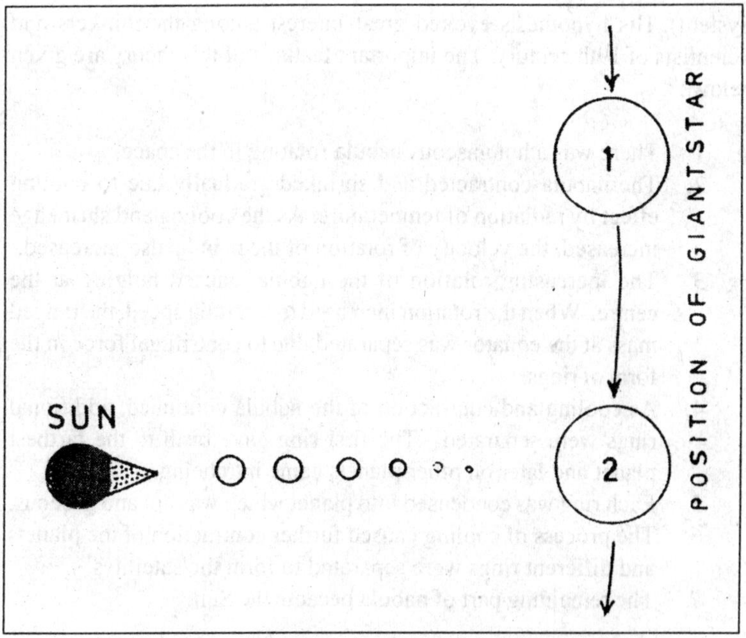

Fig. 2.2 : Planetesimal Hypothesis of Chamberlin and Moulton.

Criticism

Kant's gaseous hypothesis was the first scientific explanation of the origin of the earth. But this hypothesis was criticized on two grounds.

(i) He could not explain the source of primodial matters which formed the gaseous mass of cloud (Nabula); and

(ii) He failed to explain the angular momentum adjustment between Sun and its planets. According to the law of conservation of angular momentum, the angular momentum is directly proportional to the mass of the body. Angular momentum of the planets should be lower than the angular momentum of the Sun in the light of Kant's hypothesis. But the present solar system shows opposite arrangement of angular momentum.

Nabular Hypothesis

On the basis of earlier work of Kant, the French Scientist Laplace (1794) propounded Nabular hypothesis to explain the origin of the earth (Solar

System). His hypothesis evoked great interest among the thinkers and Scientists of 19th century. The important features of this theory are given below :

1. There was a hot gaseous nabula rotating in the space.
2. The nabula contracted and shrinked gradually due to cooling effect by radiation of temperature. As the cooling and shrinkage increased, the velocity of rotation of the nabula also increased.
3. The increasing rotation of the nabula, caused bulging at the centre. When the rotation increased to a certain speed, the bulged mass at the equator was separated due to centrifugal force in the form of rings.
4. As cooling and contraction of the nabula continued, additional rings were separated. The first ring gave birth to the farthest planet and later on other planets came into being.
5. Each ring was condensed into planet which was hot and gaseous.
6. The process of cooling caused further contraction of the planets and different rings were separated to form the satellites.
7. The remaining part of nabula became the Sun.

Criticism

The Nabular hypothesis was widely accepted during the past century owing to the confirmation of different characteristics of present solar system with the idea developed by Laplace. The merits and demerits of his hypothesis are as follows :

Merits

1. All the planets are composed of same elements — because they were derived from a common Nabula.
2. The planets revolve around the sun in same direction, in same plane and in nearly circular orbits.
3. The inner molten and upper solid part of the earth is confirmed by this theory. This explains that planets were formed by cooling of hot gaseous rings separated from Nabula. Thus the upper part became solid and inner remained liquid.

Demerits

This hypothesis suffers from a number of shortcomings and does not explain the following facts :

(i) The Sun is in slow rotation but the theory indicated that the contraction of Nabula had caused acceleration in its rotation. Thus, according to this theory, the sun should have rather more speedy rotation after about 4.5 billion years of the origin of solar system; and

(ii) The absence of any kind of equatorial bulge in the sun does not confirm the explanation that due to rotation of Nabula, equatorial bulge was formed and separated in the form of rings to form planets.

The Planetesimal Hypothesis

Two American scientists, Chamberlin and Moulton (1900) announced their new theory called Planetesimal hypothesis. In this hypothesis, they attempted to overcome the objections raised against nabular hypothesis. The process of formation of solar system was explained in the following manner:

(i) There was cold proto-sun and an intruding star of considerable size moving in the space;

(ii) The intruding moving star approached the proto-sun and a tidal bulge was caused on the surface of the sun due to gravitational force of the star;

(iii) The ejected mass of the sun was scattered in the form of solid particles around the sun. These particles were called planetesimals;

(iv) The big size of planetesimal was attracted and accreted to form the planets;

(v) The accretion and aggregation of planetesimals caused the heating of the planets. With the passage of time, size of planets became larger and the increasing gravitational compression of interior produced internal heat of the earth due to colescence of increasing number of planetesimals;

(vi) The gases, held in planetesimals, were given off as the growing earth became heated by internal compression. Thus the atmosphere was formed around the earth; and

(vii) The increasing content of water vapours started to condense due to cooling of the atmosphere and consequently rainfall occurred for a longer time and oceans were formed on the earth.

Criticism

The Planetesimal Hypothesis tried to explain the origin of earth in a very scientific way but it could remain non-explanatory about following issues posed by scientists.

(i) The formation of planets by accretion and aggregation of planetesimal's nuclei is not able to explain the orderly size of the planet of present solar system. The planets increase up to Jupiter as distance from the Sun decreases. Again, size of planet decreases as distance from the Sun increases;
(ii) The assumption of cold planets and satellites at the beginning is not also confirmed with the present solar system. Some of the planets are still in gaseous and hot state; and
(iii) The geological evidences indicate that the earth was in liquid state in the beginning and the present earth has been evolved through the cooling process. The arrangement of its mass into dense core surrounded by lesser dense matter is against the theory that the evolution of the earth passed through a process of cooling to heating.

Tidal Hypothesis of Jeans and Jeffreys

British scientists Jean and Jeffrey (1925) gave a new theory of the origin of the earth, based on the gravitational pull concept of Newton (Fig. 2.3). This hypothesis, explains the origin of the earth (solar system) in the following manner:

(i) There was a hot gaseous sun and an intruding star of larger size. They were rotating on their own orbits;
(ii) The intruding star gradually approached the Sun. Due to attractional force of the star, a tide of hot gaseous mass was formed on the surface of the Sun;
(iii) When the intruding star came close to sun, the tidal bulged portion of the Sun detached in the form of cigar like filament of hot and gaseous material and it moved away with the moving star;
(iv) When the star moved far away from the Sun, the detached filament

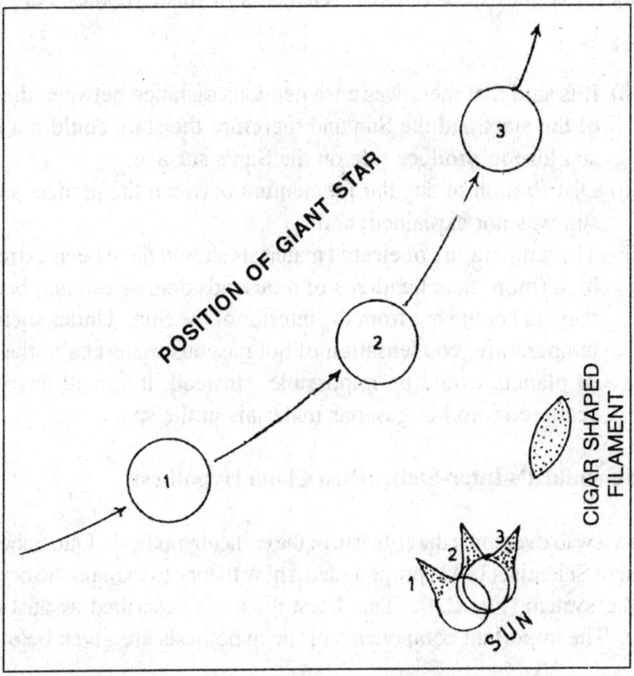

Fig. 2.3 : Tidal Hypothesis of Jeans and Jeffreys.

was once again attracted by the Sun and it started to revolve around the Sun in the form of spiral nabula;

(v) The cooling of filament broke it into masses of different shapes and size which began to contract towards nuclei, which resulted in the formation of planets;

(vi) The tidal bulging again took place among the planets due to the attraction of the Sun and thus satellites were formed; and

(vii) Through the gradual cooling of hot and gaseous and satellites, they have been from liquids to finally solids and the gases which remained outside became atmosphere.

Criticism

The Tidal hypothesis had gained wide acceptance because it explained the different aspects of present solar system. The arrangement of planets and satellites and the constituents of all the planets have been very well explained by this hypothesis. Some scientists have criticised this theory keeping in view the principles of astronomy. Demerits of this hypothesis are given below.

(i) It is said that there was a tremendous distance between the orbit of the stars and the Sun and therefore the stars could not come so close to produce tide on the Sun's surface;
(ii) Distribution of angular momentum between the planets and the sun was not explained; and
(iii) The temperature of ejected materials should have been extremely high (more than hundreds of thousands degree celcius) because they had come out from the interior of the Sun. Under such high temperature, condensation of hot gaseous materials in the form of planets would be impossible. Instead, it should have been scattered into hot gaseous materials in the space.

Otto Schemidst's Inter-Steller Dust Cloud Hypothesis

With a view to overcome the criticism of the earlier hypothesis, Otto Schemidst a Russian Scientist (1943) propounded a new theory to explain the origin of the solar system (Fig. 2.4). This latest theory is described as dust cloud theory. The important components of the hypothesis are given below:

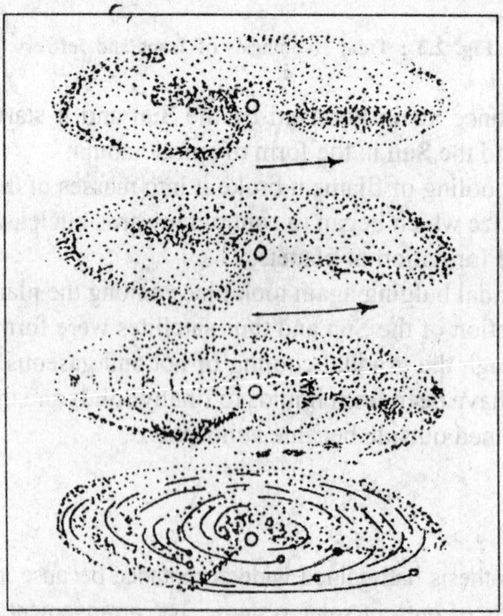

Fig. 2.4 : Formation of Disc of Dark Matter and Embroys of the Planets according to Schemidt.

(i) Dynamic dust and gas particles (Primordial matter) were scattered in the space. These particles were having their own angular momentum;

(ii) Dust particles were attracted towards the sun and they started to revolve around the Sun. The cloud of dust particles changed into form of flat disc revolving around the Sun;

(iii) Due to rotation of disc, redistribution and rearrangement of particles took place according to their size and density. Dense and big particles adjusted in middle and smaller and lighter ones shifted outward sides of the disc;

(iv) The orbits of particles intersected among themselves and due to colescence big particles became bigger in size and smaller one were extincted. The big particles became the asteroides; and

(v) Due to continuous accretion of small particles to the asteroids, their size grown up and changed into form of planets. Still some particles, were left which were not accreted in asteroids. These left particles were formed into satellites.

Criticism

This hypothesis seems to be a renewal of gaseous Nabula concept of Kant. The order of arrangement, and the variation of composition and density of

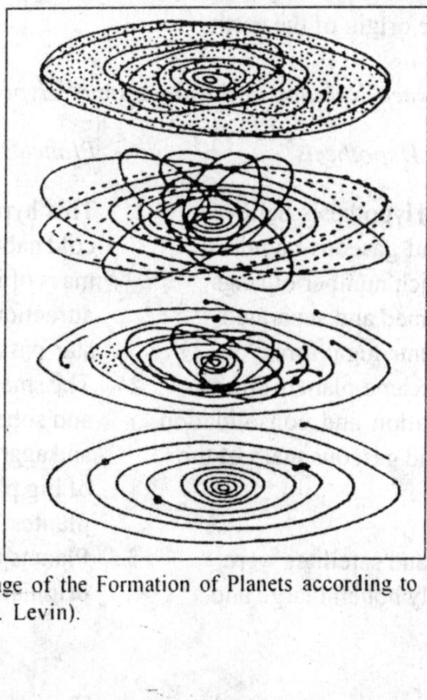

Fig. 2.5 : Final Stage of the Formation of Planets according to Schemidt (After B. Levin).

planets of solar system are well explained in this hypothesis. Lighter and smaller planets are found on either margin (inner and outer) of solar system which shows well conformity with arrangement of consolidated particles in the flat disc around the Sun (Fig. 2.5). The problem of undue distribution of amount of angular momentum of planets and the Sun is also solved because the planets were not from the Sun. Despite of above merits, the theory is also suffering from some problems as given below —

(i) This theory could not explain the source of the origin of dust clouds;and

(ii) It explains that the planets were formed by growing size of asteroids and meteorites but some scientists and astronomers believed that asteroids and meteorites were formed due to disintegration of the planets.

Comparative Study of Various Hypothesis of Origin of the Earth between Nabular and Planetesimal Hypothesis

Above discussed hypothesis are suffering from various kinds of demerits. Every successive theory has made an effort to remove the demerits of preceding hypothesis. There are some important fundamental points on which basis we can differentiate the various hypotheses from each others.

The following paras give an understanding of contrast of various theories of the origin of the earth.

Contrast between Nabular and Planetesimal Hypotheses

Nabular Hypothesis	*Planetesimal Hypothesis*
1. Nabular Hypothesis believed in hot and gaseous nabula from which number of rings were formed and separated due to centrifugal force.	1. This hypothesis believed in cold nabula (Sun) from which mass of it ejected out due to attraction pull of an intruding star passing through the Sun.
2. Rings became planets due to condensation and consolidation of hot and gaseous mass of the nabula.	2. The small planetesimals (cold and solid particles) accreted and aggregated with the nuclei of big planetesimals to form plantes.
3. Planets and satellites were originally hot and large and	3. Planets and satellites were originally cold and small and

gradually cooled and grew smaller.	gradually heated and became larger.
4. Outer most planet Pluto was formed first and others at later periods.	4. Planets and satellites were formed at the same time.
5. Earth always had an atmosphere.	5. Initially earth was without an atmosphere when she was small.

Contrast between Tidal and Planetesimal Theory

Tidal Hypothesis	*Planetesimal Theory*
1. It indicates that the molten hot mass of the sun was ejected by the gravitational pull of the intruding star.	1. It indicates that cold mass of the sun was ejected by the attractional pull of the intruding star.
2. Tidal Hypothesis indicates the formation of cigar shaped filament caused the formation of planet in a defined order. The size of planet is larger in the centre and smaller towards the ends.	2. Planetesimal hypothesis indicates that ejected cold mass of sun was scattered in not in any order and planets formed without any defined sequence in their size and arrangement.
3. According to this hypothesis the planets and satellites originally were hot and gradually cooled down.	3. According to this hypothesis the planets and satellites were originally cold and gradually heated with the collision of planetesimals around nuclei.

Contrast between Tidal and Dust Cloud Hypothesis

Tidal Hypothesis	*Dust Cloud Hypothesis*
1. Tidal hypothesis believed that the planets were formed by cooling of ejected hot materials from the Sun.	1. Dust cloud hypothesis believed that the formation of planets and satellites took place due to collision of a number of primordial matters scattered in the space.

2. Planets are formed from the Sun mass. That is why the distribution of angular momentum among planets and the Sun is not well explained.	2. Planets are not formed from the sun mass. That is why the undue variation in the amount of angular momentum among planets and the sun is justified.
3. Size and order of arrangement of planets in solar system is explained but it lacked about the density variation of the planets.	3. Size, density and order of arrangement of planets in solar system is well explained. Smaller and denser planets lie closer to the sun and big and light planets are lying farther from the Sun.

3

INTERIOR OF THE EARTH

The information and the knowledge about the internal constitution of the earth has been a mystery and matter of great contradiction. Only different kinds of guesses were made on the basis of whatever indirect evidences have been found.

Sources of Information about Interior of the Earth

We cannot peep down the internal parts of the earth by nacked eyes directly. It might be possible to get direct observation through the deep mines as well as oil drilling wells which were dug but their reaches are to very limited depth (not more than 6 km as compared to 6400 km of earth's radius). The deepest gold mine in South Africa and oil well are 4 and 6 km respectively. That is why indirect observation or sources are only ways through which we are able to have an idea about the internal composition and structure of the earth.

Indirect Sources

There are a number of indirect sources of data of the interior of the earth as given below —

Temperature

It has been observed through the mines that the temperature increases with the increasing depth beneath the earth surface. The average rate of temperature change is 1°C for 32 metres but this rate is not uniform at every depth upto 100 km depth, temperature increases at a rate of 12°C per km, between 100-400 km at 2°C per km and beyond 400 km at 1°C per km. Thus the such temperature gradient with the descent are indication of an inconcievably high temperature inside the earth. It is estimated that the actual temperature at the core of the earth would be 2000°C. On the basis of such high temperature, it is viewed by some scientists that the interior of the earth should be either in liquid or gaseous state because at 2000°C temperature rocks cannot remain in solid state.

Pressure

Along with rising temperature with depth the pressure of overlying rocks are also increasing with descent due to increasing weight of super incumbent rocks. Consequently, at the core of the earth the pressure should be very high. Thus the presence of forces of compression and expansion under the influence of simultaneously high pressure and temperature, the inference of any idea about condition of interior of the earth would be very difficult. However, some scientists argue that the increasing pressure inside the earth raises up to the melting points of different rocks and thus the core of the earth would remain in solid state despite of high temperature there.

Seismic Waves

Earthquake's waves are also source of knowledge about the interior of the earth as they travel inside earth through different passage. Most major earthquakes are caused by dislocation of rocks in the faults and different kinds of vibrations are originated from the focus (Place of origin of vibration). Of earthquake which lies in the inside of the crust. These vibrations are recorded by seismograph. These vibrations are of different kinds as given below (Fig. 3.1).

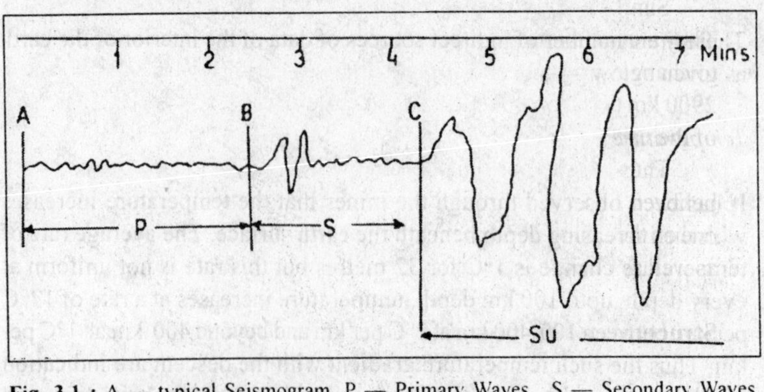

Fig. 3.1 : A — typical Seismogram, P — Primary Waves, S — Secondary Waves, Su — Surface Waves. The points A, B and C indicate the exact time of arrival of Primary Secondary and Surface Waves respectively.

1. *Primary or Longitudinal or P. Waves*

They are fastest earthquake waves in which the particles of matter move in the same direction of their propagation. They are also called as push waves and they can pass through all three mediums like solid, liquid and gaseous. They are like the sound waves.

2. *S or Secondary or Transverse or Shake Waves*

These earthquake waves are like the light waves in which motion of particles is at right angle to the direction of propagation. S-waves can travel through the solid medium only.

3. *L or Long or Surface Waves*

Such earthquake waves are slowest and marked by largest number of oscillations. They are confined only to surface region of crust and do not go into interior of the earth. Major destruction by earthquake is caused by this wave.

Moreover, these earthquake waves travel in the interior of the earth varying velocity and traverse variable courses at different depths. For instance, the velocity of P- and S-waves increases with the depth upto 2900 km. The velocity of P-waves at the depth of 100 km is about 8 km/sec and it increases upto about 13 km/sec at 2900 km depth. After the 2900 km, the velocity suddenly drops down to 8 km/sec and again increases to 13 km/sec in the inner core of the earth.

Similarly, the course or path traversed by these waves are also variable. They are not straightline but are subjected to refraction and reflection due to change in density and composition of interior of the earth. Beyond the 2900 km depth only P-waves have been recorded to pass through the core of the earth.

Thus on the basis of behaviour of these seismic waves, scientists believed that the earth planet is consisting of different layers overlying one on the other. These layers have been characterised by different Physical as well as Chemical composition.

Structure and Composition of Interior of the Earth

Above discussion regarding the indirect source of information of interior of the earth, clearly convinced the scientists, that the earth has layered structure like a onion with its shells one inside another. The density of the interior of the earth increases with the depth as indicated by concavity of P- and S-waves upward following the principle of optics. However, from the prohibition of the S-waves in the core, it is inferred that core is in liquid state. In this way, it is believed that the earth behaves as solid upto 2900 km with a core in liquid or plastic state. Thus, the earth from surface to the centre has been divided into following layers. (Fig. 3.2) :

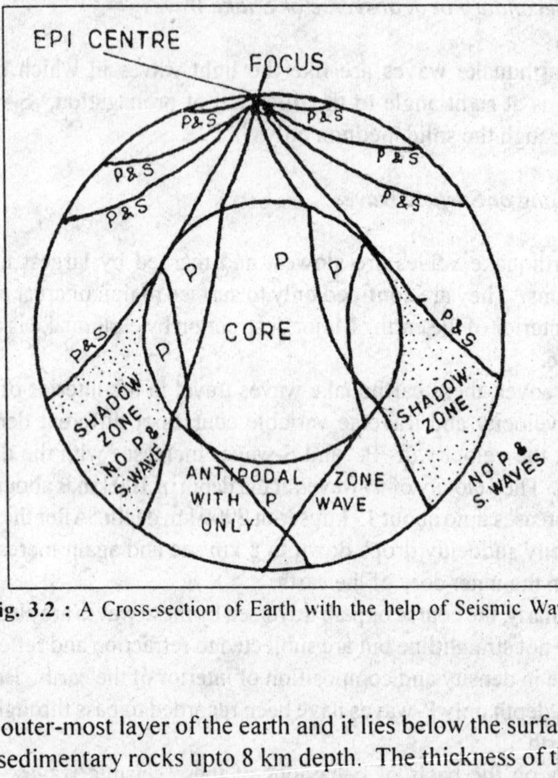

Fig. 3.2 : A Cross-section of Earth with the help of Seismic Waves.

Crust

Crust is outer-most layer of the earth and it lies below the surface capped by new sedimentary rocks upto 8 km depth. The thickness of the crust is variable between 45 to 100 km. It is greater below the continent and shallow beneath the oceans. Its density varies between 2.75 to 2.90. Chemically it is formed by mainly silica and aluminium and is also known as sial. On the basis of variation in physical and chemical characteristics of this layer from top to bottom, it may be divided as (i) outer crust and (ii) inner crust. Thus the outer crust, which is mainly under the continents, is believed to be formed of crystalline metamorphic and igneous rocks, acid in nature. According to Peter Horris, the major constituents of the outer crust are SiO^3 (Silica or quartz) 62 per cent, Al_2O_3 (Aluminium oxide)

16 per cent, iron oxides 6 per cent, Manganese oxide 3 per cent, CaO (Quiline) 6 per cent, Na_2O (Sodium oxide) 3 per cent, and K_2O (Potassium oxide) 3 per cent. Oxygen is the dominant element (47 per cent) in the crust followed by silicon. The thickness of outer crust is between 8 to 45 km.

The lower crust is believed to be underlaid by ocean and it is mainly composed of basaltic and ultrabasic igneous and metamorphic rocks like basalt and gabbro. This layer has heavier silicates, named as SIMA or silica and magnesium. It is outer silicate layer extending from 45 to 100 km depth inside the earth and the rocks are mostly in molten state.

Mantle

This layer of the earth lies between the overlying crust and underlying core. The extent of this zone is between 100 to 2900 km deep. The boundary which separates the crust from mantle is called MOHO or Mohovisic discontinuity at about 100 km depth and core is demarcated by Gutenberg or Oldham discontinuity at 2900 km depth. Mantle is formed by heavier materials like silica and magnesium, hence called as SIMA. Its density varies between 3.1 to 5.00. As a whole the Mantle holds 68 per cent of mass of the earth and 83 per cent of its volume. It has great importance in understanding various kinds of natural phenomena like plate tectonic, orogeny, continental drift etc. occur in the crust because the energy and forces involved in the process are confined to this zone.

The mantle is divisible and has been divided into upper and lower layers. The upper layer extends between 100 to 1700 km mainly under the ocean. It is an inner silicate layer composed of partly silica and aluminium and rocks are basic in nature. In this layer some rocks are of solid form and some of plastic nature or close to melting points. Its density varies between 3.1 to 4.75. While the lower mantle at 1700 to 2900 km depth is the transitional zone of mixed mental and silicates. It has wholly ultra basic rocks and its density lies between 4.75 to 5.00.

Core

The core is the innermost layer of the earth beneath the mantle. It extends from 2900 to 6400km. It is formed of heavier metallic material like iron and nickle. Hence it is also called as NiFe (Nickle and Ferrous). Its density is variable between 5 to 13. The core of the earth can also be divided into zone, i.e. outer metallic and inner metallic cores. Outer metallic has the thickness between 2900 to 4980 km around the core. This layer is either in liquid or plastic state, that is why S-wave cannot pass through the core while the inner core extends from 4980 km to the centre of the earth and formed of heavy metallic materials and it is in solid state. This is also known as Baryospher. However, the core as a whole contains the 32 per cent mass of the earth and 16 per cent of its volume and its density is double of mantle.

4

ORIGIN OF CONTINENTS AND OCEAN BASINS

After the cooling of the hot gaseous earth, the continents and the oceans came into being at first time on the earth surface. These land features are considered as first order land forms. The cause of their genesis is very much puzzled and complex. Numerous explanations and ideas have given by different scientists from time to time in the past. The present natures of earth surface, i.e. distribution of land and water and their appearance etc. have been much accounted in explaining any theory regarding the origin of continents and ocean basins.

Features of Earth Surface

The general outlook of present earth surface indicates the following important features:

(i) Land and water are unequally distributed on the earth surface. 29.80 per cent area of the earth is covered by land rather than 70.20 per cent by water;

(ii) Predomination of land in the northern hemisphere and of water in the southern hemisphere is the striking feature of the world;

(iii) Both the continents and oceans are assumed to have roughly triangular shape in their locations. Continents have their bases in the north and their apices in the south whereas the oceans show their bases in the south and apices in the north. Antarctica and Australia are having exception and they do not follow the rule of triangles. The north pole is surrounded by water, i.e. Arctic ocean. While the south pole is surrounded by land i.e. Antarctica continent. The Pacific is the largest ocean which covers about 2/3rd of ocean area of the world; and

(iv) The continents and oceans show antipodal arrangement. All the landmass, irrespective of any size, is diametrically opposite to an ocean. Patogonia and New Zealand are only two exceptions. The former is opposite to part of North China and the latter is opposite to Iberian Peninsula.

Keeping in view all these above discussed points regarding the earth, the following theories have been discussed to understand the origin of the continents and oceans on the globe.

(i) Tetrahedral theory
(ii) Wegner's Continental drift theory
(iii) Plate Tectonic theory

Tetrahedral Theory

In 1875 Lowthian Green propounded a theory to explain the origin of continents and oceans with the help of a geometrical design of tetrahedron. Tetrahedron is a shape which is formed by combination of four triangles. He developed his theory on the basis of following two assumptions.

(i) A sphere is that body which contains the largest volume with respect to surface area; and
(ii) A tetrahedron is that body which contains least volume with respect to surface area.

After making various experience, he assumed that if the forces from all directions are put on a spherical body, it would collapse and change into tetrahedral form. He came on conclusion after crushing a wrought iron tube through the experiment. This experience, he applied to the earth. He thought that various forces, used on the earth planet to contract it, would cause the collapse of spherical surface of the earth into tetrahedral form. As tetrahedron is the body formed by uniting four equilateral triangles in such a way that three are set on three edges of one triangle. The faces are planes and opposite to the coign on apex of the triangle. Thus after applying this idea to the earth, he tried to explain to antipodal arrangement of land and water surface.

Following derivations have been made in the light of the theory, regarding the earth surface (Fig. 4.1).

1. All the apices and coigns have represented to the continents (lands) and faces or planes to the oceans (Water) because if a

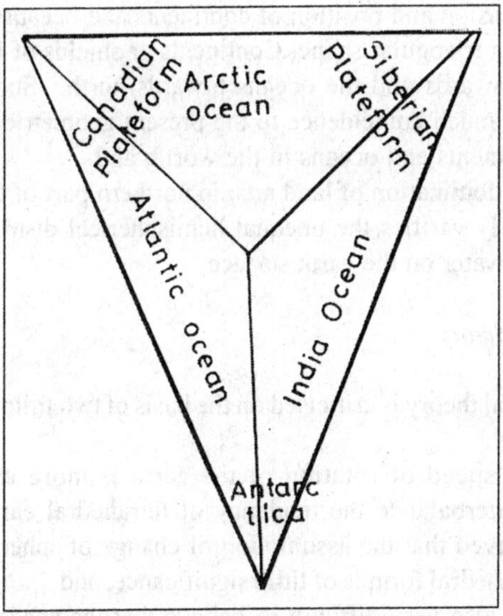

Fig. 4.1 : Distribution of Land and Water on Tetrahedron.

tetrahedron is drowned in water, the wedge or coign will be out of water and faces under the water;

2. The three important shields Laurentian, Siberian and Baltic are represented by northern edges of tetrahedron while the Antarctic coinsides with the apex in the south. The topmost face of it assumed as Arctic whereas other three faces are represented by Pacific, Atlantic and Indian Ocean.

Merits of Theory

The Tetrahedron theory of Lowthian Green, now-a-days, is having very few supporters but it will not be irrelevent to consider its main implication as it serves to focus attention on some important points in the earth's surface structure. The following facts support the validity of the theory.

(i) The arrangement of lands on coigns and waters on the faces of tetrahedron is well corresponding into the antipodal arrangement of land and water on the earth surface. Americas lie opposite to Indian ocean, Asia and Australia are on opposite of Atlantic Ocean, and Europe and Africa are antipodal to Pacific ocean;

(ii) Extension and position of continents and oceans in tetrahedron are in triangular shape. Continents or shields of landmass taper southwards and the oceans towards north. Such features are very much coincidence to the present geometrical shape of the continents and oceans in the world; and

(iii) The domination of land area in northern part of the tetrahedron clearly varifies the unequal hemispherical distribution of land and water on the earth surface.

Demerits of Theory

The tetrahedral theory is criticised on the basis of two following demerits:

(i) The speed of rotation of the earth is more enough than to counterbalance the tendency of tetrahedral earth. Thus it is believed that the assumption of change of spherical globe into tetrahedral form is of little significance; and

(ii) It is also very difficult to believe that the spherical earth with complex structure would have been contracted into tetrahedral form.

Wegner's Continental Drift Theory

Alfred Wegner (1912) propounded theory in which he tried to explain the present distribution of continents and oceans which came out as a result of drifting of continental masses into different parts, in different directions and in different periods of geological time. He was basically meteorologist and developed this hypothesis on the basis of study of Paleo-climatogists, paleobotanists, and Paleo-physists regarding the evolution of the earth.

Basis of the Theory

As being meteorologists, he found that the climate varied all over the world. The hypothesis grew out of need of explaining the major variation of climate in the past. There is clear geological evidences of climates resembling those of equatorial, tropical and arctic regions in latitudes which were removed from land where these climates prevail today.

But geological study and remnants of particular climate are found in different parts of the globe in present contrasting climatic condition. For example the remarks of sheet glaciation of carboniferous period in Africa, South India and Australia, presence of coal seams in Antarctica etc. motivated Wegner to think whether climate has changed in geological past or the continents have drifted apart in different directions in different times.

Beside, the geological, poleo-botanical and paleophysical studies have also helped in developing this theory. Resemblence in rock structure, composition and formation along the Atlantic coasts are very important basis for hypothesising the theory. Similarity in fauna and flora. Jig Saw Pit of American coasts with the Atlantic coasts of Africa and Europe, East African coast with West Indian coast and south coast of Pakistan and Australia with Bay of Bengal etc. are another basis of this theory. Curve of mid-Atlantic ridge is parallel to the curve of American coasts.

Concepts of Theory

Alfred Wegner started his theory following the Suess' classification of interior of the earth. He believed that Sial is the outer part of the earth and Sima is lower layer which underlies sial. He thought that continents are formed of sial, relatively lighter material and the ocean is formed of Sima, rather denser of basalts. He assumed that continents (Sial) are floating in Sima. The main component of this theory are as given below —

(i) All the present continents were joined together into one super continent during paleozoic geological era and it was surrounded by deep ocean. Wegner called this supercontinent as Pangea and the ocean as Panthalasa (Fig. 4.2);

(ii) During the corboniferous period (200 million years ago) the Pangea was broken up into two land masses and moved north and south direction. The northern block is called as Laurentia or Angara land, consisted of Asia, Europe and North America. The second landmass was Gondwana land consisting of Australia, Africa, India, Antarctica and South America. Between these two blocks of landmass, depression called as Tythes, existed. At present Mediterrarean sea is the relict of the Tythes sea;

(iii) During Mesosoic times, westward movement of continental masses took place under influence of Tidal force of the sun and the moon. Thus North America from Angaraland and South America from Gondwanaland drifted westwards and Atlantic ocean was formed between Americas and Europe and Africa

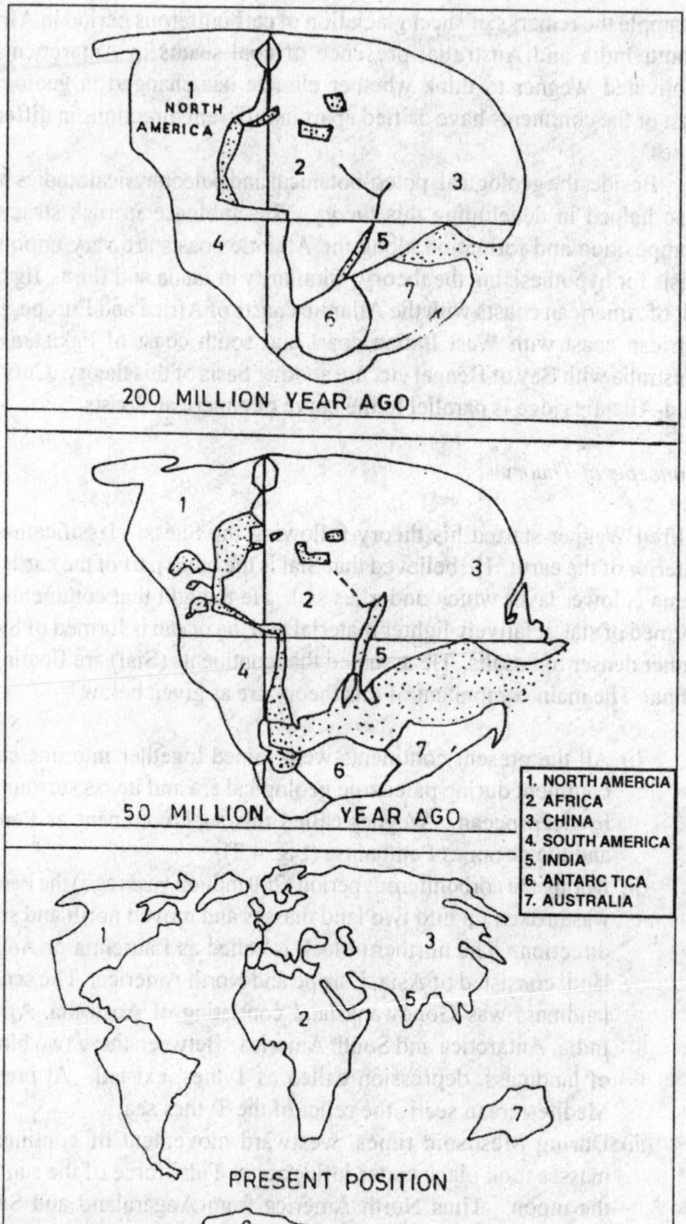

Fig. 4.2 : Phases of Continental Drift by A. Wegner.

continents;

(iv) During the tertiary times, India, Australia and Madagascar moved away northward and separated from the Antarctica continent. Indian ocean came into being after northward drifting of penisular India;

(v) Wegner has also presented the mechanism of mountain building in his theory. Alpine Range of Mountain between Europe and Africa and cordillera of North and South America along the west coast have been formed due to crumpling of the edges of drifting continental masses by resistence exerted by rocks of the sea floors;

(vi) The drifting of landmass has been the cause of formation of a number of island arcs and festoon in Pacific ocean and Atlantic ocean. At the time of westward movement of Asia and Americas, the eastern marginal edges could not keep pace and they are left behind in the form of Sakhalin, Japan, Kurile and East Indies in Pacific ocean and West Indies in Atlantic ocean respectively; and

(vii) Wegner, in colloboration with Koppen, traced the successive positions of equator and poles from corboniferous times onward with reference to the distribution of glacial, dry and humid condition. For example the relic of carboniferous forest in the form of coal beds in middle latitudes in Northern hemisphere has been used to demarcate the equational belt of corboniferous time in this area. According to him, the North pole in Silurian period was at lati. 14°N and long. 124°W, in carboniferous period at lat. 16°N long. 147°W, in tertiary period at lat. 51°N long. 153°N. Similarly the equator and South poles moved according to positions of North poles.

Force of Continental Drift

Wegner, in his theory of continental drift assumed two kinds, i.e. (i) Equatorward, and (ii) Westward movement of continental masses. For these different directional movement, differntial gravitational force and tidal force of the sun and the moon were supposed to be responsible (Fig. 4.3).

Facts Favour to the Theory

Though the continental drift theory has been criticised on many grounds, there are many evidences and facts which support the theory, still, in modern days. The evidences are as given below.

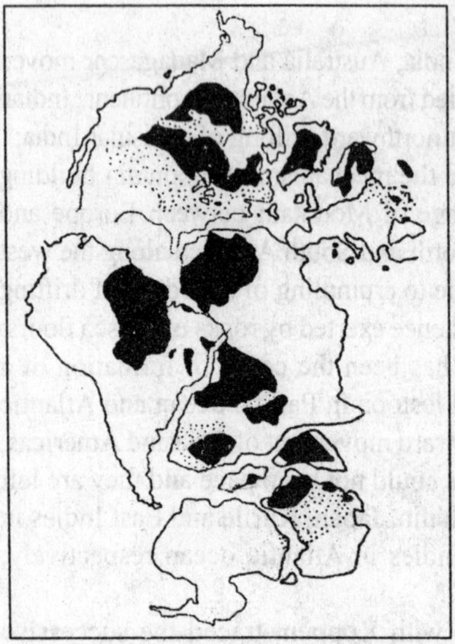

Fig. 4.3 : Reassembled continents, prior to start of continental drift. Dark pattern : areas of oldest shield rocks (older than-1.7 b.y.), Light pattern : rocks in the age range - 0.8 to -1.7 b.y. (After P.M. Hurley and J. R. Rand, 1969, *Science,* Vol. 164, p. 1237, Figure 8.)

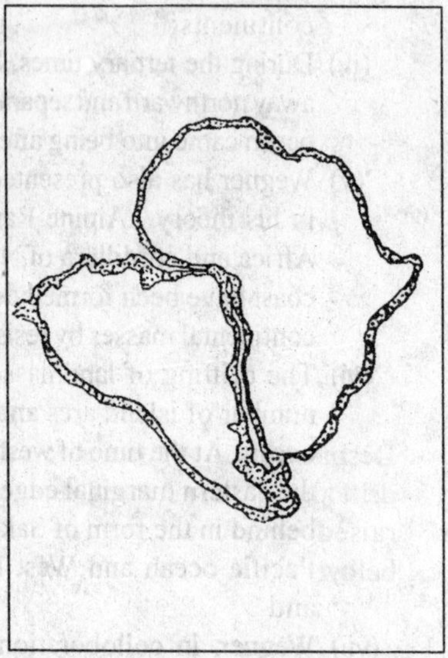

Fig. 4.4 : Jig Saw Fitting Juxtaposition of South America and Africa

(i) The most important feature which support the theory is the 'Jig Saw Fit' (Fig. 4.4). If all the present landmasses are brought together, they would be approximately, represented as Pangea. For example the eastern coast of Atlantic can be refitted with the western coast. Australia can be refitted in Bay of Bengal;

(ii) The similarity of structure, lithology, fauna and flora and minerals along the both coast of South Atlantic ocean are important testimonials to the drift theory of Wegner;

(iii) The marks of ice sheet spreading and glaciation in Africa. Indian peninsula, Australia etc. in present tropical region indicate that they were drifted from landmass which was under glaciation during corboniferous period;

(iv) Geodetic survey has shown that the Greenland is drifting westward at a rate 6 cm per annum. This fact also supports the theory of Wegner;

(v) The Caledonian and Hercynian mountain folds of eastern coast of North America and Western Coast of Europe are the same range extending across the Atlantic ocean. Such continuation of mountains of same period revealed that once both sides were united; and

(vi) The presence of glassoporis flora in South India, Australia, South Africa, Falkland and Antarctica is indicating that all these lands were once united.

Criticism

Despite of many geological and geographical supports to the continental drift theory of Wegner, a number of objections against it have been raised by scientists. The objections against this theory are given below—

(i) The theory is not capable to explain the force responsible for such kind of drifting of continental mass. Wegner has assumed differential gravitational and tidal force caused the displacement. The tidal force invoked by Wegner to account for supposed westerly drift of the continent would need to be 10,000 million times as powerful as it is today to produce required effect. If it would have such value, the rotation of the earth should have been stopped in one year. Similarly the differential gravitational force, which was contended by Wegner for equatorward drift, is not adequate for such effect. It could not produce appreciable movement in such Viscous substratum;

(ii) The Wegner's continental drift theory has been criticized on the ground of the date of occurrence. He did not explain about any drift before paleozoic time. He did not give any explanation how did pangea remain together till the hour of drifting began and why did not this process start till that date;

(iii) The direction of drifting of landmass, as assumed by Wegner, has been also questionable. F.B. Taylor deduced an outward radial movement from two poles. According to him a ring of mountains are found along the outer edge of drifted blocks. The Alpine Himalayan chain, island arcs of East Asia, the western cordillera of North America and the range of Handuras and West Indies are forming the ring in northern hemisphere.

J.W. Evans has shown the evidences of eastward and southward, westward from America towards the centre of Pacific; and

(iv) This theory also presents the contradictory views about the Sima. Initially Wegner assumed that continents are floating on the Sima freely but later on he contended that mountains have been formed by crampling of edges of continents due to resistance caused by Sima on the drifted blocks.

Plate Tectonic Theory

Plate tectonic theory is a new concept in geology to explain tectonic movements like earthquakes, volcanic activities, formation of mountain, sea arcs, sea floor spreading etc. on the earth. It is based on the assumption that the plates are formed of crust and some part of mantle. They have the thickness between 50 to 150 km. They are of both oceanic and continental crust. Plates are floating in underlying hot liquid low viscous material extending from the base of lithosphere up to 700 kms depth. There are six major plates all over the world (Fig. 4.5). The zone, in which plates are floating, is called as asthenosphere.

Movement of the Plates

All the plates are not in stationary positions but they move into different directions at different rates. Their rate of motion varies between 2 to 20 cm per year. There are three ways in which plate's motion takes place. First, convergent in which the plates come closer from different directions. The second is divergent where plates move apart while the third is parallel in which plates move parallel to one another. The nature of plate motion is associated with particular types of tectonic phenomena.

Plate Boundaries and Earth Movement

There are three types of plate's boundaries on the basis of direction of their movement with relation to each other. They are : (i) convergent plate boundary , (ii) divergent plate boundary, and (iii) transform plate boundary. Along these boundaries various kinds of tectonic features are formed. They are discussed in the following lines (Fig. 4.6).

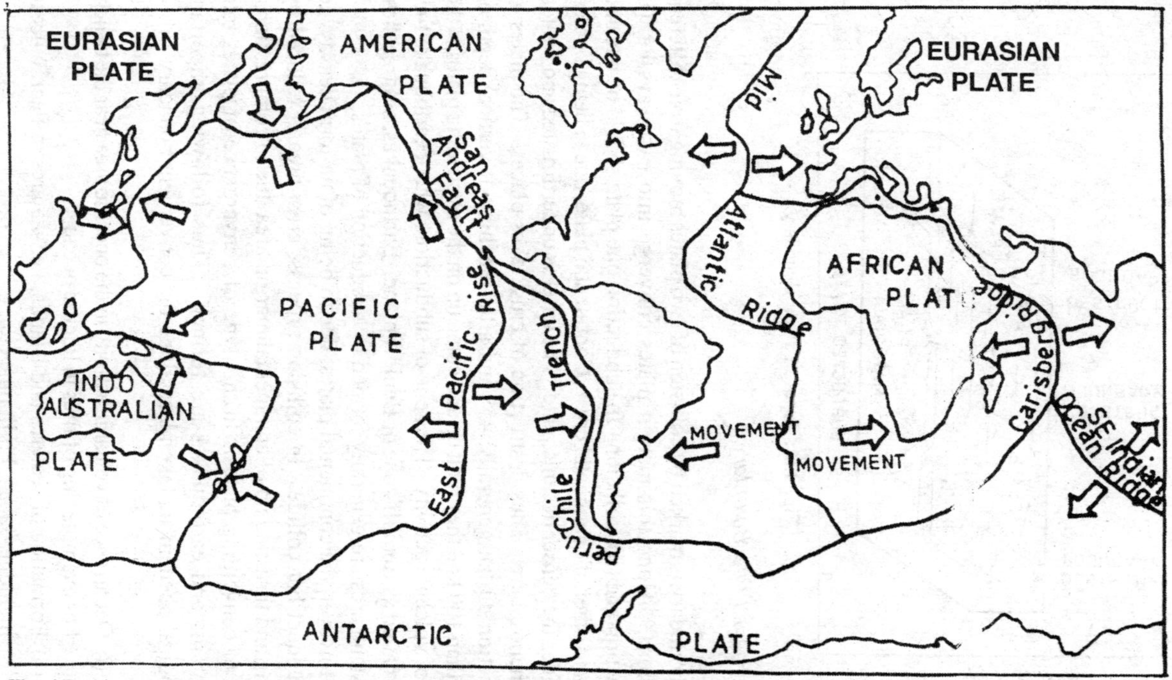

Fig. 4.5 : Six principal plates were proposed by Le Pichon in 1984. The plates are growing along spreading ridges indicated by arrows moving in opposite directions that is diverging plate material is being consumed in trenches, where the paired arrows are converging.

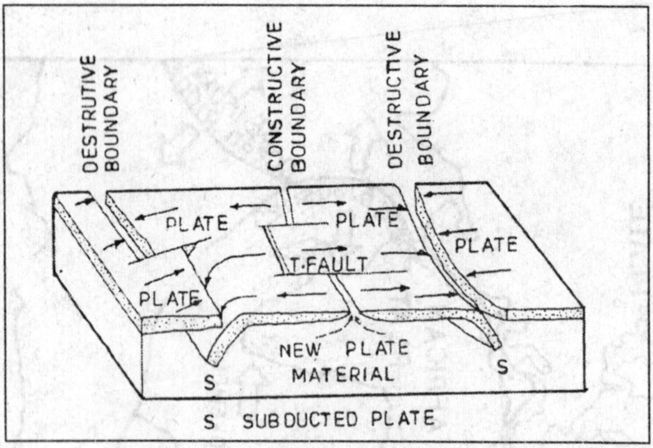

Fig. 4.6 : Types of Plate Boundary.

Convergent Plate Boundary

Such boundary is marked by collision of two plates moving from different directions. Along this margin plates converge into two ways i.e. (i) subduction and (ii) collision. In subduction one plate should be oceanic crust and other of continental crust. Continental plate, due to being lighter, overrides the denser oceanic plate. The latter is underthrusted and melted in asthenosphere. Thus destruction of crust takes places. The plates of subduction dip at, generally, 45° and after reaching at vertical depth of more than 100 km, they melt up and come up on the overriding plates in the form of volcanic activity. The zone of subduction, where earthquake foci lie, is called as Benioff zone or Benioff plane. Earthquakes, sea trenches, and island arcs are generally formed by subduction of plates.

However, the continental plates, due to being of uniform density, do not subduct but collide. In collision process, crust area is reduced by folding and thrusting up of continental margin area due to compressional force of converging plates. In this type of convergence of plates fold mountains and earthquakes are formed. Thus following important geological phenomena take place along the convergent boundary.

(i) Occurrence of volcanic activity in the oceanic area near the trench. For example ring of fires in Pacific ocean;
(ii) Formation of sea arcs and islands, i.e. sea arcs in Pacific ocean including Japan, Philippine etc.;
(iii) Earthquakes occurence; and
(iv) Formation of fold mountain and sea trenches.

Divergent Boundary

The plate's divergent boundary is one in which two plates are moving apart in opposite direction from a common line. Due to moving away of the plates, a rift is formed along this boundary or margin. The width of rift or fracture is widening as long as plates move. Through this passage of fracture, molten materials are upwelling and deposited both sides of the fracture. Such margins occurs usually in ocean and along the mid-oceanic ridge. Along the mid-oceanic ridge especially in Atlantic ocean, due to deposition of new ejected molten materials, new oceanic floor is created. The creation of new oceanic floor is described as ocean floor spreading. Hence the divergent margin is also called as construction margin as new crust is formed. The following events are associated along this margin.

Table 4.1 : Summary of Plate Tectonic Movement and Tectonic Phenomena on the Earth's Surface.

Margins	Boundaries with reference to the nature of plates		
	Ocean-ocean plate	Ocean-continental plate	Continent-Continent plate
Convergent	1. Trench Formation	1. Subduction	1. Formation of fold mountains
	2. Island Formation	2. Volcanic Eruption	2. Earthquakes
	3. Earthquakes	3. Formation of Trench	
		4. Earthquakes	
		5. Young mountains	
Divergent	1. Fissure eruption	—	1. Formation of Rift valley
	2. Ocean floor spreading, Ridge Formation	—	2. Inland sea
	3. Earthquakes	—	3. Volcanism
			4. Earthquakes
Conservative (Parallel)	1. Transform fault	—	1. Transform fault
	2. Ridge and valley fracture		2. Earthquakes
	3. Earthquakes		3. No volcanic activity

(i) Formation of rift or fracture;
(ii) Upwelling of molten materials from the mantle;
(iii) Formation of ocean floor spreading;
(iv) Occurrence of fissure type volcanic eruption.

Transform Margin or Conservative Margin

All the plates from two sides (both converging and diverging plates) move parallel to each other and slide along common margin. Such margin is called as Transform margin because transform fault is formed. Plates, in this process, may join ridge to ridge, a ridge to trench or a trench to trench. A ridge-trench transform will always change length with time. A trench to trench transform may either lengthen, shrink or remain constant depending on which of plates, that form the subduction system, is the downgoing plate and the transform which join two ridge axes will change in length with time. The following events are taking place along the transform plate margin.

(i) *Formation of Transform Faults* : Along the edge of parallel movement of plates (Conservative boundary) a fault is formed in which the blocks of rock move or displace horizontally along the fault plane at very low dip. It is also called as strike slip fault. In such fault, usually scarp does not occur.
(ii) *Earthquakes* : When two plates slide along the edge of each other, a series of vibrations are set inside the crust. Due to this type of vibration, usually, a shock is felt on the earth surface in the form of earthquakes.

5

MATERIALS OF THE EARTH CRUST

The cooling and solidification of the hot and molten earth formed a solid thin layer around the earth. This outer layer is commonly known as earth's crust or sial. The earth crust is consisting of about 98 chemical elements. Out of them, eight are dominant and account for 98.59 per cent of the earth crust. These elements are oxygen, silicon, alumunium, iron, calcium, sodium, potassium and magnesium in descending order of their proportions.

However, these elements, in the crust, do not occur in their elemental form but mix with each other to form compound and substances with a definite chemical and physical characteristics. These substances are called as minerals. Thus minerals may be defined as the natural abiotic substance consisting of one or more elements bearing with definite chemical and physical properties. These minerals combined together in the crust and formed a new substance which is called as rock. There are about 2000 minerals involved in the formation of rocks of the crust. But the most of the rocks are formed of six important minerals like feldspar, quartz, pyroxenes, amphiboles, mica and olivine. In brief rocks may be defined as an aggregate of number of minerals found in the earth crust. It is of both biotic and abiotic nature.

Contrast between Rocks and Minerals

Rocks	*Minerals*
1. Rock is an aggregate of minerals.	1. Mineral is an aggregate of elements.
2. Rock is of both inorganic and organic nature.	2. Mineral is an inorganic natural substance.
3. Rock is not having fixed chemical formula.	3. Mineral has its fixed chemical formula.

Classification of the Rocks

Moreover the study of rock in geology is a separate branch which is called as petrology. Rocks are of various nature and characteristics. Some may be soft like shale or hard like granite. Some may be acid or basic and some may be organic or inorganic origin. They also vary extensively in their mode of formation. Thus rocks may be classified into different groups based on different criteria. Most accepted and well recognised classification of rocks is based on the mode of formation. On the basis of mode of formation, rocks may be classified into following three groups.

(i) Igneous Rocks
(ii) Sedimentary Rocks
(iii) Metamorphic Rocks.

Igneous Rocks

The word igneous is derived from the Latin word 'ignis' which means fire. Thus the rock which is formed by cooling and solidifying of the molten materials is called as igneous rock. It is also known as primary rock, because the first rock formed by cooling of hot earth surface. Granite, basalt and gabbro are typical examples. Due to cooling process of hot magma or lava, a number of crystals are formed in this rock. Igneous rock may be grouped into different categories on the basis of their distinct characters.

Classification of Igneous Rocks

Igneous rocks have been divided on the basis of different criteria like chemical composition, texture, place of occurrence and crystal size. But the most common classification is based on the place of their location or occurrence. There are mainly two types of igneous rocks as given in (Fig. 5.1).

(i) Extrusive igneous rock.
(ii) Intrusive igneous rock.

Extrusive Igneous Rocks : The rock formed by cooling of lava on the earth surface, is described as extrusive igneous rock. In the formation of such

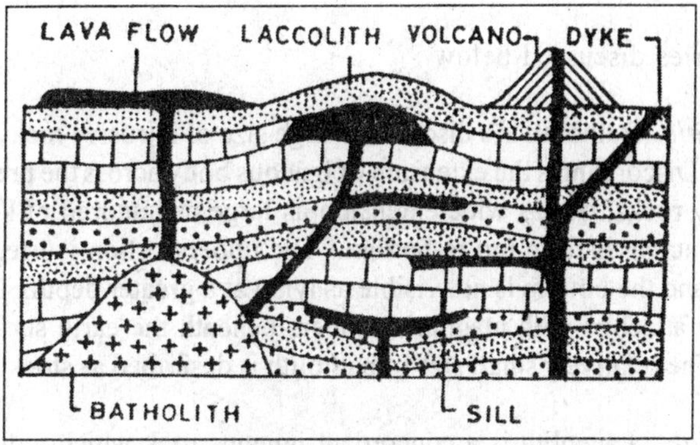

Fig. 5.1 : Various types of Igneous Rocks.

rocks, the rate of cooling of lava is rather faster due to being exposed to atmosphere. The fast rate of cooling results in formation of fine crystals. Thus the extrusive rocks show glossy appearance. Basalt is the typical example of such rocks.

Intrusive Igneous Rocks : Intrusive Igneous rocks are formed by cooling of magma below the earth surface at various depths. The rate of cooling of magma is rather very slow. Due to slow rate of cooling, the crystals of rocks are of coarse grained, i.e. granites. The intrusive rocks may occure at various depths and form a distinct characteristics in their texture, composition and crystal sizes. On the basis of depth of occurrence this rock may be further subdivided as (i) hypabyssal and (ii) plutonic igneous rock. Hypabyssal intrusive rocks are found at shallow depth. The crystals are coarse grained. Gabbro is the most suitable example of such rocks. While the plutonic rocks occur at greater depth of the earth. The name plutonic is derived from Pluto, the god of under world. The crystals of the rock is rather more coarser than the hypabyssal one due to very slow rate of cooling of magma at the greater depth.

Igneous Bodies

The intrusive igneous rocks are found in different shapes and sizes below the earth. Their nature, location and forms are determined by the intrusion of magma across the overlying rocks beneath the earth surface. The intruded magma cooled down and formed different igneous rock mass of different dimensions beneath the surface. Such rock bodies are known as igneous bodies. They may be of different categories discussed below :

Batholith : Batholith is a discordant large size of intrusive rock at greater depth. Discordant is the extension of igneous body across the pre-existing country rocks. It has wide extension as large as hundred of kms. The width increases from top to bottom. The upper portion is irregular and rough and the bottom is not visible as lying at a greater depth. Its base is usually attached with magma chambers beneath the earth surface (Fig. 5.1). The relatively small size of batholith is described as stock.

Laccolith : Laccolith is a concordant igneous rock which is found in a chamber formed between two layers of the rocks beneath the earth surface. It is a dome shaped body. The upper surface is circular and base is uniform. This igneous body is formed due to cooling of intruded magma in an anticlinal chamber of rocks. This anticlinal chamber is formed due to upward movement of magma. Such bodies are usually occurring at lower depth from the earth surface (Fig. 5.1).

Lopolith : Lopolith is concordant igneous body found between two layers of the rocks. Its shape is just reverse of laccolith. When magma is accumulated in synclinal depression formed between two horizontal layers at the time of unsuccessful volcanic erruption (Fig. 5.2).

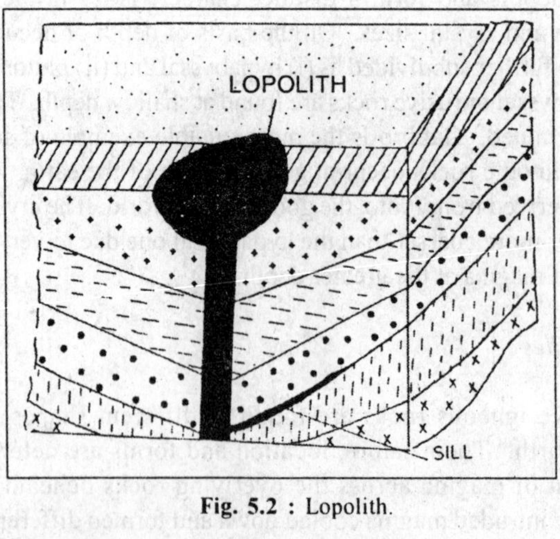

Fig. 5.2 : Lopolith.

Phacolith : It is a thin crescent shaped of concordant igneous body occurring along the syncline and anticline of folded rocks beneath the earth surface (Fig. 5.3).

Dyke : Dyke is a discordant igneous bodies lying with vertical or nearly vertical attitude, across the country rocks like a hard pillar. Generally the volconic vents convert into dyke after the closing of eruption and cooling and solidifying of magma within the vent.

Sill : Sills are the tubular concordant igneous bodies found parallel to the structure of country rocks. They are formed when intruded magma flowed into space between two beds of the rocks and became solid as it cooled down (Fig. 5.4).

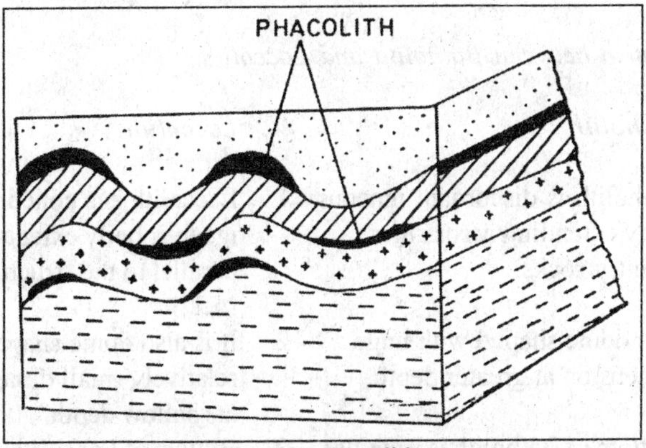

Fig. 5.3 : Phacolith.

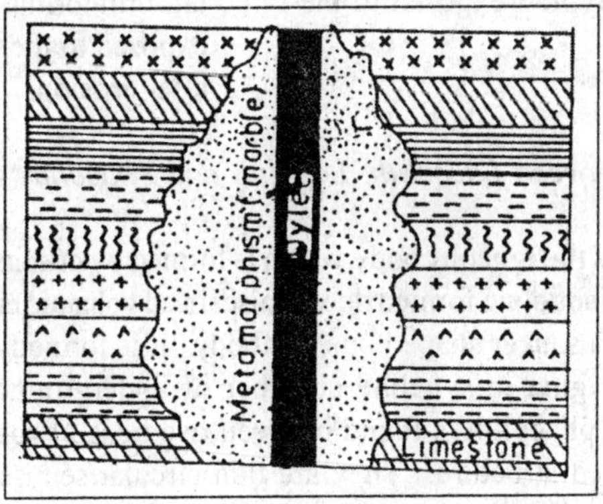

Fig. 5.4 : Illustration of contact or Thermal Metamorphism.

Characteristics of Igneous Rocks

Igneous rocks have various kinds of peculiarities on which basis they can be differentiated from the other rocks. The following characteristics of igneous rocks are given below :

(i) They are crystalline rocks. Various sizes of crystals occur in different rocks due to differential rate of cooling of the magma / lava;

(ii) They are non-fossiliferous rocks. Fossils are almost nil, if they were availiable in any case, might burnt out due to excessive heat; and

(iii) They are mineraliferous rocks. Different kind of minerals are concentrated in the rocks.

A Comparative Study of Igneous Bodies

Distinguish between Batholith and Laccolith

Batholith	*Laccolith*
1. Batholith is discordant igneous body extending across the country rocks.	1. Laccolith is a concordant igneous body extending parallel to the structure of the rock.
2. It is dome shaped with huge dimension at greater depth.	2. It is also dome shaped but of relatively small dimension at shallow depth.
3. Bottom of Batholith is irregular and usually non-exposed to the surface.	3. Bottom of Laccolith is of very uniform formation and exposed to the surface after erosion.

Distinguish between Laccolith, Lopolith and Phacolith

The laccolith is the igneous body which is formed by cooling of magma in the chamber of anticline formed due to miscarried volcanic eruption. While the lopolith is a saucer shaped igneous body. It is formed due to cooling of intruded magma in syncline chamber of underground folded rock. Moreover, the phacolith is found between two beds of rock of crest and trough of folded structure. They are thin circular/semi-circular bodies without uniform base.

Sedimentary Rock

When the igneous rocks or any other rocks expose to the atmosphere, they are subjected to disintegration and decomposition of weathering process. The wheathered materials are removed by different agents of erosion, like running water, glacier and wind, from their original place to another place. The carried or transported materials are of different sizes and collectively called as sediments. Thus the sediments are deposited at suitable place. With the passage of time due to increasing pressure of sediments, the sediments are converted into a compact and hard rock. Such rock, which is formed by deposition of sediments at suitable place by agents of erosion, is called as sedimentary rock.

Process of Formation of Sedimentary Rocks

The sedimentary rocks are not formed in one day or two days but it undergoes different stages before complete formation of rock. The following process are involved in the formation of sedimentary rocks.

(i) Sedimentation : The small particles of broken exposed rocks are carried away by rivers, glaciers and winds. They deposit the carried materials at suitable places. For example river deposits the sediments enroute, along the foothill, in flood plain and ultimately in the sea or lake where it discharge its water. Glacier deposits the materials at the place where it melts and reduces its carrying capacity to considerable extent. Thus the process of settling down of sediments is called as sedimentation.

(ii) Compaction : The thickness of sediment increases due to repetition of deposition of sediments of every year. The increasing thickness and pressure of overlying sediments result in compaction of underlying sediments. With the passage of time different layers are marked between the new and old sediments. This discussed process is called as compaction.

(iii) Cementation : Some cementing stuff like silicon, calcium carbonate and oxide of iron are dissolved in the water. These materials abide the loose particles into one and the compact sediments became very hard. Such process of binding the particles of sediments is called as cementation.

(iv) Stratification : After the completion of above discussed process, many thin or thick layers of sediments at various time are formed. The thin layer is called as laminae and thick layer is known as strata. Thus the formation either laminae or strata is described as stratification.

Classification of Sedimentary Rocks

The sedimentary rocks may be classified into different forms on the basis of following criteria :

(i) Mode of formation
(ii) Place of formation.

Type of Sedimentary Rocks on the Basis of Mode of Formation

There are mainly three types of sedimentary rocks on the basis of mode of formation.

(i) *Mechanically Formed Sedimentary Rocks*

When the sedimentary rocks are formed by deposition of sediments produced by breaking of exposed rocks by agents of erosion. The sediments are categorised, on the basis of diameter of the particles, into different groups like clay, silt, sand, gravel etc. according to increasing diameters of the grain size. Thus the following rocks are the mechanically formed sedimentary rocks.

(1) *Sand stone :* The sedimentary rocks which are formed by sand particles.
(2) *Clay :* The sedimentary rocks which are formed by clay particles.
(3) *Conglomerate:* The rocks which are formed by cementing of gravel size particles.
(4) *Shale :* The rocks which are formed by deposition of silts.

Organically Formed Sedimentary Rocks : Such sedimentary rocks are formed by deposit of sediments produced by decomposition of organic matters, like skeleton of animals and remains of plants in the water. There are three types of organically formed sedimentary rocks (i) Calcarious rocks (ii) Silicious rocks, and (iii) Carbonaceous rocks.

Calcarious Sedimentary Rocks : When the sedimentary rocks are formed by deposition of calcium corbonate derived from the skeleton and remains of animals and plants respectively. For example the skeletons of coral polyps produce calcium corbonate. Limestone is the typical example of calcarious sedimentary rock.

Silicious Sedimentary Rocks : Animals like sponges and rediolaria leave behind them complex skeleton of silica which formed the silicious sedimentary rock.

Corbonaceous Rocks : The decaying plants leave behind carbon and other elements due to their decomposition under the pressure of sediments. Thus after the accumulation of carbon content, a corbonaceous sedimentary rock is formed like coal and peat.

(ii) *Chemically Formed Sedimentary Rocks*

Chemically formed sedimentary rocks are formed by deposition of chemical elements precipitated from the solution. Salt and gypsum are the best example of such rocks.

Type of Sedimentary Rocks on the Basis of Place of Formation and Agents of Transportation of Sediments :

On the basis of place and agents of tormation of sedimentary rocks, it can be classified into various forms.

(i) Argillaceous Sedimentary rocks.
(ii) Aeolian Sedimentary rocks.
(iii) Glacial Sedimentary rocks.

Argillaceous sedimentary rocks are such which are formed in the water. They may be terrestrial origin and the sediments may be brought by river, sea waves or decomposition of marine organism in the water. For example limestone, shale etc. They are also known as lacustrine sedimentary rock when formed in the lakes, riverine when formed in the river valley route and marine sedimentary rock when found in the sea water.

Aeolian sedimentary rock is one which is formed by deposition of sediments brought by winds. Loess is very suitable example of aeolian sedimentary rock.

While the *glacial sedimentary rocks* are formed by deposition of sediments brought by glaciers. Till plain is typical example of glacial sedimentary rocks.

Characteristics of Sedimentary Rocks

Following characteristics are associated with Sedimentary Rocks.

(i) They are fossiliferous rocks. Remains of animals and plants, in some cases, are embedded in these rocks.

(ii) They are non-crystalline rocks.

(iii) They are stratified rocks. Different stratas are developed by deposition of sediments year after year.

(iv) Minerals are rarely found in such rocks. Coal and Petroleum are solely concentrated in these rocks.

Metamorphic Rocks

Sedimentary and igneous rocks may be altered or metamorphosed by a variety of processes. Metamorphism means, literally, a change of form. There are mainly three processes by which rocks are metamorphosed. First process called as thermal in which when country rocks come into contact with ejected magma, the country rocks are melted with excessive heat. Melted rocks cooled, solidified and recrystallined and changed in their physical and chemical forms. The second process is related with the pressure of overlying rocks. Due to mountain building movement rocks are folded down to a considerable depth. Thus pressure of overlying rocks causes change in chemical and physical properties and a new rock is formed. Third is the chemical solution process in which different chemical substances (gases) dissolved in underground water reacted with country rocks beneath the earth surface. Due to certain reactions chemical and physical form of certain rocks are changed. Thus the metamorphic rock may be defined as one which has been formed by alteration of chemical and physical characteristics under the influence of high pressure, high temperature and chemical solution.

Kinds of Metamorphic Rocks

There are mainly two kinds of metamorphic rocks on the basis of the factors and area involved in metamorphism of the rocks.

(i) Contact or Thermal Metamorphic Rocks : At the time of volcanic eruption through the vent of volcanoes a super heated magma is ejected out. The rocks which are in contact with this passage of magma, they are melted. Again they may be cooled down, solidified and recrystallized. Thus the metamorphic rocks which are formed under such process is called contact or thermal metamorphic rocks. For example limestone and shale are converted into marbles. The area of this process is very limited to few km distance from the contact point (Fig. 5.4).

(ii) Regional or Dynamic Metamorphic Rocks : Such rocks are formed due to dynamic nature of different rocks. At the time of earth movement the sedimentary rocks are folded up and down due to compressional forces. The down folded rocks move down to a such depth where temprature is very high. In addition to, the increasing pressure of overlying rocks causes the melting, regrouping and recrystallization of the rocks. Such rocks are formed in an extensive region. That is why, this rock is called as Dynamic or Regional metamorphic rocks (Fig. 5.5). for example shale is converted into schist, coal is converted into anthracite and anthracite into graphite.

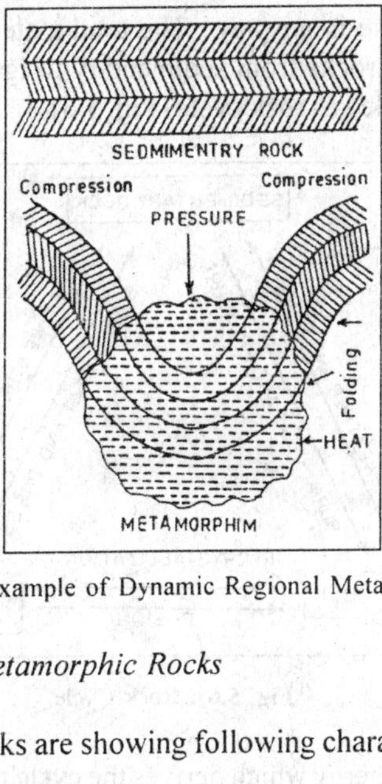

Fig. 5.5 : Example of Dynamic Regional Metamorphism.

Characteristics of Metamorphic Rocks

The Metamorphic rocks are showing following characteristics :

(i) They are having foliation in which the mineral particles are aligned in a strip. The foliation resembles the stratification but often in wave line.

(ii) Cleavages are also common in this rock. They resemble the foliation but are independent of direction of stratification.

(iii) They are crystalline rocks. A number of minerals are concentrated in them. These rocks are known as storehouse of minerals.

(iv) They are non-fossiliferous rocks.

Rock Cycle

The primary rock (igneous rock) has been weathered and eroded by agents of erosion. The eroded materials are deposited in sea, lakes or some other places and consequently sedimentary rocks are formed. The sedimentary as well as igneous rocks have undergone alteration and subsequently metamorphosed. Again, the exposed rocks of any category, are eroded and deposited. Some rocks due to earth movement get contact with superheated magma and melted up. They are ejected up in hot magma or lava and through the cooling and solidifying process they change into igneous rocks. Thus the materials of earth crust are changing their places but they do not loose in anyway. The total materials remain the same. Thus the change of rocks from sedimentary to igneous or metamorphic and *vice versa* is called as rock cycle (Fig. 5.6).

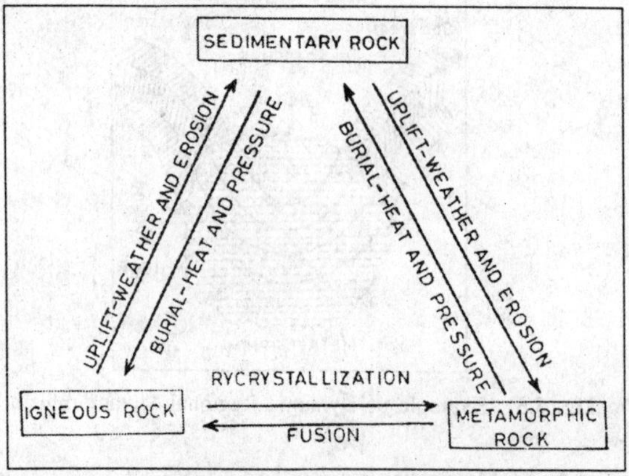

Fig. 5.6 : Rock Cycle.

However, the energy which derives the cycle is from the two direct or indirect sources. Sun is the main source which provides the energy to the external forces (agents of erosion and weathering) to sculpture the exposed rocks on the earth surface. The second is the geo-thermal energy which are originated due to disintegration of radioactive elements like uranium, plutonium and thorium beneath the surface. Through this energy, rocks are melted and converted into magma by which igneous and metamorphic rocks are formed.

6

EARTH MOVEMENT

Our earth is unstable. Every second some of disturbances are taking place in any part of the earth. The earth has undergone various changes and alteration over a long geological history. The various distinct landforms appear today at whatever places they were not in the past at these places. At the place of lofty mountain Himalayas, there was geosyncline, called as Tythes sea before billions of years ago. In the beginning there were neither mountains, plateaus nor plains. Before 120 million years ago, the isolated continents were united as pangea. But now they are lying far away from each other. Shaking of the earth and its disastrous effects are felt everyday along the weak zone, in various parts of the earth. Submergence of coastal forest of western coast of India near Bombay and sinking down of west coast along fault line and emergence of east coast of India are many examples of the disturbances of the earth. Volcanic eruption and ocean floor spreading also deserve attention to be cited as earth's crust disturbances. All kinds of disturbances on the earth surface is caused by internal forces hidden in the crust. Thus the changes which are taking place on the face of the earth due to the internal forces within the crust is describes as earth movements.

Causes of the Earth Movements

The speculation of land forms on the earth surface in different ages of geological times and at different places enforce the geologist and geomorphologist to think over the genetic forces and processes responsible for their origin. It is believed that the forces of earth movement rest in the crust. Now questions arise how the forces are originated? There are two ways by which forces of earth movement came into being. First, from time to time certain sections below the earth surface become heated, probably, by heat given off by radioactive elements. This causes expansion which brings about a series of events resulting in the folding of adjacent sections of the rock at the surface. Secondly, the elevated parts of continental regions are continually denuded and the eroded materials are transported and deposited in the sea. Consequently, the continents become lighter and the seas become heavier. Thus a slow adjustment takes place by the pushing down the oceanic segment and forcing up the land segment. Due to this isostatic adjustment, convectional as well lateral forces are originated in the earth's crust.

Type of the Earth Movement

The earth movements, which are experienced on the earth surface by people, are of varying magnitude and intensity. Their effects may appear in different duration of time. They are differentiated on the basis of whether they are lifted or depressed parts of the surface or whether they are subjected to lateral stretching or compressions. Earth movements may be classified into mainly two groups on the basis of time which is taken by them to appear on the surface.

Sudden Earth Movements

When the effect of the earth-movement appears on the surface suddenly or abruptly and they finish in short time. For example earthquakes and valcanoes. By these movements heavy loss of property and life is caused due to sudden occurrence.

Slow or Secular Earth Movemnt

This type of earth movement continues for longer, more than span of our lives. The effect of these movements is felt very slowly. Such movement is called secular or slow movement. Raising up east Indian coast by 15-30 metres, sliding down of western coast of India, submergence of forest near Bombay, formation of Himalays from the floor of Tythes are some Indian examples of secular earth movements. The slow of secular earth movement has been classified into two forms.

(i) Epirogenic Earth Movements.
(ii) Orogenic Earth Movements.

Epirogenic Earth Movement

The vertical force which causes the upliftment or subsidence of some segment of the earth surface. When a part of the earth is uplifted, called as upwarping while the subsidence of the earth is called down warping. Continents and oceans are formed due to such movement. So it is also known as epirogenic or continental building earth movement.

Orogenic Earth Movement

The orogenic word is derived from greek word 'Oro' which means mountain. This movement is also known as mountain building earth movement because the various type of structural mountains are formed by this movement. It is a lateral movement of force causes stretching and tension in the rocks. There are two kinds of such movements on the basis of direction of movement of the force —

(i) Compressional Force
(ii) Tensional Force

Compressional Force

When the lateral force inside the crust move from opposite direction to a common point is called as compressional force of earth movement. Due to compression, the rocks are deformed and their area are squeezed. Consequently, a wave like structure, called fold, is formed (Fig. 6.1). The upfold is described as crest or anticline and the downfold is described as trough or syncline. The nature of folds vary from region to region due to variation in the amount of compression exerted on the sides of the folds.

Kinds of Folds

Every fold has two limbs and these limbs decline at different angles. On the basis of dip or angle on which the limbs incline, folds may be recognised into various groups (Fig 6.1).

1. Symmetrical folds
2. Asymmetrical folds
3. Isoclinal folds
4. Monoclinal folds
5. Recumbent or overfolds
6. Overthrust or Nappe folds
7. Tensional Force

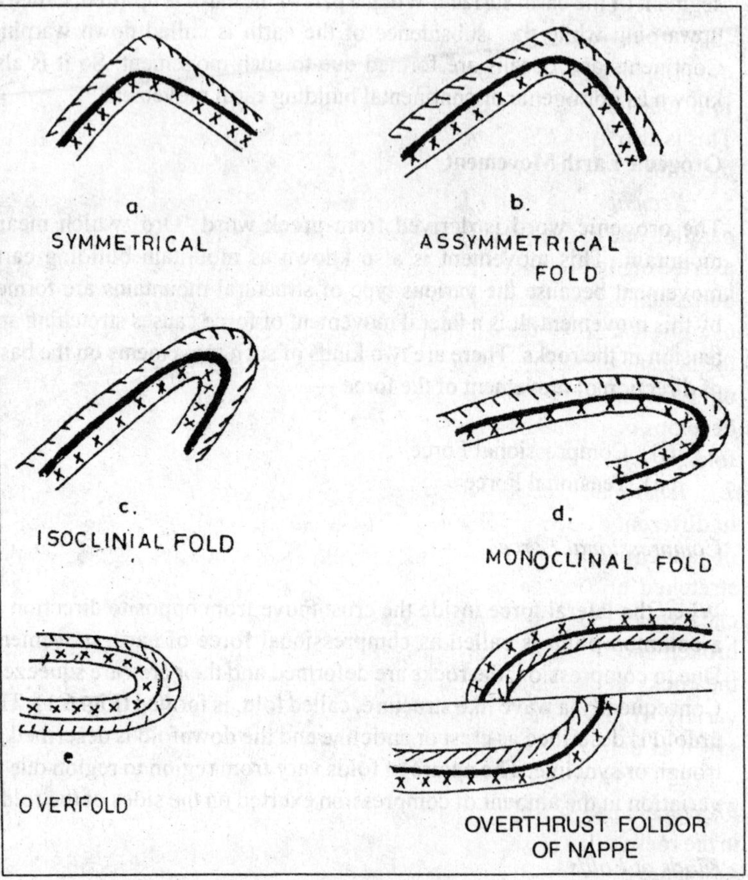

Fig. 6.1 : Groups of Folds.

1. *Symmetrical Folds :* When the two limbs of the fold dip at equal angle due to compression of same amount of them. Such folds are called as symmetrical folds.
2. *Asymmetrical Folds :* It is the type of folds in which the limbs of folds dip at unequal angles. One limb inclines steeply than the other limb.
3. *Isoclinal Folds. :* In this type of folds, the compression force is very strong and the two limbs of folds dip in the same direction and become parallel. The location of their axial plane is neither vertical nor horizontal.

4. *Monoclinal Folds :* Monoclinal fold is a modified form of a symmetrical fold. One limb of the fold dips at right angle while the other at acute angle. This is also formed by imposition of unequal compression on the limbs of the folds.
5. *Recumbent or Overfold :* Sometimes, two limbs of the folds become parallel, one lies above the other, due to excessive amount of compression exerted on them from opposite sides. Their limbs overturned and become horizontal in their axial plan.
6. *Overthrust or Nappe Folds :* When the compression force exceeds the bearing capacity of the rocks, the folds become very acute and break up. After breaking up, one limb thrust over the other. Thus a thrust fault develops in a recumbent folds. Such fold is called as overthrust or Nappe fold.
7. *Tensional Force :* When the endogenic lateral forces moving aside in different directions from a common point, is described as tensional or tangential force or earth movement. In this movement, rocks are stretched and strained upto a certain limit controlled by elasticity capacity of the rocks. When the strain become stronger, the rocks are broken and a gap is formed in the rock. This kind of gap or opening in the rocks is called as fracture or joints. By the effect of this force, variety of landforms may emerge out on the earth surface. They are given as below :

Fractures : Due to action of tensional force, a number of openings develop in the rocks of given region. This kind of opening in the rocks is described as fracture.

Fault : The broken blocks of rocks move into different directions with varying magnitude, depending upon the intensity of acting force, along the line of fracture. Such dislocation of blocks of rock along the line of fracture is called as fault (Fig. 6.2). A typical fault have a number of important components. Fault plane, hanging wall, strike, hade etc. are common. These parts of fault have been described briefly below.

1. *Fault Plane :* It is the plane along which displacement of broken blocks of rock takes place.
2. *Heave :* Horizontal displacement of blocks in a fault is called Heave.
3. *Throw :* The vertical distance between two displaced blocks, of fault along the fault plane.

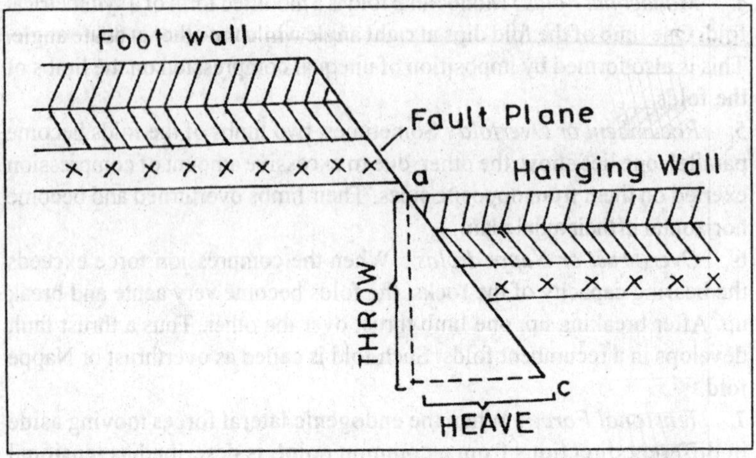

Fig. 6.2 : A Normal Fault and its Components.

4. *Hanging wall :* The block of fault which lies above the fault plane.
5. *Footwall :* The block of the fault which underlies the fault plane.

Kinds of Faults

On the basis of magnitude of displacement, direction and dip of the fault, it may be classified into different following forms (Fig 6.3).

1. Normal fault
2. Reverse fault
3. Step-like fault
4. Tear fault

Normal Fault : A normal fault is one in which the hanging wall moves downward in comparison with the adjoining footwall along the fault plane. It is formed by tensional force. The important characteristics, by which it may be recognised, is that the horizontal distance (Heave) between two blocks on either side increases as the vertical displacement (Throw) increases.

Reverse Fault : A reverse fault is one in which the hanging wall moves upward in comparison with the footwall. The horizontal displacement

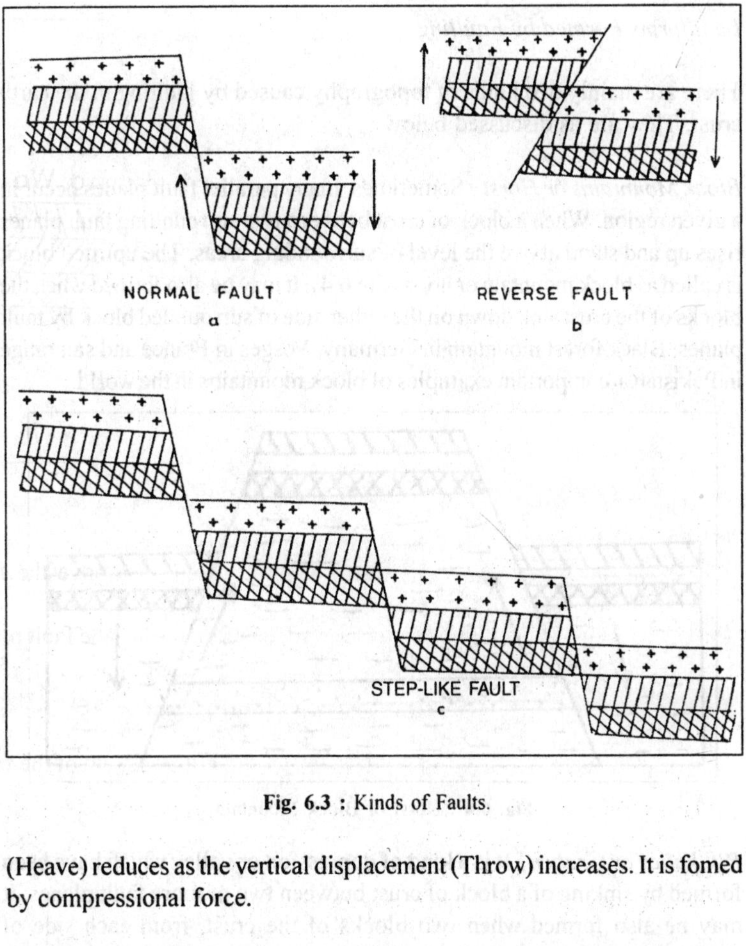

Fig. 6.3 : Kinds of Faults.

(Heave) reduces as the vertical displacement (Throw) increases. It is fomed by compressional force.

Step-like fault : In a region of a number of parallel fractures, numerous faults are originated. These faults are usually having same dips and strikes and direction of movement. When the successive wall of these faults are down thrown, a typical steplike features are formed in the region. Such feature is called as step-like fault.

Tear fault : Tear fault is one in which blocks of rock move horizontally and one overrides the other. This is formed due to compressional force at the time of complex mountain building process. The fault's dip is very gentle and scarp is almost nill.

Landforms Formed by Faulting

There are mainly two kinds of topography caused by faulting in the earth crust. They are as discussed below :

Block Mountains or Horst : Sometimes, many parallel fault planes occur in a given region. When a block of crust between two surrounding fault planes rises up and stand above the level of surrounding areas. The uplifted block is called as block mountain or horst (Fig 6.4). It may be also formed when the blocks of the crust sink down on the either side of surrounded block by fault planes. Black forest mountain in Germany, Vosges in France and salt range in Pakistan are important examples of block mountains in the world.

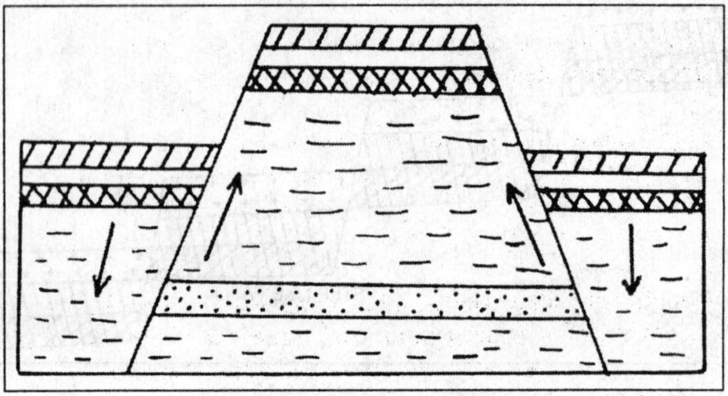

Fig. 6.4 : Horst or Block Mountain.

Rift Valley or Graben : It is a kind of depression or valley which have been formed by sinking of a block of crust between two or more fault planes. It may be also formed when two blocks of the crust, from each side of bounded blocks, rise up (Fig 6.4). Dead sea, Death valley (California), Rhine Rift valley, Mid-land valley (Scotland) are some examples of rift valley of graben (Fig 6.5).

Sudden Earth Movements

There are two kinds of sudden earth movements :

(i) Volcanism (ii) Earthquake

Volcanism or Volcanoes

Volcanism is a process in which the hot rocks either in the form of liquid

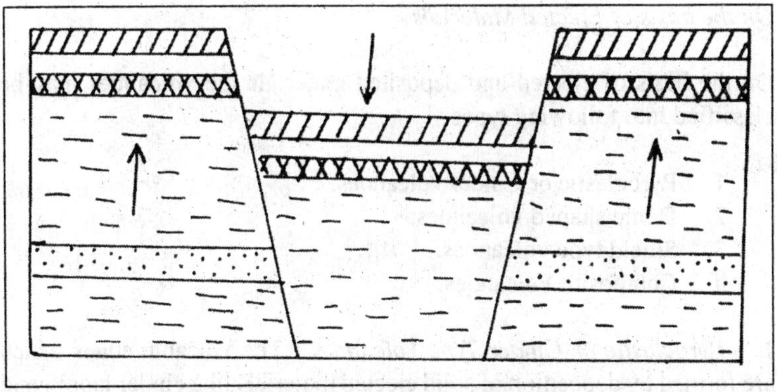

Fig. 6.5 : Rift Valley or Graben.

like magma or in the form of solid small pieces are ejecting out on the earth surface through a vent or passage of the crust. Beside, streams gases and smokes are also coming out at the time of volcanic eruption. Thus the ejecting out of these materials from beneath the earth surface through the Vent is known as volcano.

Classification of Volcanoes

The volcanoes may be classified on the basis of nature of eruption and ejected materials into different forms.

On the Basis of Nature of Eruption

There are two kinds of valcanoes on the basis of nature of the eruption, as discussed below :

1. *Central Type Volcanoes :* When the volcanic eruption takes place through a vent with a great force and the ejected materials are deposited around the vent to form volcanic cones. Such volcanoes are kown as centrally typed volcanoes.

2. *Fissure Type Volcanic Eruption :* When the magma or hot materials from within the crust are coming out on the surface through the fractures or cracks in the crust. In such process the flow of ejection is slow and the ejected materials spread over the wide area like a thin sheet called as Volcanic shields. Deccan plateau of India is formed by deposition of magma through fissure eruption.

On the basis of Ejected Materials

On the basis of ejected and deposited materials the volcanoes may be classified into following types.

1. Pyroclastic or Cinder volcanoes.
2. Dome shaped volcanoes.
3. Shield type volcanoes.
4. Composite Volcanoes.

1. *Pyroclastic or Cinder Type Volcanoes :* The volcanic cones which are formed by deposition of solid ejected materials like cinder lappili and ash through the vent of the volcanoes, from beneath the earth surface. Such cones are having rather more heights and steep slopes. Such volcanoes of greater height with steep slopes are called as cinder or pyroclastic volcanoes (Fig 6.6).

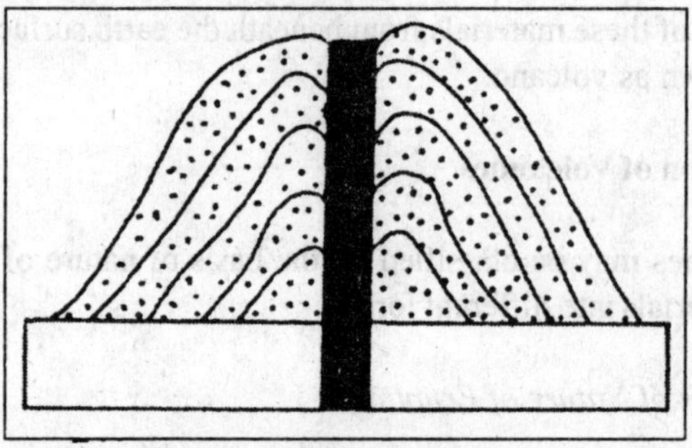

Fig. 6.6 : Cinder Type Volcano.

2. *Dome Shaped Volcanoes :* Sometimes, very viscous magma comes out in volcanic eruption. The viscous material does not spread over a larger area but piles up in small area around the mouth of vent. After the deposition of such viscous magma a dome shaped volcano is formed (Fig 6.7). It is having narrow base and rather more height with steep slope. The dome shaped volcano is also known as acid lava volcano.

3. *Shield or Basalt Type Volcanoes :* The fluid basis magma spreads over a larger area after the explosion through the volcanic vent due to its low viscosity. Due to high rate of fluidity the ejected and accumulated magma does not pile up to the considerable height. Base is broader and the height is rather low. Such type of volcanic formation is known as

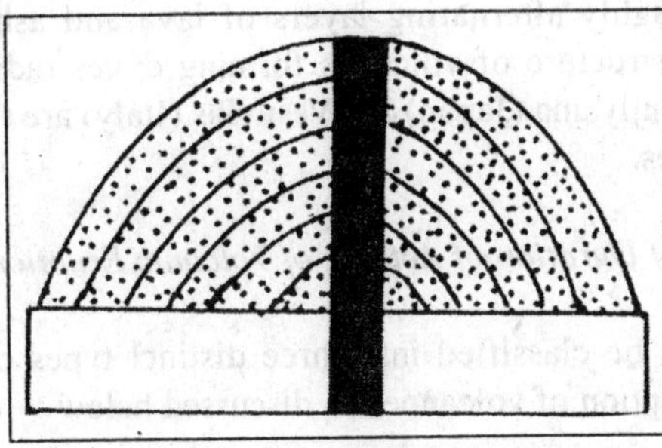

Fig. 6.7 : Dome Shaped Volcano.

shield or basaltic volcano. It is plateau rather than mountain. All howalian valcanoes are of this type. Usually volcanic formation in oceans and seas are also of the same type, rarely, rise above sea level (Fig 6.8).

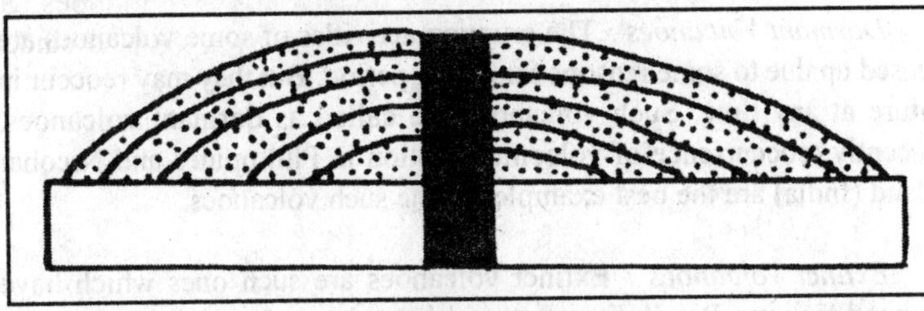

Fig. 6.8 : Shield Volcano.

4. *Composit Volcanoes :* Rare volcanic eruptions with an individual material in isolation occur in the world but they are the combination of different types of ejected materials. When a volcanic cone is formed by alternate accumulation of Pyroclastic (solid) materials and magmas ejected out around centre of the vent, called as composite volcano (Fig 6.9). It is

Fig. 6.9 : Composite Volcano.

formed of roughly alternating layers of lava and ash. Magma may penetrate the structure of volcanos forming dykes radiating from the central vent. Fujiyama (Japan) are Vesuvius (Italy) are the examples of such volcanoes.

On the Basis of Duration of Activity of Volcanic Eruption

Volcanoes can be classified into three distinct types on the basis of duration of eruption of volcanoes as discussed below :

1. *Active Volanoes :* The volcanoes which show continuous eruption are known as active valcanoes. There are about 500 active volanoes all over the world.

2. *Dormant Volcanoes :* The eruption activities of some volcanoes are ceased up due to some reasons for a long period. But they may reoccur in future at any time. Such volcanoes are called as dormant volcanoes. Recently reoccurrence of volconic eruption in Philippines and Nicobar island (India) are the best examples of the such volcanoes.

3. *Extinct Volcanoes :* Extinct volcanoes are such ones which have stopped for ever. Possibility of reoccurrence is not found there.

Materials Ejected from the Volcanoes

Three types of materials are coming out through the volcanic eruption in different parts of the world.

1. *Pyroclastic Materials :* Pyroclastic Materials include all types of solid fragmented materials come out through volcanic eruption. Bomb, Cinder, Lappilli, ash and volcanic dusts are common pyroclastic materials according to the size of diameters of particles in descending order.

2. *Lava and Magma :* Lava or magma are the main constituent of volcanic eruption. When the magma reaches on the earth surface is called as Lava. There are two kinds of magma.

(i) Acid, and (ii) basic, depending upon the content of silica. The acid magma is more viscous with high percentage of silica while the basic is non-viscous with low proportion of silica.

3. *Vapours and Gases :* Steams and gases of different types are ejecting out at the time of volcanic eruption. Carbon dioxide, hydrogen, sulphur dioxide, carbon mono-oxide etc. are important volconic gases.

Causes of Volcanism

Formation of volcanoes takes place due to the ejection of hot solid, liquid and gaseous materials from beneath the earth surface. The most important matter, which is to be explored is the cause of formation of magma and its ejection after the breaking of overlying rocks. There are different explanations, regarding the origin of volcanism in the world. On the basis of the factors involved in such process, the causes of valcanoes may be grouped into following categories.

Percolation of Water : Water percolates down from the bottom of water bodies like sea and oceans in larger proportion to underground. Temperature in the interior of the earth is high as it increases at a rate 1 °C for 165 mt depth from the earth surface. When the percolated water gets contact with hot rocks, converts into steams and vapours. These steams and vapours make endeavour to come out and along the weak zone of the earth crust, wherever outlets are available. Thus due to great force of these steams, some molten materials like magma begin to eject and form the volcanoes on the earth surface. More than 90 per cent volcanoes are found along the continents and ocean's margins.

Tectonic Earth Movement : At the time of mountain building movement, the crustal rocks are deformed into form of folding and faulting. In the case of faulting, dislocation of rocks takes place. Dislocation and displacement of rocks along the fault planes cause the release of super incumbent pressure of overlying rocks inside the earth crust. As we know, temperature beneath the earth surface, relatively, at 100 km depth is very high due to disintegration of radioactive elements. But under high temperature rocks are not melted due to increasing pressure of overlying rocks. The pressure increases the melting points of the underlying rocks. Thus, when pressure of overlying rocks is released by dislocation of the rocks in tectonic movement, the underlying rocks are melted and magma is formed in the upper mantle. This magma ejects out through the weak zone of the earth crust to form the volcanoes.

Plate Tectonic Movement

Plate tectonic is a new concept in geology which explains well the cause of genesis of volcanic eruption. It is believed that the whole earth is divided into six major plates which may consists of either, or both, continental and oceanic crust materials. The plates are having thickness of 60-100 km, consists of both crust and upper mantle. Plates may move away from one another, may converge and slid past one another. When the two plates are moving for away, a rift is formed. Along this boundary of divergence baslatic magma flows out and spread over as ocean floor spreading and new oceanic crust is formed. Thus a quite or fissure eruption type volcanoes are formed along the mid oceanic ridges due to plate movement. When two plates move to each other and converge, over riding of plates take place. Along the boundary of convergence, the heavier plate formed of SIMA undergoes the lighter plate of sial. The undermoving plate reaches upto depth of 100 km and it is melted there due to high temperature inside the earth. Consequently magma is produced and it may flow out through the weak zone of the earth crust. The boundary of convergence is also known as zone of subduction or Benioff. Thus the convergence and subduction of plate is also very important cause of volcanic eruption. (Fig 6.10)

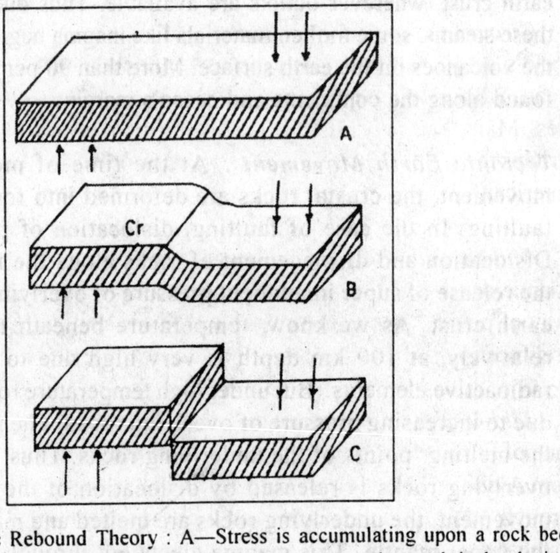

Fig. 6.10 : Elastic Rebound Theory : A—Stress is accumulating upon a rock bed. B—The bed has been deformed to some extent and a crack is developing. C—Along the crack the two fractured blocks rebound to their original position of no strain.

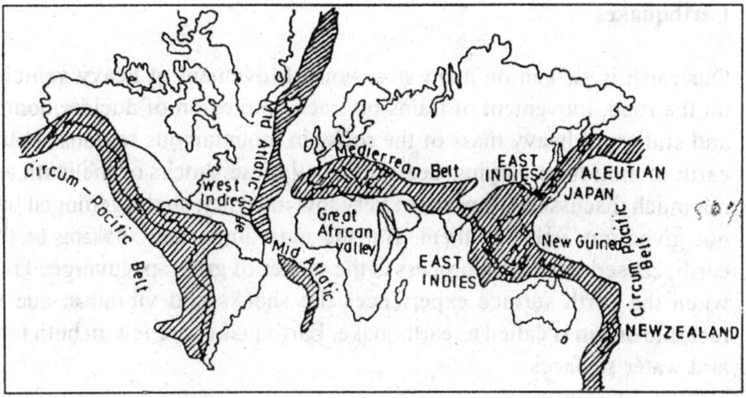

Fig. 6.11 : World Distribution of Earthquakes and Volcanoes.

Geographical Distribution of Volcanoes in the World

Keeping in view the causation of volcanoes, we may recognise three important distinct volcanic regions in the world (Fig 6.11).

1. *Circum Pacific or Zone of Subduction Region :* Larger number of world's active volcanoes are concentrated along the margin of Pacific ocean. Valcanoes are found in a chain here. It is also known as the ring of fires. Main cause of high concentration of volcanoes in this region is perhaps the subduction of plates in oceanic trenches along the Asiatic and American coasts.

2. *Mid-Oceanic Regions :* Along the mid-oceanic ridges especially in Atlantic ocean fissure type volcanic eruptions are common due to moving away of plates. Magma is of basaltic type and volcanoes of high elevation rarely occur.

3. *Mid-Continental Regions :* Mid-continental regions include the weak zone of Europe and Asia along the mountain chain of tertiary era. It starts from Alps fold mountain in Europe and reaches up to islands of S.E. Asia through Afghanistan, Baluchistan, India and Indo-Burma border. Crustal dislocation during mountain building movement is the common cause of occurrence of volcanoes in this region.

Earthquakes

Our earth is shaken on many occasions. Movement of heavy vehicles on the road, movement of trains on track, explosion of nuclear bombs and sliding of heavy mass of the rocks in mountainous regions set the earth in motion to varying scales. But all these shocks of the earth are not much discussed. They are of very low intensity, so the geologist had not given any value to them. But the most important shaking of the earth, caused by natural process is the matter of great speculation. Thus when the earth surface experiences the shocks and vibration due to tectonic origin is called as earthquake. Earthquakes are felt on both land and water surfaces.

Causes of the Earthquakes

As we put or throw a stone in water bodies, a series of waves originate from the stike point of the water and moving outward side in decreasing intensity with increasing distances. Similarly, when rocks are broken up inside the crust by any reason, a series of vibrations are originated and move upwards and downward in the crust. When these vibrations reach to surface, earthquakes are felt. Besides, there are some other ways by which earthshocks are established. The following causes may explain the origin of the earthquakes.

Faulting : When the rocks are broken up inside the crust and a fault is formed along this fracture or break point. Due to dislocation of broken rocks, vibrations are originated and earth is shaken. Majority of earthquakes occur due to faulting.

Volcanic Eruption : At the time of valconic eruption, hot magma is erupted with great force violently due to force of ejected materials, the earth is shaken and earthquakes are originated. Generally the earthquakes belts are found along the belt of volcanoes.

Elastic Rebound : When the orogenic force acts upon a rock from above and below, a strain is caused. The rock is strained upto a level, when the elasticity of it, is greater than the strain. When the elasticity ends, the rock is broken up. The broken blocks of rock make endeavour to regain their original place before strain like a rubber. Thus through the movement of blocks of rock to get their original place, earth is shaken and earthquakes are originated (Fig 6.10).

Plate Tectonic Movement : Different plates of the earth crust are moving in different ways. Due to variation in the direction of movement, three kinds of plate, margins have been recognised. They are :

(i) Margin of convergence or subduction or Beni-off zone.

(ii) Margin of Divergence.

(iii) Margin of lateral movement.

Along these three boundaries various kinds of tectonic movements, take place. Faulting and quite type volcanic eruptions are common along the zone of divergence. Folding and violent volcanic eruptions occur along the subduction zone while the transform fault or tear faults are caused along the margin of lateral movement. Thus along all these boundaries, the earthquakes are originated. But the maximum earthquakes occur along the zone of subduction.

Important Elements of the Earthquakes

Focus : It is place of origin of earthquake's waves. It lies usually beneath the earth at varying depth.
Epicentre : The place on the earth surface where first shock is felt. It lies vertically above focus.
Earthquake Waves : They are also known as seismic waves. There are three types of waves, i.e. (i) P. Waves (ii) S. Wave (iii) L. Waves. (Their properties have been discussed in chapter 3).
Isoseismal : It is an imaginary line, drawn on the map, connecting the places of uniform intensity of earthquakes. Intensity of earthquakes are measured by seismograph on Richtar Scale.

Effects of Earthquakes

Earthquakes are very disastrous natural phenomena. Damages are caused into following ways :

(i) The earthquakes have caused deformation on the earth surface. Formation of thrust fault, upliftment and subsidence of part of the earth are common features produced by earthquakes. For example Assam earthquake of 1897 caused the formation of a fault with 10.6 m x 19.3 m size. Similarly, a number of faults, originated by earthquakes, have been observed at the mouth of Mississippi river.

(ii) Earthquakes have caused the loss of lives and properties by destruction of cities and towns. Buildings and some other human structures, road, railways, communication lines etc. are collapsed and men and women are buried under large debris and rubbles of collapsed structural materials of the buildings. Destruction of Tokyo (Japan) and Russia are the recent examples of havoc and damages of life and properties caused by different earthquakes which occurred in first half of 1985.

(iii) Economic structure of earthquakes affected area is badly affected. Industrial as well as agricultural economies are damaged by seismic waves. Destruction of industries, breaking up of communication system by disturbing the communication line, bending of water pipelines, collapsing of irrigation networks of canals and tubewells are many aspects of economic loss caused by earthquackes.

Earthquakes of Latur (Maharashtra) in 1994, may be cited as an example of such type of destruction.

Area of the Earthquakes

Like the volcanoes, the earthquakes are also concentrated into three important geographical regions as given below :

(i) *Circum Pacific Region :* Along the eastern coast of Asia and western coast of Americas most concentration of earthquakes are found. The origin of earthquakes are, probably, due to volcanic eruption and convergence of plates.

(ii) *Mid-Atlantic Region :* Along the mid-Atlantic ridges, a number of earthshocks have been recorded due to divergence of plates and outflow of basaltic magma from beneath the earth surface.

(iii) *Mid-Continental Region :* This is very important zone of the earthquakes. Due to being a weak zone of the earth crust, recent mount aim building movement and convergence zone of the plates, every year, many severe shocks occur in this region. This region extends from Alps region in Europe passing through Iran, Afghanistan, Baluchistan, Himalayan region to East Indies.

7

MOUNTAIN BUILDING OR OROGENY

Mountians simply refer the uplifted or elevated land with reference to the surrounding areas. They differ in their origin, structures, arrangements of the rocks, location, size and geological periods. They are characterized with the association of peaks, valleys intermountain plateaus and plains and glaciers and rivers. The mountains may be classified into various categories on the basis of different criteria.

Geographical Classification

Mountain may be classified into following groups :

(a) *Cordillera :* It is a series of more or less parallel ranges of mountains together with intervening valleys and plateaus. For example cordillera of Andes. They are formed in different periods by different processes.

(b) *Mountain System :* It is a group of mountain ranges or mountains which originated in a single period with the same process.

(c) *Mountain Range :* The group of long and narrow ridges which are formed during the same period and by same process. For example Himalayan Mountain ranges.

(d) *Mountain Ridge :* It refers a long, narrow, high hills sloping at different angles on the either sides.

Classification of Mountains on the Basis of Process of Origin

On the basis of mode of formation, mountains may be grouped as follows:

1. Fold Mountains
2. Block Mountains
3. Volcanic Cone
4. Relic Mountains

Fold Mountains : Such mountains are formed due to crumpling and squeezing of sediments of geosyncline between two stable lands. The collision of ocean-continents and ocean-ocean plates have resulted in the formation of fold mountain, i.e. Andes, Rockies, Alps, Himalayas etc. They are also classified as (i) young fold mountain, (ii) mature fold mountain, and (iii) old fold mountains.

Block Mountains : Block mountains are formed when a block of land between two faults is uplifted due to tensional forces. Black forest mountain is an excellent example of block mountain.

Volcanic Cone : Such mountains are the result of accumulation of volcanic materials, i.e. Cinder, Pyroclastic dust, lava etc. around the vent.

Relic Mountains : These mountains are also known as denuded mountains. They are formed when a mountains or plateaus are reduced to low height after continuous erosion. They are characterized by smooth peak or top. Aravali Hill is an example of relic mountain.

Classification of Mountains on the Basis of Periods of Origin

1. Pre-cambrian mountains
2. Caledonian mountains
3. Hercynian mountains
4. Alpine mountains

Pre-cambrian mountains refers the mountain which originated during pre-cambrian period. Such mountains occur around old land masses or shields. Laurentian Mountains, Scandinawian mountains etc. are the examples.

Caledonian mountains include Applachian system, mountain of Scottland, Aravli etc. have been originated in silurian and devonian period.

Hercynian mountains came into being during permian and permo corboniferous periods. Altai, Tien Shan, Mountains of Dzungarian basin, Tarim basin etc. are the Hercynian mountains.

The Alpine mountains, however, which were uplifted from the ocean beds or geosyncline during the tertiary periods are included in Alpine

Mountains, Rockies, Andes, Alps and Himalayas are the examples of Alpine mountains.

Theories of Orogeny of Fold Mountains

By definition, fold mountains are the uplifted structure which originated by foldings and thrusting of the beds of sedimentary rocks found in the geosyncline or continental margin regions. Structurally overfold, Nappe, and overthrust and faulting are also common in complex folded mountain system. Various theories have been propounded by geologists to explain the mountain building movements during different geological periods. However, every movement includes lithogenesis (formation of weakzone either as continental margin or geocyncline) orogenesis (compression and folding) and the gliptogenesis (upheaval and isostatic equilibrium through denudation and sub-surface flow of magma from beneath ocean to continental crust). The following theories of mountain buildings are discussed below -

1. Geosyncline Orogen Theory of Kober,
2. Thermal Contraction Theory of Jaffreys,
3. Continental Drift Theory of Alfred Wegner,
4. Thermal Convection Current Theory of A. Holmes,
5. Radio Active Theory of Joly,
6. Continent Sliding Theory of Daly, and
7. Plate Tectonic Theory.

Geosynclinal Orogen Theory of Kober

Kober is an important propounder of theory of mountain building based on the contraction of the earth. He is considered as contractionist scholar. His theory is based on the analysis of various kinds of land forms, like mountains, stable landmass, oceans and inter mountain valleys and plateaus, distributed throughout the world in various geological periods. He contended that the earth is divided into two geological structures, i.e. (i) Orogens and (ii) Kratogens. Orogens imply the weak zones on the earth where foldings, earthquakes and volcanic eruption are very common. While the Kratogens include the old rigid stableland masses which surround the orogens. Important stable land masses are Russian, Canadian, African Siberian, Chinese, Indian, Australian Antarctica and Brazilian plateforms or shields. They are considered as the foundation stones of present continents. Fracture, tension, formation of block mountain and rift valley etc. are important geological phenomena in Kratogens.

Kober observed the present land form systems and found that there are two systems of mountains in N-S and E-W direction in orogens regions. The first is great Alpine system which extends from Gibralter as far as Sunda Islands. It is originated due to squeezing out of sediments of Tythes geosyncline bounded by Eurassia foreland in the North and Afro-Indian foreland in the south. The second system, follow N-S direction, includes the submarine ridges in Atlantic, Indian and partially in Pacific ocean. These two systems are connected by great loops in Aleutian-Alaskan arc, Sunda arc and Antilles arc. On the basis of analysis of global distribution of landforms, and their geo-tectonic characteristics, Kober recognised eight morphotectonic units found in the world. They are : (i) India and Australia, (ii) Africa with the part of Atlantic and Indian ocean, (iii) Eurassia, (iv) the North Pacific continent, (v) the South Pacific continent, (vi) North America, (vii) South America, and (viii) Antarctica.

Development of the Thesis of Mountain Building Movement

Kober thought that the earth has undergone the cooling and contraction since the stage of star. The contraction of the earth led to the origin of force of movement of blocks of the earth. Relationship between the orogens and Kratogens is the main component of the theory. His contention is that the fold mountains came into being at the place of a broad and extensive shallow depression bounded by at least two rigid landmasses (Kratogens) along two sides. This depression is described by him as a term of 'Geosyncline'. According to Kober a mountain (fold mountain) is evolved through various geological processes or events in a given area (Fig 7.1).

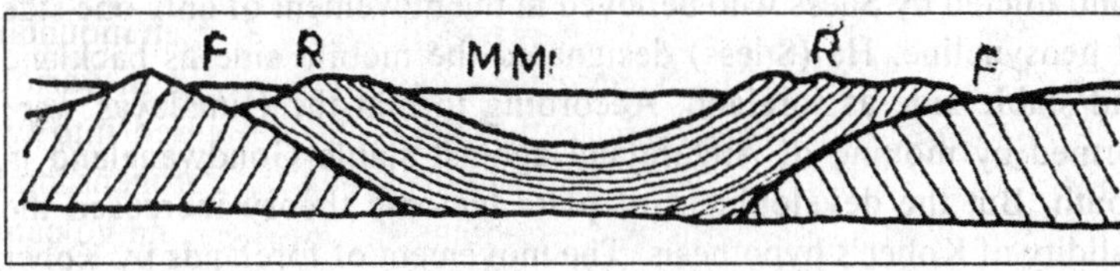

MM = Median Mass, R = Ranges of Mountains, F = Fore Lands

Fig. 7.1 : Geosyncline Model of Mountain Building of L. Kober.

Mountain building process begins with the formation of geosyncline by subsidence or foundering of some parts of landmasses through tectonic movement. The geosyncline is filled up by the deposition of sediments brought by various agents of erosion like rivers and glaciers from the surrounding landmasses. With the passage of time the thickness of sediments becomes very high and subsequently the deposited sediments are subsided under the pressure of increasing weight of overlying thick sediments. In this way, geosyncline is assumed to have a great thickness of sediments on its floor.

Kober again, mentioned that every geosyncline has two edges on opposite side. These edges are described as Forelands by him. These two forelands move towards each other at the differential rates under the influence of compressive force originated due to contraction of the earth. The sediments are squeezed out and folded up into ranges along the forelands of geocyncline. These ranges along the forelands are called as Randkettens. The remaining part between two randkettens is described as Median Mass. The median mass, sometimes, disappears when the compressive forces are very severe and strong. Two forelands might be brought in contact with one another. In this situation, whole sediments of geosyncline is squeezed up and tectonically complicated randketten without median mass is formed. This typical randketten is called as Narbe. Thus the mountain ranges, with a number of intermountain plateaus, longitudinal valleys, and plains, came into being through the process mentioned above. This type of explanation of mountain building is known as Geosyncline Hypothesis of mountain building (Fig 7.1).

However, this theory of mountain building is criticized with reference to the force of movement of forelands. It is objected that the forces originated by contraction of earth is not sufficient to cause the formation of such lofty and extensive mountain system like Alps, Himalayas, Rokies and Andes on the earth surface. This theory is also contradicted by Suess who believed in the movement of only one side of geosyncline. He (Suess) designated the mobile side as backland and stable side as foreland. According to him the Himalayas were formed by moving of Angaraland against stable Gondwanaland in South. But the development of plate tectonic theory increased the validity of Kober's hypothesis. The movement of forelands by Kober coincide with the mobility of two plates along the boundary of convergence. No doubt, the Kober's theory is one of the scientific and rational hypothesis of mountain building movements.

Thermal Contraction Theory of Mountain Building by Jeffreys

Jeffreys is one of the important exponent of contraction theory of the earth. He viewed that the evolution of mountains, fault, thrusts, foldings, continents and oceans took place due to contraction of the earth since its origin. Contraction of the earth's different shells from surface to interior (core) took place at different rates due to variation in the rate of coolings. The reduction in the velocity of earth's rotation through geological periods is also assumed as a factor of reduction of radius of the earth. It is believed that some 1600 million years ago, the earth rotated in 0.84 of our present day of 24 hours and in still earlier period the rotation period was even smaller. During last 1600 million years the contraction in equatorial circumference of the earth may have been about 18 kilometres. Thus Jeffreys calculated, mathematically, the reduction or shortening of the crust to the order of 200 kms and the surface area reduction is of $5 \times 10^{16} cm^2$.

According to Jeffreys, it is through that the earth is consisting of different concentratic shells from surface to core. The interior of the earth below 700 kms from the surface may have undergone no appreciable change in temperature and volume due to greater heat inside the earth. But outer 700 kms recorded a loss of temperature at different rates in different shells due to radiation of the heat. The rate of cooling and contraction of a over lying layer is more than the layer immediately below it. Thus the reduction in volume of outer layer is rather high than the layer immediately below it. The reduction in volume of outer layer would not fit on the underlying inner layer which may have rather larger volume. But at the depth of 700 kms, the rate of cooling and contraction might have reached to a condition in which overlying layer would be fitted well on the underlying layer. This layer recorded no tension due to contraction in volume. Hence it is known as No-strain-zone. Below the No-strain layer, again unadjustment occurred between two adjoining layers. Rate of cooling and contraction are decreasing towards the centre. Consequently a overlying layer will be smaller than the underlying layers and an unadjustment would occur between them due to variation in their volumes. This implies horizontal stretching and thining of overlying layer with a view to adjust on the underlying layers. Hence the overlying layer will be probably fractured due to tensional force.

Moreover, the upper layer of the crust has cooled down at the maximum level and no further cooling is possible, while the inner layers, continue to cool and contract. Thus the outer crust with larger size would not fit on the inner contracting region or layers. Adjustment between them may probably occur by crumpling and crushing the outer crust through the compression force. Consequently buckling and folding occurred on the earth's surface. Under the gravitational force the compressed surface above no-strain zone move down into fractured and thinned layer zone below the no-strain zone. Consequently outflow of materials from substratum took place and the upfolds were lifted up further to get a great height.

He further stressed that rate of cooling beneath the ocean is faster than the continents. This led to the fact that the compressional forces caused by cooling and contraction, moves from ocean to land, hence the formation of fold mountains have taken place along the continental margin parallel to the coast. Circumpacific fold mountains are the typical examples.

Periodicity of mountain building is also described by Jeffreys. According to him the mountain building forces, i.e. compressional and tensional are continuously accumulated unless the strength of rocks is exceeded. Fracture and Crumpling, i.e. mountain building, will commence and continue until the forces are weakend. Mountain building is followed by a period of quiesscence in which forces are accumulated. Sooner or later mountain building may reoccur. Thus he believed in five orogenesis mountain building periods on the basis of scientific study of mountain systems on the earth originated during geological history.

Thus Jeffreys tried to explain the process of formation of mountains with a great reasoning and scientific and mathematical explanation. However, this theory is very much criticized on the following grounds :

(i) Jeffreys used very complex mathematical calculation which is very difficult to understand for non-mathematical scholars;

(ii) His calculation of the shortening of the surface area and the radius of the earth's crust by contraction is the half of the known estimated shortening of earth crust by folding compression and thrusting in orogenic belts. But theoretically both must be equal;

(iii) This theory is also questioned about the distribution of mountains on the earth surface. Since cooling and contractions on spherical earth would have taken place equally from all sides on every place. Therefore, according to the thermal contraction hypothesis the mountains should have been universally distributed in different directions and should not have had a specific location as tertiary mountains of circumpacific and Mediterranean locations;

(iv) The orogenic forces generated by cooling and contractions are not sufficient to produce a great earth panning mountain ranges but rather a large number of comparatively minor puckers or minor folds; and

(v) This theory also fails to coincide with the recent most valid geological theory of Platetectonic, Paleo maganetism, ocean floor spreading etc. in which drifting of continents and oceans have been highlighted.

Continental Drift Theory of Mountain Building by Alfred Wegner

Alfred Wegner propounded the theory to explain the present locations of continents and oceans on the earth with the help of lateral drifting of different crustal parts in different directions. This hypothesis is described as continental Drif theory. He developed and postulated the theory in 1921 with the help of paleo-climatology, paleo-ontology, paleo-botany and geophysics etc. desciplines. He studied and found many similarities of fossils, geological structure, coastal geometry — between east and west coast of Atlantic ocean. Mid-Atlantic ridge also shows parallelism in its direction with the Atlantic coasts. Thus he assumed that during carboniferous period, all the continents were united together as single supercontinent, called as pangea. This pangea was assumed to be constituted of Sial. Pangea was surrounded by ocean (Pacific) called as Panthalusa which floor is formed of SIMA. Due to variation in the density between SIAL and SIMA, the pangea (SIAL) was floating into SIMA.

He further assumed that the pangea was broken up and moved in south (equatorwards) and west direction. The movement of continental blocks towards equator depends upon the relation of centre of gravity of moving or floating mass and the centre of boyancy of same mass. While the westward movement is ascribed to tidal force of the sun and the moon on the earth surface.

These two movements of continents were obstructed by the resistence of the heavier sima in which formers were floating. This resistence led to the crumpling, rapturing and folding up of continental margins in the form of mountains. Rockies and Andes mountains along the western coast of north and south America were formed according to Wegner due to west-ward drifting of continents of America from Eurassia and Africa. While the formation of Alps-Himalayan system is sought to be formed by drifting of continent, towards equator when the tertiary equator laid approximately along the line of Alpine folding.

Wegner has also explained the process of origin of island arcs and festoon. East Indies, West Indies arc of southern Antilles between Tierra del Eugo and Antarctica etc. have come into existance due to differential rate of drifting of mainlands. They owe their formation to the westward movement of Asia and Americas from which they have been cutoff by their attachment to the solidified floor of the oceans.

Drifting theory of Wegner is found more valid and accepted after the development of Holmes thermal convectional current theory and plate tectonic theory. But some scientists are doubtful about the forces of drifting of blocks of continents. They viewed that tidal forces of the sun and the moon described by him, have not been sufficient to cause drifting apart of continents. It required 1000 times more than present gravitational force. If it was so, the rotation of the earth would have been ceased out. Similarly, the differential gravitational and bouyancy force of moving mass also appear inadequate for the purpose of drifting equator wards. Theory is also criticized on the issue of resistence caused by SIMA on the floating SIAL for the formation of mountain like Rockies-Andes and Alps-Himalayan systems. Because he has already assumed that SIAL is floating in SIMA freely.

Mountain Buildings by Thermal Convection Currents by Arthur Holmes

In 1921, an explanation, based on the circulation of thermal convection currents inside the earth, for the formation of mountains has been given by Arthur Holmes. According to Holmes, the crust of earth includes the Sial and upper part of SIMA. Below this, substratum lies. It is consisting of thermally fluid or glassy part of lower layer. Beneath the ocean the upper sialic parts do not occur with exception in a fragmentary parts in Atlantic and Indian ocean, substratum is found as continous both beneath the continents and oceans.

Holmes assumed that substratum as well as parts of crust consist of different kinds of radioactive elements which add heat to the interior especially substratum of the earth. Amount of radioactive elements and addition of temperature vary with equator and poles due to variation in thickness of crust. Thickness of crust is greater at the equator than the poles. Similarly, radioactive elements also varies beneath the ocean and the continents. Consequently, the temperature gradient also varies from place to place in the crust.

It is believed, further, that the earth crust's stable condition will prevail until a certian critical temperature gradient is exceeded. This critical temperature gradient is 3°C per km. The presence of radioactive elements and their disintegration increase the temperature gradient which led to the origin cf convection current circulation and the excess heat is lost through the conduction of radiation from the earth's crust. Convectional currents also help in scaping out the heat from the interior of the earth. The average loss of heat is 60 calories per sq. cm. area of the earth surface. The circulation pattern of convective currents will depend upon two factors (i) the varying thickness of substratum from Equator to poles, and (ii) the varying thickness and radioactive elements of crustal cover from continents to oceans.

However, due to variation in the thickness of substratum from equator to poles, two kinds of convectional thermal currents are originated. Ascending of thermal currents takes place beneath both continents and oceans at equator and descending and convergence of currents take place at poles beneath both continent and oceans. Due to variation in radioactivity of the continental masses and oceans, temperature is rather higher beneath the continents than the sub-oceanic masses because the former has more radioactive elements due to greater thickness. Due to these two ascending and descending currents, various kinds of landforms are originated. The ascending current moving upward along with crustal matters and reaching at lower level of crust where they begin to diverge towards polar regions north and south direction due to variation in temperatures between two latitudinal zones. The divergent currents dispalce the continental mass apart laterally. Hence the crust above the ascending current is raptured and broken off. A depression, called as Newsea, is produced, Volcanic activity may also occur as hot materials, sometimes, move with upward moving currents. For example, mid-Atlantic ridge is the place of diverging crust and fissure eruption as ocean floor spreading is very common. The formation of Tythes sea is also an excellent example of formation of depression above the ascending thermal currents in equatorial regions (Fig 7.2).

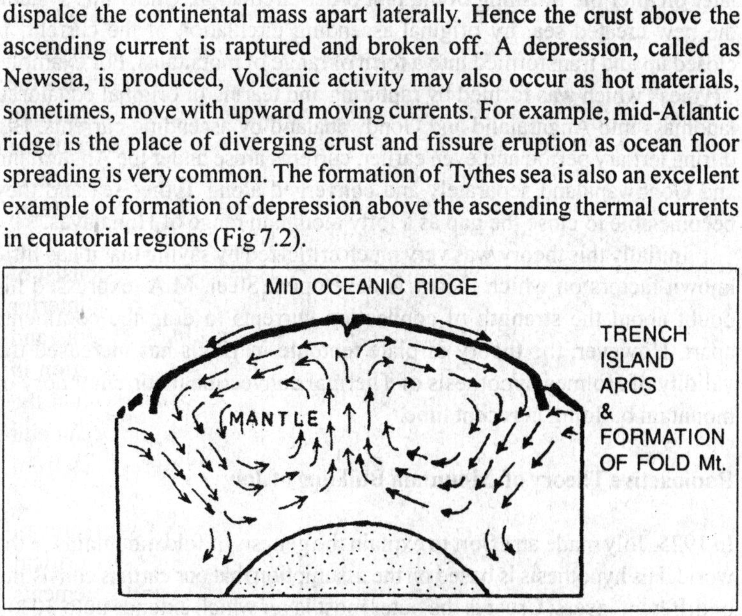

Fig. 7.2 : Covection Current Hypothesis of A. Holmes.

While the converging currents both from ocean and continent move closer. The brought materials are accumulated near the continental margins or shelves. Due to compression of converging currents, the shelves area is subsided and moving downwards as the currents begin to descend down towards the substratum. With the passage of the time, through the sedimentation and subsidence, the shelves are transformed into a geosyncline. When the circulation becomes faster and severe, the forelands come closer and sediments are squeezed out in different folds. As it is believed that this circulation system is not continuous and fixed location phenomena. It may occur at any point and finish when loss of heat is greater than production. It is slackened when the upward current is supplied with cold materials and the descending currents with the hot materials. Due to weakening and wanning of ascending circulation system, the supply of material to the geosyncline (place of descending currents) is also reduced. Overlying pressure is reduced aand the folded materials of geosyncline are uplifted in the form of mountains.

Holmes, further described that when the original system (First system) of circulation of currents dies away, its place is taken by a highly irregular series of currents. A new system originates in divided blocks of continents later on after the finishing of the first order circulation. Under this system, the new created sea, by original ascending circulation of the current, is closed up and transformed into a form of range of mountains. For example, "Tythes" which was formed by rapturing and tearing of original equatorial landmass into Angaraland and Gondwanaland by ascending currents. But during tertiary period and even earlier, currents arose under the Angaraland and Gondwanaland separately and converged along Tythes sea and they become able to close the gap as a lofty mountain range of Himalayas.

Initially this theory was very much criticized by saying that it has little known factors on which theory is developed. Steer, M.A. expressed his doubt about the strength of convection currents to drag the continents apart. However, the theory of plate tentonic in 1960s has increased the validity of Holmes hypothesis of Thermal convectional current theory of mountain building in recent time.

Radioactive Theory of Mountain Building of Joly

In 1925, Joly made an effort to explain the genesis of fold mountains in the world. His hypothesis is based on the assumption that our earth is consisting of different layers. Crust is the outer most layer which extends upto 30 km depth. It consisits of SIAL beneath the continents and SIMA under the oceans. Beneath the crust, substratum (SIMA) is found. The temperature increases with the depth due to disintegration of radioactive elements like uranium, thorium etc. inside the earth. Continental masses or sialic crust have more radio active elements than the SIMA, so temperature beneath the continent is greater than the sub-oceanic SIMA.

Joly viewed that temperature inside the earth increased due to the disintegration of radioactive elements found in both SIAL and SIMA. The production of heat by radioactivity is rather very slow but it became very high after its accumulation over a long period of millions of years. He estimated the temperature equivalent of 1050°C at 30 km depth and 1150°C at the lower level of SIMA (Substratum). The temperature of substratum is equivalent to melting point of basalt. According to him such high temperature (1150°C) would have been produced in 33 to 50 million years. Transfer of heat would have not taken place from SIMA to SIAL beneath continents and from lower SIMA to upper SIMA beneath oceans.

Moreover, the substratum became liquidified under such high temperature and consequently several very interesting changes occurred on the earth's surface. First the volume of substratum increased up and subsequently radius of the earth and the distance of crust from the centre would be necessarily increased. Secondly, due to decrease of density of substratum, the continental masses subsided into sima and the level of sea water would be raised up with reference to surrounding continents. Continental margins would be submerged under sea water and be converted into geosynclines by deposition of sediments. This process is described as transgressia of sea (Fig 7.3a).

Thirdly, on the ocean bottom due to absence of continental crust, the ocean floor is broken up in fracture and through which the molten materials reached on the floors and islands (volconic) came into formation. Many such islands are distributed in Atlantic and Pacific oceans.

When the loss of heat increased through fractured ocean bottom, the substratum again cooled down. Solidification of the substratum would reoccur, and its density must have increased proportionately (Fig 7.3b). Conversely, the radius of the earth would be reduced and the subsided continents in the molten state would be again jumped up due to increase of density of substratum. Sea level fell down and the water retreated back leaving behind the continental margins. Ocean bottoms also reduced and exert pressure along the continental margins which would be squeezed in between neighbouring oceans. This squeezing would necessarily exert compression upon the compratively soft, sediments laid down in the transgressional seas and so these sediments would be strongly folded.

This is the first step of formation of mountains through lateral compression. The folded mountain is raised up further, as a whole due to isostatic recovery of deeply folded mountain on account of complete solidification of substratum. It is second face of mountain buildings.

Joly also highlighted mountain building periodicity in his theory like Kober, Jeffreys and Holmes. According to him mountain building takes place during the period of solidification of substratum after the end of

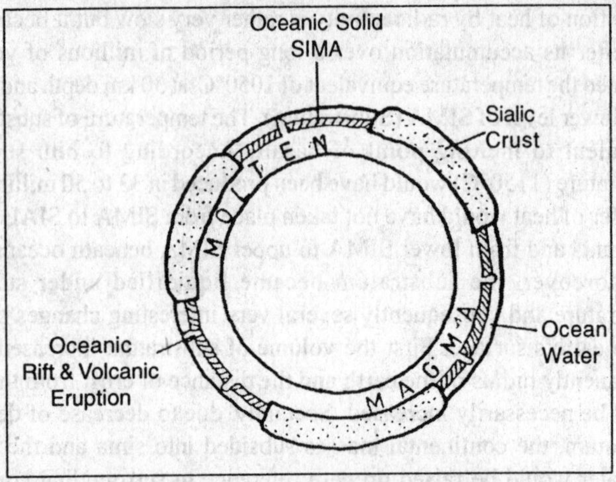

Fig. 7.3 (a) : Enlarged Earth during Molten State of Sima.

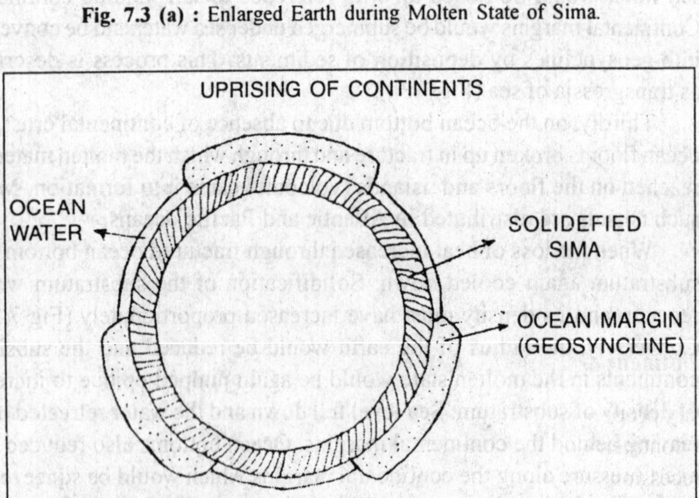

Fig. 7.3 (b) : Shortened Global radius due to resolidification of Sima (Substratum) and Contraction of Volume.

linquidification. The complete period between two solid states of substratum is described as revolution. He estimated the time taken for melting the SIMA is supposed to vary between 33 to 60 million years. His mountain building periodicity is equally spaced through out the geological times. His mountain building should have been cyclic and regularly recurrent. He also assumed that the great mountains faced the greatest ocean. According to him, during the molten state of substratum the continents, moved westward under tidal effects. It changes the position of continents and oceans as former occupied the position of latter.

This theory is very much valid due to its scientific and reasoning based explanation. It deals with the formation of geosyncline and continental drifting envisaged by "Alfred Wegner" and "Holmes". However, the theory is criticised by different scholars on some important grounds.

(i) Joly is criticized for the assumption of structure of the earth. Jeffreys is against his crust of 30 km thickness. He believed in 16 km thick crust as determined by seismology.
(ii) Joly's assumption of cyclic and regular reoccurence of mountain buildings is doubted by some authors. Authorities agree with the number of orogenic periods revealed by geological history of the earth but they are unjustifiable to conclude that all these revolutions are equally spaced.
(iii) His dictum the greatest mountains face the greatest oceans is also not accepted. He thought that the greater oceans exerted stronger lateral pressure against continents and greatest mountain, were formed on the either side of oceans. But the empirical evidences indicated the validity of this concept only along the pacific coasts. It is not well tested and significance with reference to Atlantic ocean.
(iv) Westward drift of continents is also criticized. On the basis of mathematical calculation it can be said that there is no well known force adequate to cause such displacement.

Continent Sliding Hypothesis of Mountain Building of Daly

Daly propounded a hypothesis of mountain building in 1926 on the basis of continent sliding and the gravity of the earth. He did avoid the tidal, precessional or other such forces which are involved in causing continental drift in different theories given earlier.

Daly started his hypothesis with the assumption that there are three riding landmasses, i.e. (i) Polar (North), (ii) South polar and (iii) Equator landmass. These landmasses were separated by depression of north mid latitude furrow (Tythes sea) South mi-latitude furrow and the Pacific ocean. These water bodies covered 1/2 of the world area.

The rigid landmasses (domes) were sloping to the depressions. He also assumed that the primitive crust was formed over the molten earth and all these ancient land masses and furrows are formed of this crust. The density of continental crust is rather higher than the substratum which is formed of glassy basalt.

He further moved with the description that in the furrows and the Pacific, the sediments brought from different surrounding regid landmasses were deposited. Due to increasing the thickness of sediments, with the passage of time, the pressure of overlying sediments and water increased to a considerable extent. Consequently, the heavy material beneath furrow moved to the root of rigid landmasses of lighter materials. Rigid domes uplifted more and their size increased. In this way the weight of sediments became very high and due to compressional force of continents and high gravity force (as ocean bottom is closer to the centre), the layers of rocks are inclined and bent down. When the overlying pressure exceeded the resistence capacity of underlying rocks of geosynclines, the rocks are broken up into blocks due to rupture. In this way the continental dome slided down beneath the broken block of the geosyncline rocks. Due to compression of sliding continental mass, the geosyncline beds were folded up as mountains. Further, the continental block sunk into the substratum and subsequently the former is melted and volume of the substratum expanded. Hence, the folded mountain were raised up further which is

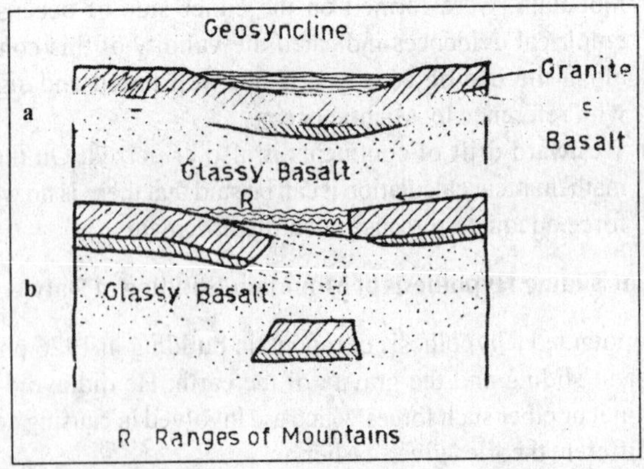

Fig. 7.4 : Formation of Mountain by Crumpling of Geosynclinical Rock Layers and Sliding of Broken Block of Earth's Crust [From R.A. Daly Dur Mobile Earth (1926)].

described as second phase to mountain buildings (Fig 7.4).

However, the Daly's hypothesis is more explanatory and accepted to the geologists especially on the basis of distribution of fold mountians in north-south and-east west direction along the Pacific coast and Tythes sea respectively. But it is criticized on various grounds, by different scholars.

(i) His theory is criticized with respect to the structure of the earth. He believed that density of crust is higher than the substratum which is against the seismological evidences that the density of the earth increases with the depth; and

(ii) His assumption of furrows (ocean) as geosynclines, is also not accepted. Because, a shallow and narrow depression with sedimentation and subsidence is declined as geosyncline. So the deepest ocean like Pacific could not be considered as geosyncline. He is not able to explain about southern furrow.

Plate Tectonic Theory

Plate tectonic theory is widely accepted concept of mountain building movement in recent years. Much have been discussed in Chapter of Origin of Continent and Ocean Basin. In this theory it is believed that mountain, trench, island arcs, and volcanic ridges are formed along the boundaries of convergence and divergence.

Along the boundary of ocean and continental plate convergence, the lighter plate of the continent moves over the denser oceanic plate. This is known as subduction. The oceanic plate melts into substratum and increases in volume. Consquently island arcs, oceanic trenches, volcanic eruptions usually take place along the boundary. The thick sequence of geosynclinal sediments along the continental margin is compressed and deformed into narrow belt of fold mountain ranges.* The descending slope of oceanic plate, after melting, led to the formation of instrusion of Magma in continental crust, trench and island arcs. Formation of Rockies and Andes Mountains are the best example of mountain formation along the ocean-continent boundary.

While along the convergent boundary between continental plates, a high complex mountain system is produced. When two continental plates move from either side of a geosyncline like Tythes sea, the geosynclinal

* One-sided geosyncline along the continental margin which opens towards ocean is called as geosyncline.

sediments as well as oceanic crust or sediments of geosyncline squeezed into narrow zone. Consequently, geosynclinal sediments, or the forelands as described by Kober, are deformed into ranges of the mountain. The oceanic part (median mass) descends into substratum under gravity force due to overlying pressure and it melts into form of magma. This magma flows beneath the deformed ranges and consequently they are uplifted as a very high and complex mountains.

Contrary to this, the boundary between converging plates of oceanic structure is known for presence of ocean trench and island arcs or islands. One plate of relatively lighter density overrides the denser one of opposite side. Along the points of subduction deep sea trenches are formed. When underlying plate reaches a depth of 100 km, it melts and results in metamorphism of the rocks, volcanic activity and the earthquakes. The volcanic activity adds the piling up of the volcanic materials. When the piled or accumulated materials appear above sea water is described as island arcs. Japan is an excellent example of island arcs of volcanic origin. Similarly, ocean ridge is formed along the divergent boundary of two plates in which ocean beds are raptured and magma comes out as fissure eruption. After the accumulation of magma, mid-oceanic ridge like Atlantic ridge is formed.

8

PROCESS OF GRADATION

Naturally the surface of earth is very irregular, consisting of different kinds of landforms like mountains. Plateaus, plains, rivers, lakes and so on. These contrasting landforms on the earth surface are, directly or indirectly caused by endogenic forces lying inside the earth's crust. Simultaneously the external forces work on the surface with an objective to remove the irregularities or roughness on the surface of the earth. The important forces are rivers, glaciers winds and sea waves. They are continuously involved themselves in cutting down the highlands and depositing the transported sediments at lowlands with a final target of levelling up the ground. Thus the process which tend to bring the surface of land to a common level is called as process of gradation.

The process of gradation involves two kinds of works, i.e. (i) degradation, and (ii) aggradation. Degradation refers to the cutting down of highlands with a view to bring it at surface level. It is also called as work of denudation. While the aggradation means the deposition of carried materials by the agents of denudation at low lying area with a view to level up depression to common surface level.

Works of Denudation

As the definition indicates, under the works of denudation, the rocks are broken up into smaller particles and the broken fragments are carried away from their original place to another place. Thus the denudational works may consist of wearing and tearing of rocks and transportation of broken rock debris. Breaking of rock takes place into two ways : (i) weathering, and (ii) erosion.

Weathering

The exposed rocks to weather on the surface of earth are subjected to disintegration and decomposition by physical and chemical factors. The broken materials do not move from their original place except in sloping terrains where fragmented parts move down the slope under the influence of gravity and accumulate at foot of highlands. This kind of vertical movement and accumulation of broken materials and called as mass movement and talus or scree respectively. Hence weathering may be defined as the process of distintegration and decomposition of exposed rocks at or near the surface *in situ* by natural agents like temperature, water, gases and biological organisms.

Type of Weathering

Keeping in view the factor involved in the process of either disintegration or decomposition of rocks, two general types of weathering are normally recongnized. First, physical weathering which involves rock disintegration without any change in chemical constituents of rocks. Secondly, chemical weathering in which mineral consitituents may suffer from either decay or alteration by such agents as water, oxygen or various organic acids.

Physical (Mechanical) Weathering : Physical weatherings are caused by a number of factors like change in temperature, freezing of water and melting of ice in the pores of the rocks. Crystallisation of salts and action of animals and plants.

Change in temperature factor is dominant in desert regions where temperature in days increases and declines in the night to a considerable extent. Thus the rocks are expanded in the days and contracted in the nights owing to increase and decrease in temperature respectively. By the repeating process of alteration in the volume of rocks in day and night, the rocks are weakened and broken. The change in temperature also causes the peeling off the rocks into different thin horizontal layers.It is due to the fact that the rocks are bad conductor of heat. The rate of heating of over lying layers of the rocks is faster than the underlying layers of the rocks. This unequal rate of heating with depth results in unequal expansion of rocks of different layers. Hence the adjustment between overlying and underlying layers of rocks is disturbed and consequently thin layers of rocks are peeled off from the exposed rocks. The peeling off the thin layer of the rock is called as exfoliation.

Freezing of water and melting of ice in the pores or openings of rocks are important agents of weathering in cold region where the temperature reaches below freezing point in night and above freezing point in the day. Thus the water, in pores or fractures, is frozen in the night and melted up in day. Consequently, the volume of water increases in the nights causing extra stresses on the both sides of the pores and in day the volume of water decrease due to melting of frozen water, causing release of the pressure exerted on the both sides of the pores. Thus the alternating process of increasing and decreasing volume of water led to the weakening and disintegration of rocks usually in cold climatic region. This kind of process of distintegration of rock is known frost wedging.

Crystallization of salts causes the disintegration of the rocks when the dissolved salts like calcium corbonate and calcium sulphate and sodium chloride in water solidify in the rock after the drying up of water. Thus crystallized salts set up a disruptive effects in the rocks and consequently rocks are weakend and distintegrated.

Animals and plants are also responsible for breaking up of the rocks in different ways. The penetration of roots of trees and shrubs through the crackes of the rock to a considerable depth weaken and dislodge the different blocks of the rocks. Burrowing animals and insects also disintegrate the rocks as they dig the rocks for the formation of their habitats. Man himself, is the greater breaker of the rocks. He causes the disintegration of rocks by the activities like deforestation, ploughing the soils, digging the mine, the wells, explosions of nuclear devices and so on.

Chemical Weathering : Water, oxygen and carbon dioxide are the important chemical elements which are responsible for the decomposition of rocks in different ways. On the basis of chemical elements which are very actively involved in the chemical reaction in different kinds of rocks, chemical weathering may be classified into different categories.

Oxidation

Water is week acid. It gets oxygen from the atmosphere when the water combined with oxygen passes through the rocks having iron content, the oxide of iron is formed. Thus the rocks are decomposed and weakened through the chemical reactions between oxygen and iron in the them. This process of weathering is called oxidation.

Carbonation : When the rain water equipped with CO_2 moves through corbonate rocks like limestone, the rock are dissolved into solution form in the water. Through this solution action rocks are weakened and decomposed. This process of disintegration and decomposition of the rocks is called as corbonation.

Hydration : When a certain type of minerals like gypsum expand as they absorb the water passing through them in the rocks. Thus the expansion of voume of mineral constituents of rocks causes an additional stresses in the rocks. This process led to the disintegration of concerned rocks. Such process of weathering is known as Hydration.

Solution : Some soluble minerals, like salts are dissolved into water and they are removed from the rocks. Thus after the removal of rocks constituent minerals the rocks are weakened and disintegrated due to change in their chemical composition, such process is called as solution.

Erosion

Erosion is the displacement of weathered fragments of rocks. During the time of travel or transportation of broken materials, they are further broken into finer particles by the process of attrition, abrasion and solution action of erosion.

Process of Denudation

Cutting down of the rocks by different agents of exogenetic forces and the transportation of these broken materials by them are commonly described as process of denudation or erosion which takes place on the earth surface. The weathering process helps to the erosional works made by various agents to considerable extent. But the former is different from latter in such a way that in former case, the rocks are disintegrated and decomposed *in situ* while in the case of latter the transportation work is prevalent along with the wearing and tearing of rocks. In other words in the process of erosion the broken material are transported from one place to another place by agents of erosion. Thus the work of erosion mainly can be divided into, (i) breaking of rocks, and (ii) transportation of eroded materials. Moreover, important agents of erosion are running water, glaciers, winds, sea waves, and underground water. The works of each agent is discussed below individually.

Running Water (River)

River is one of the important agents of erosion on the surface of the earth especially in humid region. When rainfall occurs, some parts of it evaporate to atmosphere, some parts percolate down to underground and remaining flow in the form of run off through some narrow minor stream channels.

After the meeting of a number of minor streams, a big stream is formed which is described as river. Usually the rivers originate from highlands like hills, mountains, and plateaus and they end into seas or lakes. Thus through the development of their courses from mountain to sea, they do the work of erosion (wearing and tearing of rocks), transportation of eroded materials and deposition of carried sediment, at appropriate place.

Process of Erosion by Rivers

The ability of river, to sculpture the country rocks, is controlled by number of factors, i.e. velocity, load, discharge and slope of the river. Steeper slope, higher velocity and more load of river may perform greater amount of erosion along the river course. In addition to nature of rocks whether hard or soft and chemical composition of water also determine the erosional capability of river. However, under different sets of condition the rivers do the work of erosion through different process discussed below:

Hydraulic Process : Hydraulic process refers to the breaking down of rocks under the impact of continuous flow of water in river valley. Water enters the cracks, joint and pores and weakens the rocks and dislodge them from original place due to influence of moving water.

Abrasion : Through the hydraulic process, river breaks-up the country rocks and carries them along its channel. The fragmented particles are dragged along the bottom and sides of river valley and they may cuase sculpturing of the river bed and the dragging materials themselves. Thus the breaking up of attacking and attacked rocks, by the flow of river water, due to friction between them along the valley of river is described as Abrasion process of river erosion.

Attrition : The fragmented particles of the rocks, which are transported by river, get the chance to collide with each other in the way of river flow. By the collision among the transported fragments enroute of river, they are broken up into smaller sizes. Such type of breaking of rocks by river's erosion action is described as Attrition process.

Solution : Water is a weak acid and it gets associated with a number of chemical solvents like carbon dioxide from the atmosphere during the rainfall. Thus a number of mineral constituents of rocks and the rocks themselves get dissolved in the water combined with CO_2. Hence, the process by which the rocks are eroded in the form of solution action is called as solution process of river erosion.

Transportation by Rivers

The river also does the work of transportation along with the erosional work. The fragmented materials produced by both weathering and erosion are carried away from their original place to another place. The carrying capacity of the river is determined by its velocity, volume of water and the size of the loads. If the velocity of river is doubled, its transportation power will be increased by 64 times and when the volume of water is increased two times, the transportation power will increase by two times also. Bigger size of transported materials retards the carrying capacity of river as compared to small and fine size of particles.

Process of Transportation

The transportation activities, made by river, are taking place into following ways.

By Traction : The big size of particles like boulders and gravels are dragged by river along the bottom and along the sides of the river valley because the river fails to lift up such heavy particles. Such type of transportation process is described as traction.

Saltation : Some particles, having medium size texture, jump up and down on the bottom of valley like frogs during the transportation process of the river. This type of transportation process is known as saltation.

Suspension : The particles of relatively smaller sizes are transported by river in such way that they are hanging in the mobile water of the river. Such type of transportation process is designated as suspension.

Solution : It is selective type of transportation process made by river. A few rocks, which are soluble in water like salts, limestone etc. dissolved in the water of the river and are displaced in the form of solution in the water.

Erosional and Depositional Work Done by Rivers

The longitudinal cross-section of river valley represents a very smooth curve from the source region to the mouth. It is flattened at to near the mouth and slightly rising upstream. The river represents differential nature of denudational and aggradation works along the whole length of the river course.

In the upper stream area, the source region, usually have very steep slope resulting higher velocity, dominant erosional activity as especially down cutting of valley. In this part of river valley, depositional work is almost absent. A number of landforms like V-shaped valley, gorge or canyon, waterfall and rapids and pot holes are formed by the erosional work of the rivers. Each landform with reference to their formation process is discussed below.

Gorge or Canyon

The river in the mountainous regions, flows through very narrow channels. Due to steeper slope and greater velocity, deepening of valley is continued to a great extent and a narrow and deep valley resembling the I-shaped, is described as gorge and canyon. Himalayan rivers have cut a number of gorges along the Siwalik range. They are formed due to down cutting and development of pot holes and plunge pools. The depth and slope of gorge depends upon the nature of the rocks. When softer rocks like shale is bounded by resistant rocks like granite and basalts from two sides, the valley will be more narrow and deeper.

Water Fall

A number of resistent rocks lie in the path of river valley. Harder rocks are eroded with rather slow rate as compared to adjoining softer rock. Resistent rocks stand stable and river water falls down from greater height of resistant rocks into the softer rocks. This type of falling or plunging down of water is called as waterfall and rapids. Thus the waterfalls usually occur into three geological conditions, i.e. (i) harder rocks overlie the softer rocks horizontally, (ii) the tilted rocks should be dipped towards up stream, and (iii) the harder and softer rocks should occur alternatively with vertical locations (Fig. 8.1). Under the above discussed, conditions a number of waterfalls are formed by rivers in their youth stage in the world. For example Niagra waterfall (U.S.A.), Victoria fall on Zembezi river in South America and Jog or Gersoppa in India.

Rapids : When a band of resistant rocks occur very close to one number

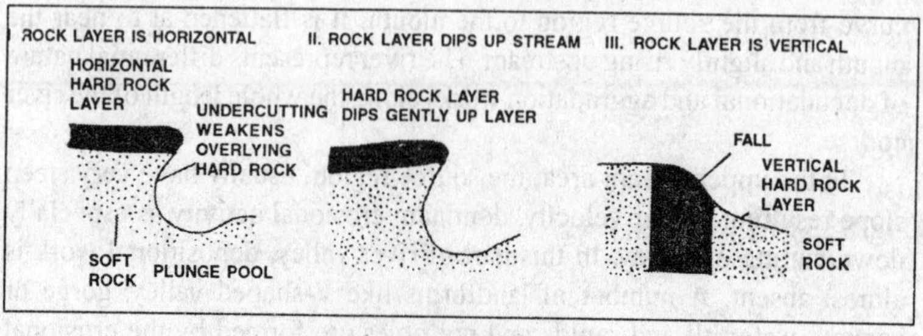

Fig. 8.1 : Waterfall.

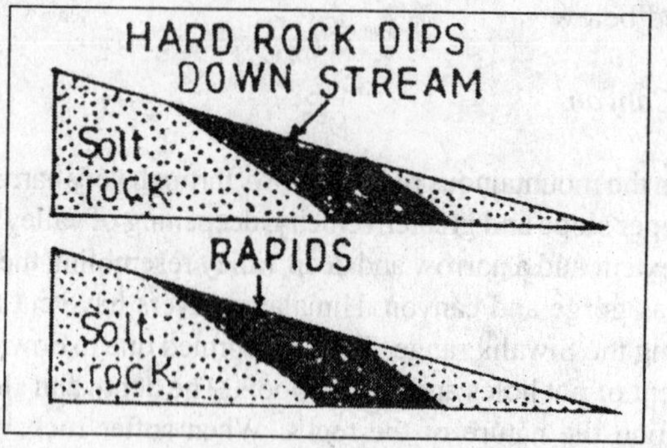

Fig. 8.2 : Formation of Rapids.

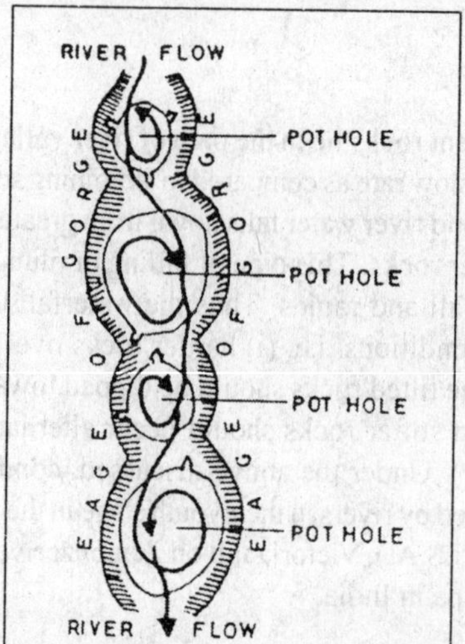

Fig. 8.3 : River Pot Holes and Rock Gorge.

in the way of river valley, the river erodes the softer rocks easily and the resistant rocks remain standing. Thus the water of the river flows through step like valley falling down from one step to another. Such type of falling water from relatively lower height is described as rapids. They are very dangerous for ship sailing in the river. The rapids may be also formed when the resistant rocks dip towards downstream (Fig. 8.2).

Pot Holes : When some of the fragmented rocks are cought up in crevices or joint of river valley and these blocks of rocks make whirling motion along with the water. Due to abrasion action of the rocks, a hole is formed on the bed of river. Such recorded depression in the bed of river are described as pot holes. The diameters of pot holes may vary from few cms to metres and depth sometime reaches 7 to 8 mts when the pot holes become larger and deeper known as plunge pool. These pot holes or plunge pools may remain for longer time if they are formed of resistant rock and it may remain for shorter time if they are formed by softer rocks (Fig. 8.3, 8.4).

Moreover, the river descends down from the highland to the plain. In the plain region the velocity, gradient, erosional and transportational capacity of river are reduced to a considerable extent. A number of tributary rivers join the mainstream and consequently the discharge of water and loads of river also increase. Lateral erosion and depositional works are also taking place and down cutting is relatively slowed down due to decline in gradient of the river. Under this situation of river, different kinds of erosional and depositional features are formed in middle and lower courses of river. These landforms briefly discussed below.

Alluvial Fan or Cone : When the river enters into plain area from the mountain, sediments of coarse sized transported by river are deposited along the meeting point of mountains and plain. Such type of depositional feature along the foot of the mountain is known as alluvial fan or cone as they are having appearance like cone or fan. When a number of cones combine together, is described as piedmont plain (Figs. 8.5, 8.6).

River Meanders and Oxbow Lakes

River in the plain area gets more and more loads by different tributaries on the one hand and carrying capacity of river declines on the other hand due

Fig. 8.4 : Formation of Pot Hole.

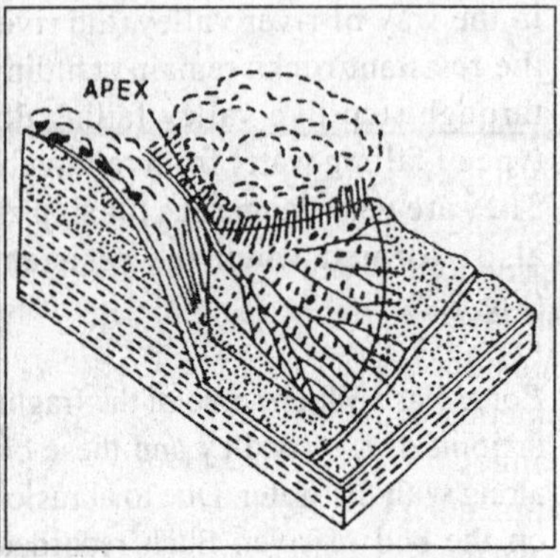

Fig. 8.5 : Alluvial Fan.

to change in the gradient from steeper to gentler. Due to decrease in the transportation capacity of it, the deposition of sediments along the sides and on the bed of valley takes place. When the accumulation of loads on the one side becomes greater than other side, the river beds towards the lower side by making erosion and a small turn develops in the river course. With the continuation of this process, this turn develops further and a meander is formed in the course of river like S-shape. The S-shaped river course is described as meanders of the river (Fig. 8.7). When a river course bends, the strong current follow the outer bank of meander. Thus outer bank is undercut and a outer steep bank of meander is formed. This steep bank is called as clift of the slope. Simulaneously, along the inner bank of the bend, deposition of sediment, takes place and this deposited inner banks known as slip of slope : Thus the formation of cliff and slip of slope cause the curve of meander to be enlarged and the neck of its to be narrowed. During the flood period when discharge of water is increased, the river follow the straight course and cut it off the meander from the neck. Thus the detached meander loop changes into oxbow lake (Fig. 8.8)

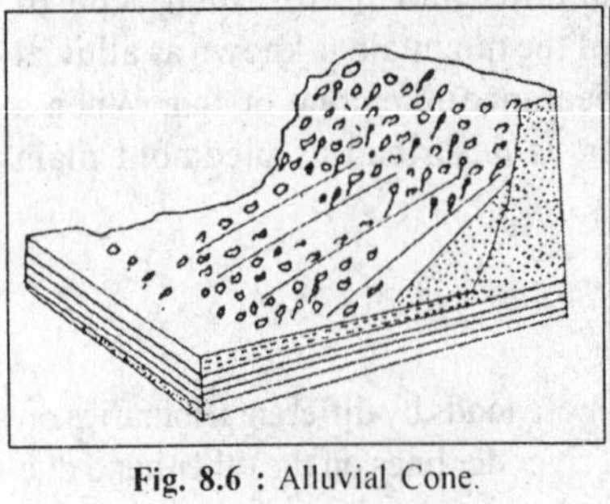

Fig. 8.6 : Alluvial Cone.

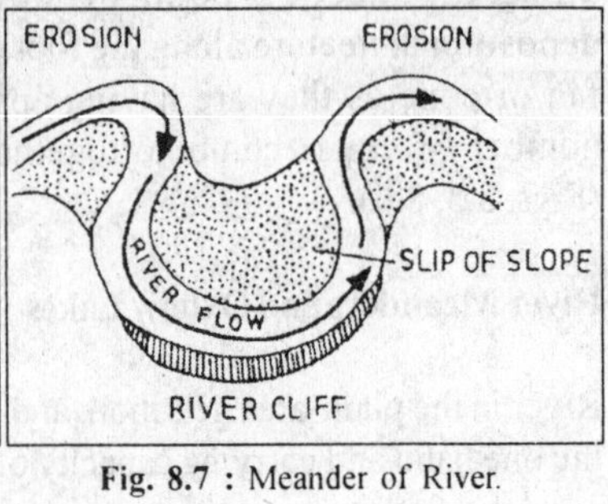

Fig. 8.7 : Meander of River.

Floor Plain and Levees

During the monsoon season the existing river channel is not enough to adjust the excess rain water in it and the water flows over the banks of the river and spread in a wide area on the either sides of the river channel. During the dry season when the water retreats to its channel, deposits of thin layer of fine sediments are left behind on the flooded area. This depositional plain on the both sides of river is described as flood plain. Sometime the hill like deposi..ion takes place on the either sides of the river channel by slack water (flood water). Successive floods, add more sediments to these ridges, so that they grow upward untill they are well above the level of flood plain. This ridge like embankment along the sides of river in called as levees (Fig. 8.9). Deposition also takes place on the river bed itself, so that the level of river bed rises up. Thus the river becomes liable to break up the levees and cause disastrous floods.

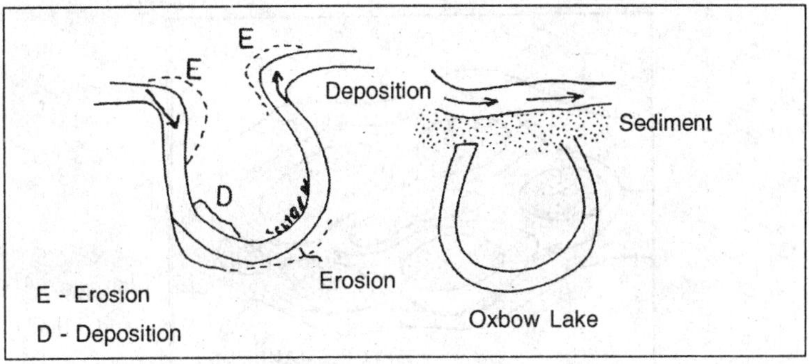

Fig. 8.8 : Formation of Oxbow Lake.

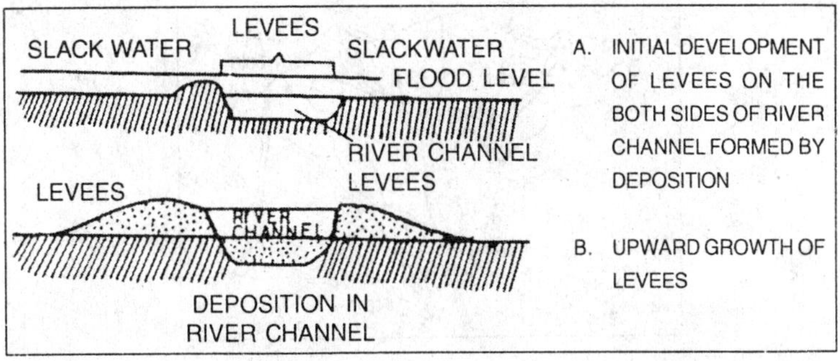

Fig. 8.9 : Flood Plain and Levees.

Braided River : In the old stage of the river when the river becomes broader, deposition becomes dominant due to reduction of gradient to considerable extent. Deposition of sediment in the river bed occur into form of sand bars. Due to inability of river to remove the bars, a number of minor streams bifurcated and flow between the bars like distributaries. Such river with sand bars and distributories in old stage is called braided river (Fig. 8.10).

Delta

The river, originating from highland, ultimately joins the lake or sea. At this stage when river meet the sea, the velocity of river is reduced to almost zero and the transported loads are deposited at the mouth of river forming a triangular strips like the Greek letter delta. Due to this reason the plain at the mouth of river is called delta (Fig. 8.11).

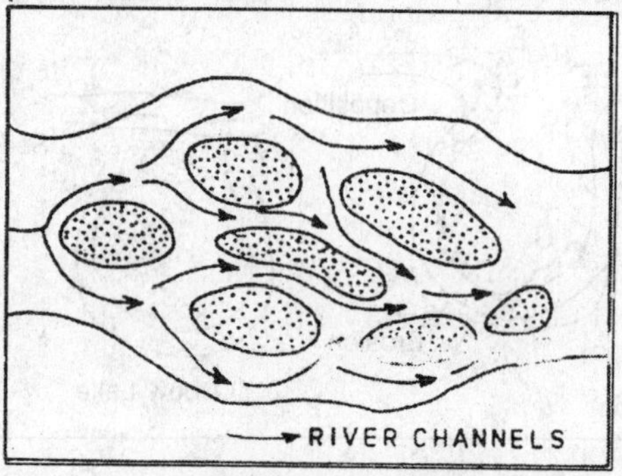

Fig. 8.10 : Braided River.

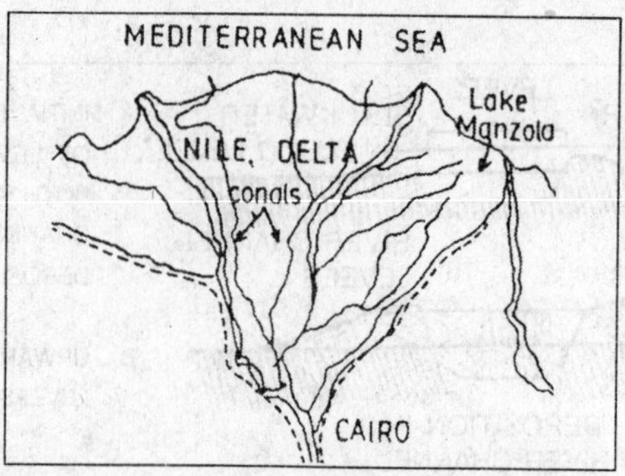

Fig. 8.11 : Formation of Deltas.

Moreover, the formation of delta is a time taking process. It gradually extends seaward under the suitable conditions. Gentler coastline, absence of active tide, wave and ocean current and larger amount of river's loads are the favourable condition for the better development of delta over the time. A number of minor streams are flowing over the delta. These streams are formed when the main channel of the river is choked up by deposition of sediments and the water of river bifurcates into different minor channels formed over the delta. Such minor streams bifurcating from the main river are known as distributories. Sunderban delta of Ganga-Brahmaputra rivers, delta of Nile river etc. are the important examples delta found in the world.

Cycle of Erosion

When any part of the earth is uplifted, a number of agents of denudation attack the elevated part of the earth's surface. This elevated region has to pass through different stages to attain the grading level on the surface. The rate of denudation depends upon the nature of rock forming land-forms, strength of the agent of erosion has been continued. Thus W.M. Davis has propounded a dictum regarding the development of a landform on the earth surface. The dictum is that a landform is the product of structure, process and time. According to him, the agent of erosion makes endeavour to reduce the highland into rolling plain level or peneplain. Such massive change in landform occurs through a well defined sequence or stages of process of erosion. These sequences are described as cycle of erosion by W.M. Davis. According to him the initial raised relief is fully reduced at the completion of this cycle and the surface is levelled down by that time. Thus usually three stages like youth, maturity and old age, have been recognised by geomorphologists for explaining the evolutionary history of landforms by different agents of change.

Rejuvenation of Rivers

The cycle of erosion is supposed to be completed when the highland is levelled down to a peneplain and the down cutting of the river is reached to its last limit and further depending is ceased. The level, beyond which the further down cutting by river is restricted, is described as the base level of the river. This base level is controlled by the level of sea water or lake water where the river disposes of its water ultimately.

Moreover, sometimes, the sea level falls down due to climatic change or the some portion of up stream in uplifted due to tectonic forces. This change either in the sea level or in the relief of upper part of river valley disturbs the base level of the river and the process of erosion is renewed before completing the cylce of erosion. Thus the recycling of erosion by river, due to change in the base level, is called as rejuvenation of river. However, the cycle of erosion is rarely completed and it is merely theorerical rather than practical.

Landforms Formed by Rejuvenation of the River

Alluvial Terraces

The rejuvenated river cut its new channel within the pre-existing channel. A new flood plain is also formed on the both sides of the new channel of rejuvenated river but below the old flood plain formed by the old river. When the river is rejuvenated again and again, a number of flood plains lying one below the another, exhibiting a step like structures, are formed. Such type of landform is described as alluvial terraces (Fig 8.12).

Incised Meanders

The rejuvenated river is not able to adjust itself in the meanders of old river and hence it cuts new channel within the pre-existing meanders. Thus the new carved channel in the old meander is described as "incised or entrenched meander".

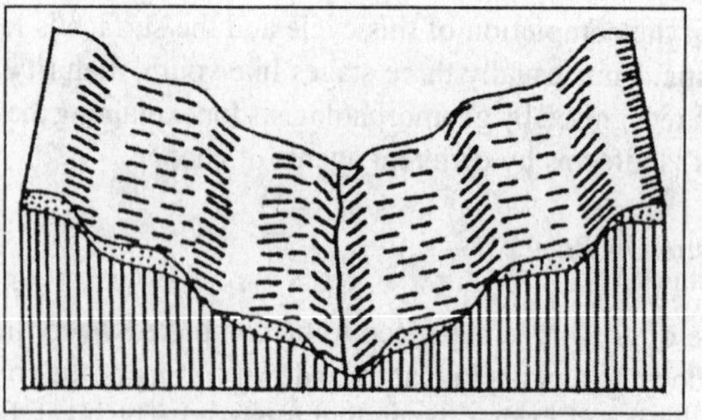

Fig. 8.12 : Example of Paired River Terraces.

Development of River Valley

A river valley from its source to the mouth has to pass through different stages and gradients. In the begining or youth stage the river valley has very steep slope and erosional works are dominant and depositional functions rarely occur. Deep valley like gorges and canyons are formed. Waterfalls, rapids and potholes are not uncommon in this stage of the river. At the mature stage, the slope is gentle and widening of valley takes place. Both erosional and depositional works go together and balanced to each other. A number of erosional like features oxbow lakes, meanders, rock benches etc. and depositional features like alluvial fans, flood plains, levees etc. are formed at this stage of river. Lastly, at the oldstage, gradient becomes almost zero and only depositional work is prevailing and a depositional feature like delta is formed at the mouth of river where river meets the lake or sea. Thus the longitudinal profile of river valley shows very smooth curve from head to mouth. The slope of valley increases slowly towards the heads of the valley and relatively flat near the mouth. Such type of river profile is described as 'graded profile' (Fig 8.13).

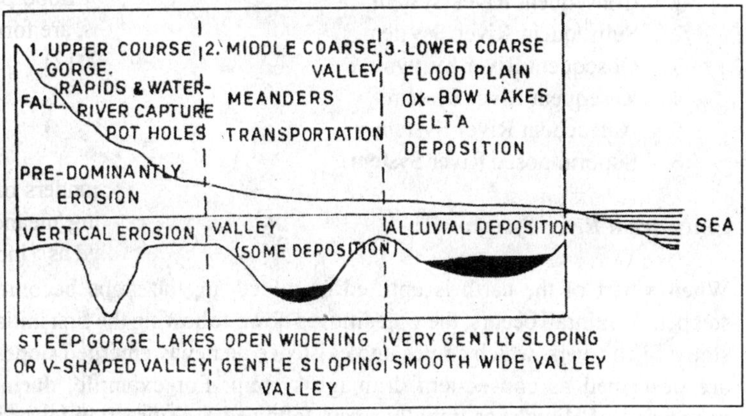

Fig. 8.13 : Development of River Valley.

Drainage Basin, Type and Pattern

From an uplifted or watershed area a number of rivers originate in different directions following the nature of slopes. With the passage of time, a river is joined by various tributary rivers between its source and mouth. The river becomes a large one in terms of discharge and loads and consequently a large area may supply water and loads to the respective rivers flowing through it. This region of supply of water and loads to a river and its tributaries is called as drainage or river basin or river's catchment area.

The river begins in the form of small stream as rill and gradually enlarges in length, breadth and depth as it moves down the hills and enters into plain areas. It is extended and integrated through the process of differential rates of erosions and depositions following the structure, nature and dip of the underlying rocks. A large number of small tributaries meet the main stream at different points at different angles on either side of river channel. Consequently a very complex network of river channels develop in a given river basin. This complex system of drainage is called as drainage system or order.

The rivers show different orientation and integration with the geological structure and the dip of underlying rocks over which they flow. Direction and depth of rivers and nature of integration of a number of tributaries with the main river vary in space and consequently following types of river system have been recognised by geomorphologists.

1. Consequent River System.
2. Subsequent River Ssystem.
3. Obsequent River System.
4. Resequent River System.
5. Antecedent River System.
6. Superimposed River System.

Consequent River System

When a part of the earth is uplifted or folded up, the slope becomes steeper. As rainfall occurs, the water moves down following the first initial slope. Such rivers, which comes into existence on newly emerged slopes, are described as consequent drainage system. For example, during orogenesis of Himalayas initially rivers followed N-S direction towards Indo-Gangetic depression. Volcanic cones, folded structures and emerged submarine floors are ideal regions of consequent river system.

Subsequent River System

As the consequent river moves ahead from mountain to plain, a large number of tributary rivers join the former stream at different points. These tributaries also follow geological structure and the existing slopes on the either side of consequent rivers. They are usually lying transverse to the course of master consequent. Thus the rivers transverse to the consequent due to geological, structural and tectonic causes are called as subsequent rivers. In folded structure the land surface is thrown into various ranges and a large number of intervening and longitudinal valleys which are transverse to the consequent river. Arrangement of soft and hard rocks alternatively also help in the development of subsequent rivers.

Obsequent River System

When the river flows in direction to the opposite flow of consequent river is known is obsequent river. It is also a consequent river system because of following the direction of slope on emerged land. But due to opposite direction to the pre-existing consequent river, it is described as obsequent. Such rivers may originate on the opposite side of watershed. For example, Himalayan region where Ganga and Yamuna follow N-S direction while the numerous streams occur on the northern slope following S-N direction. Sometimes, obsequent tertiary tributaries join subsequent rivers in a given river basin due to local change in the slope of the surface and dip of rocks.

Resequent River System

In this river system, river flows in the direction of master consequent river. Reconsequent river originates at late age than the main consequent river. It is a post-peneplain river which comes into being after completion of first cycle when inversion of relief is created. Anticlime is set at the place of old syncline and *vice versa*.

Antecendent River System

The river which cuts its valley without taking into consideration the geological structure and tectonic upheaval of land across its course, is known as antecedent river system. In this system rate of down cutting by the river exceeds the rate of upliftment. Many rivers like Sindh, Satluj, Chaghra etc. are important antecedent rivers in Himalayan ranges. Rivers originated initially from inner Himalaya in N-S direction. With the passage of time, many other ranges came into existence in southern parts and the pre-existing rivers maintained their channels across the southern ranges irrespective of their upliftment during orogenesis process.

Superimposed River System

Some rivers do not follow the slope and geological structure of underlying rocks. When a soft younger rocks overly the old crystalline or hard rocks with contrasting dip and slopes. The river develops over the new younger overlying rocks following slope and structure. Gradually overlying rocks are eroded down up to the hard underlying rock, with different geological structure. The river cut the hard rock at very slow rate and maintain its original course irrespective of dip and structure of underlying rocks. Such river is described as superimposed drainage system. The radial drainage of lake district in N-W England is superimposed on the existing surface rock.

Drainage Pattern

The chief river, with its tributaries from different sides, form a definite geometrical design in a given river basin area. This network of main and tributary rivers is described as river pattern. Geomorphologists have broadly recognised following river pattern in the world.

1. Dendritic pattern.
2. Trellis and rectangular pattern.
3. Parallel drainage pattern.
4. Radial pattern.
5. Centeripetal pattern.
6. Annular pattern.
7. Herring Bone pattern.

Dendritic Pattern

In an area of uniform structure and resistance of rocks, consequent river and its tributaries make a network of valleys like a tree. In upper part a large number of tributaries join the main consequent stream and the middle course remains almost without any river on either side. At the mouth again numerous distributaries and subdistributaries diverge in various directions. The subsequent (tributaries) rivers usually form acute angles at the confluence points. Between the valley of consequent river and tributaries inter-fluves appear like a crests or ridges. Thus a drainage pattern analogous to a tree is known as dendritic pattern. It is very common in all climatic and geomorphological regions all over the world (Fig 8.14).

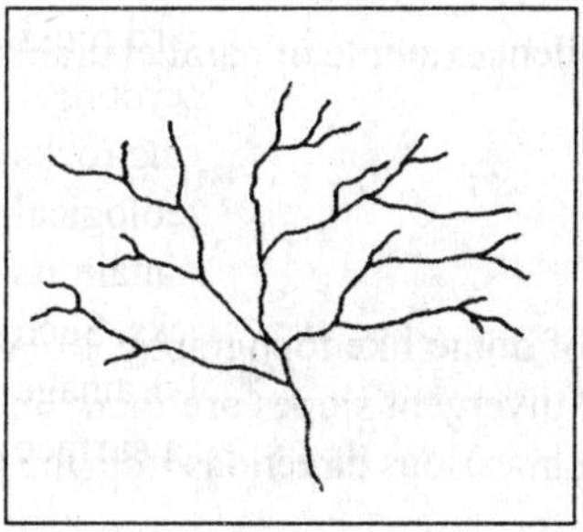

Fig. 8.14 : Dendritic Drainage Pattern.

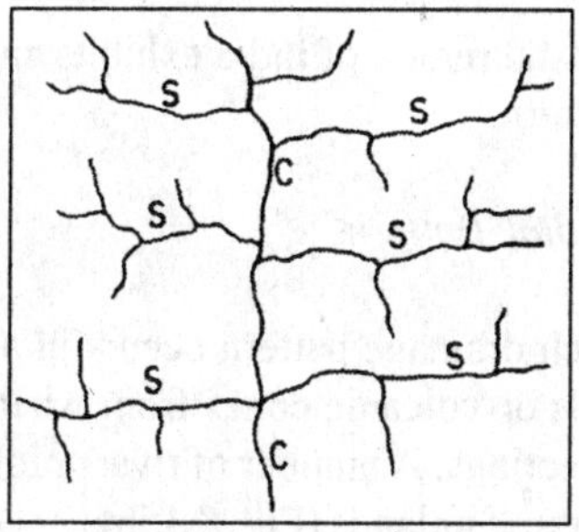

Fig. 8.15 (a) : Trellis Drainage Pattern.

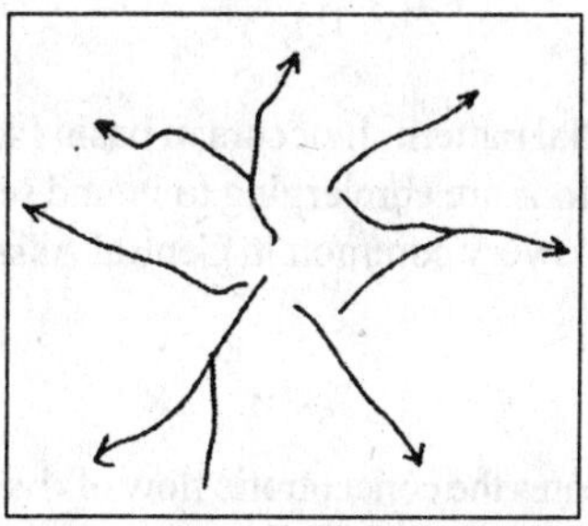

Fig. 8.15 (b) : Radial Pattern.

Trellis and Rectangular Pattern

This type of drainage pattern exhibits a rectangular geometric design of the network of consequent and its tributary rivers. In this pattern the consequent river follows the dips and the subsequent rivers are parallel to strikes. Alternation of parallel hard and soft rocks transverse to the consequent valley is the ideal condition for the formation of trellis pattern. Such drainage (Fig. 8.15a) pattern is common in folded structure and coastal areas. When the space between subsequent rivers is wider, the pattern is described as rectangular whereas the space between them is narrower it is called as trellised pattern.

Parallel Drainage Pattern

A large number of rivers are forming parallel course in a given area. This pattern of river valleys is described as parallel pattern. It develops in the same geomorphological and geological conditions as required for trellised pattern. Such pattern is usually common in narrow coastal region. Western coastal region of India exhibits an excellent example of parallel drainage pattern.

Radial Pattern

Such drainage pattern occurs in areas of dome like topography, isolated hills or volcanic cones from where the divergent slopes are found in all directions. A number of rivers radiating in various directions from the top of these uplands (Fig. 8.15b).

Centripetal Pattern

It is an opposit to radial pattern. It occurs in basin like region where rivers from different directions are converging to inland seas, lakes or structural basins. Such pattern is very common in Central Asia (Fig. 8.16).

Annular Pattern

Annular pattern indicates the concentratic flow of rivers around the uplands. It, usually, develops when hard & soft rocks are alternatively arranged in concentratic form in a given dome like structures (Fig. 8.17).

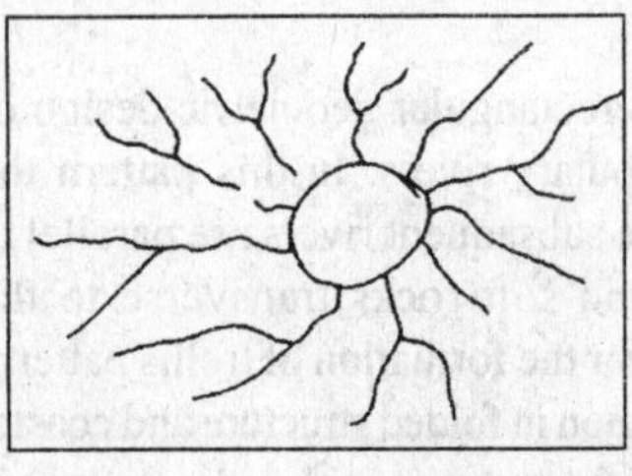

Fig. 8.16 : Centripetal Pattern.

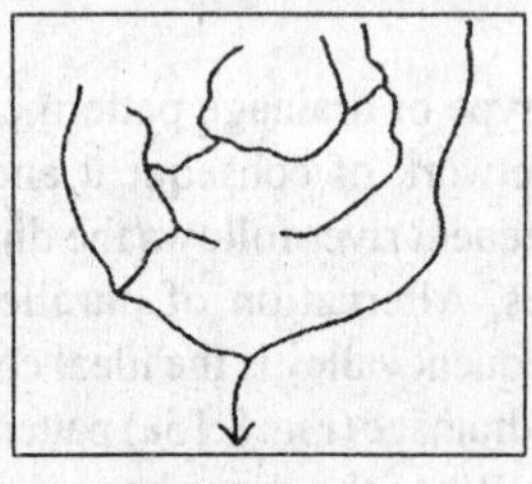

Fig. 8.17 : Annular or Ring like pattern.

Herring Bone Pattern

It is a rib like pattern which is formed by joining of a number of parallel tributaries from both sides of the consequent river. Tributaries converge the main stream at right angle. Such pattern of drainage occurs in the mountainous region where the valley has steep slopes. For example, valley of Kashmir is very important example of Herring bone drainage pattern. Moreover, when the tributaries meet the main river at acute angles, pinnate pattern is formed.

River Capture

The growing rivers widen and extend its valley through the process of headward erosion of water divides and lateral erosion of interfluves between two rivers. The rivers on the different sides of waterdivide do the work of erosion at different rates depending upon the amount of precipitation, gradient and nature of rocks. Gradually, the most powerful stream cuts back the water divide and interfluve at faster rate and consequently water divide or barrier between them is removed. The weaker river, usually lying at relatively higher level and the water of this river begins to divert and flow into stronger river by detaching the lower course from the upper portion of the captured river. This process of diversion of flow of water from any river is called as River capture, piracy or beheaded in geomorphology [Fig. 8.18 (a) & (b)]. The capturing or robbing river is called as pirate river and the captured stream which water is cut off from the head water, described as beheaded stream. The point of capture or point of diversion of water is known as bend of elbow. Some part of beheaded river valley immediate below the elbow point becomes dry due to cutting off from the head water. This part of valley is defined as wind gap and the remaining beheaded river is called as misfit river because the valley becomes very narrow unfit to the pre-existing original valley. There are three types of river capture.

(i) River capture through Head Water Erosion.
(ii) River capture through Lateral Erosion.
(iii) River capture through Inter-section of meander.

River Capture Through Head Water Erosion

When two rivers are flowing along the opposite side of a watershed, the stronger river cut back the watershed by headward erosion. After sometimes, the head of the weaker river on the opposite side is cought up by the stronger one and ultimately the water of coughtup river starts to flow into the stronger river.

River Capture Through Lateral Erosion

When two or more rivers flowing into any area, they must maintain some

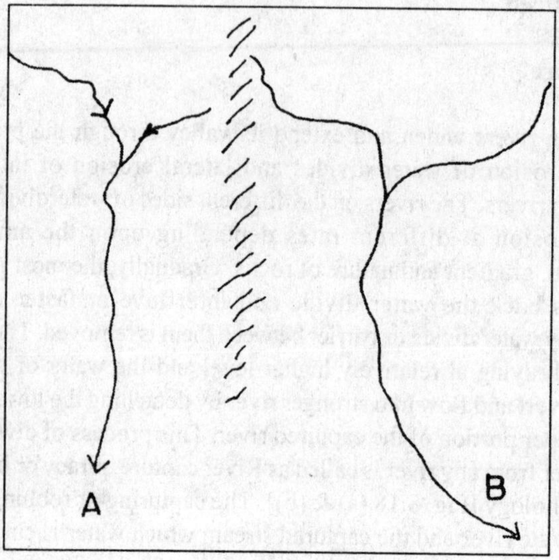

Fig. 8.18 (a) : Rivers A and B flow on either side of Watershed

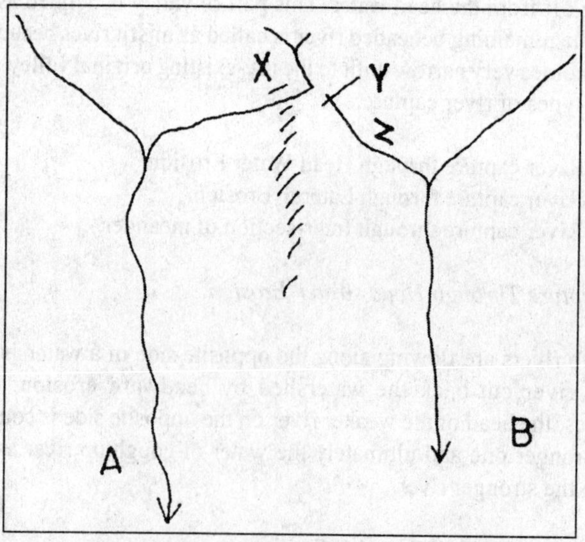

Fig. 8.18 (b) : River B is Captured by A
X — Elbow bend, Y — Water gap, Z — Pirate or Beheaded River or Misfit stream.

spatial gap (Interfluves) among them. Through the lateral erosion especially in mature stages, the interfluves between two rivers is reduced and narrowed. After sometime, the courses of the river join to each other. Thus the water of weaker and shallower river is drawn by stronger, steeper and deeper river. This process is described as piracy or capture of river by lateral erosion.

River Capture Through Inter-Section of Meander

During the mature and old stage, the rivers make a number of meanders due to declining in the slope and lateral erosion and deposition of sediments enroute. Sometimes, the meanders of two closely spaced rivers are intersected with each other. The strong river catchs up the water of weak river. It is known as the capture of river by inter-section of meanders. It follows the same process as in the river capture by lateral erosion.

9

GLACIERS

Definition and Meaning

Precipitation takes place in form of snowfall in higher latitudes and high altitude regions in the world. Snow accumulates after the snowfall and gives rise to snowfield. The snowfield lies above the snow-line. The snow-line is defined as a lowest level above which perpetual snow is found. Due to increasing weight of accumulated overlying snow, the loose snowflakes convert into hard solid crystalline ice. The underlying compact solid ice is called as Neve or Firn. When the ice becomes so thick that the lower layer becomes like plastic and consequently moves outwards or down hills under the influence of overlying pressure of accumulated snows. Thus the mass of ice moves either lateral or down hill, is called as glacier.

Kinds of the Glaciers

Glaciers are widely spread in the area where temperature lies below freezing point. Glaciers are found in every continents except Australia. The source of the glaciers, i.e. snowfield occurs above snow-line. The height of this line varies from region to region. It is at over 5000 mt. in equatorial region and at the sea level in polar region. Glaciers are found at different localities and spread into different forms on the basis of their occurrences, and dimension. They may be grouped as : (i) Valley glaciers, (ii) Continental glaciers, (iii) Piedmont glaciers.

Valley Glaciers

At sufficiently high altitude both in high and low latitudes, glaciers are formed due to low temperature and high orographic precipitation. They are long and narrow because they occupy previously formed river valleys and they move down slope from colder climate to the warmer climate where glacier begins to melt. Due to their location in mountainous region, they are also called as mountain or alpine glaciers. Such glaciers are commonly found in Himalayan ranges and Alps mountain. Siachen (72 km), Gangotri (25.5 km long) are important Himalayan glaciers.

Continental Glaciers

In Arctic and polar regions, the termperatures are low enough and snows accumulate over wide areas due to wide spread snowfalls. As a result, the areas of many thousands of sqaure of kms are buried under gignatic plates of ice. During the glacial period, large part of northern Europe, Asia and northern half of the North America were covered by thick sheet of ice. This extensive type of ice mass is called as continental glacier of ice sheets. Now the whole of the Greenland and Antarctica continent are covered with ice sheets. They are several thousands of metres thick and occupy a total area of about 9 million sq. kms.

Piedmont Glaciers

When the ice are accumulated along the foothill is known as piedmont glacier. It is formed when a number of valley glaciers descend down upon the glaciated plains and unite together and form a thick ice sheet covering plain and foothills. Such glaciers are generally found amidst the continental glaciers where mountains occur. For example, the Bering glacier of Alaska.

Denudational Works of Glaciers

Like the other agents of denudation, it also does the work of erosion, transportation and deposition. Different kinds of landforms are formed by these works of the glaciers especially in cold regions. The works of glacier are discussed below.

Work of Erosion

The mobile blocks of ice catch up the rock fragments found in the way of glaciers and these cought up materials work as tools of erosion by glaciers. The following processes are involved in the erosional works of the glaciers.

Frost Wedging : In the neighbourhood of glaciers, the fractures, cracks and pores in the country rocks are filled by water formed by melting of ice. The filled water refreezes in the opening and expands in volume causing excessive pressure against both sides of openings. Again, the frozen water melts due to increasing temperature by any reason and pressure of freezing and melting of water and ice respectively, the rocks are weakened and shattered off into different fragments. Such process of wearing and tearing of the country rocks is called frost wedging.

Plucking of Quarrying : The frost wedging process is followed by plucking process. The plucking process is one in which the .broken and shattered rock fragments are cought up by mass of ice and carried away from their original place to other place by glaciers.

Abrasion : The glaciers erodes their valleys with the help of plucked materials with them. The carried materials like pebbles and gravels, caught up firmly within the mass of ice, rub themselves against the valley floor and sides of the valley and consequently both the pebbles and gravels, and valley floor are grooved, polished and straited due to frictional effect between these two. Such process of wearing and tearing of rocks by any agent of erosion is called as abrasion.

Transportation

As discussed earlier, the broken or weathered rocks are cought up by mass of ice within it and they are plucked and transported from their parent rocks. These carried materials and firmly caught up till the time when the glaciers begin to melt. After the thawing of ice the caught up materials are dropped down and deposited at suitable places.

Deposition

The materials, removed by valley glaciers through plucking, scratching or abrading, are dropped down and deposited enroute and at its terminal point. These glacial deposited materials are unsorted, unstratified and angular unlike the sediments deposited by stream. Such kinds of all glacial deposits are called as "Glacial Drift or tills".

The deposition of glacial derived materials takes place at different locations along the valley glacier. According to mode of occurrence, the glacial deposits may be grouped as following (Fig. 9.1).

(i) *Terminal Moraines :* The deposits of glacial origin materials at the end of the valley glacier due to melting of front of the glacier.

(ii) *Lateral Moraines :* Debris also tends to accumulate along the edge of glacier as it flows down the valley. After melting of ice, the material remains as trace like deposit of till along the side of glacier is called a Leteral moraines.

(iii) *Medial Moraines :* The glacial origin materials are also deposited at the junction of two glaciers and it continues down the central part of valley. Such deposit of till is known as Medial Moraine.

(iv) *Ground Moraine :* When the glacial deposit is found on the bottom of glacial valley is called as 'Ground Moraine'.

Fluvio-glacial Deposition

At the end of glacier, the ice melts into water and a stream flows from the glacier's front. Fine particles of sediment are carried away by the stream and deposited after the terminal moraine. The deposited materials are sorted and stratified and they are collectively known as fluvio-glacial deposit. Such deposits are of several types as given below (Fig. 9.2).

Outwash Plain : Stratified and sorted sediments are occurring down the valley beyond the terminal moraine. These materials are deposited by streams, originated from the end of the valley glacier after the melting of ice. These deposits form an extensive plain having the characteristics of flood plain. Such land form formed by fluvio-glacial deposit is called "outash plain".

Kettle Hole : Kettle holes are said to be formed by due to enclosure or burial of block of ice within glacial drift and with the passage of time the arrested blocks of ice are melted and consequently a number of hollows are formed on the outwash plain. Such hollows are described as kettle holes. They are either circular or elongated in shape and vary in size and dimension considerably.

Kame Terraces : They are ridge like glacio-fluvial deposits found usually in valley of glacier, forming moraines and an outwash plain. They are isolated or collection of windly mounds formed by deposition of pebbles and gravels by sub-glacial streams in the region of eskers.

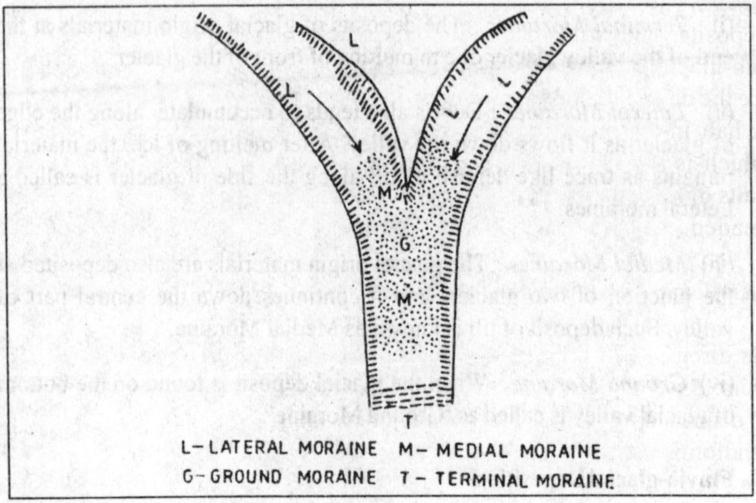

Fig. 9.1 : Glacial Deposits.

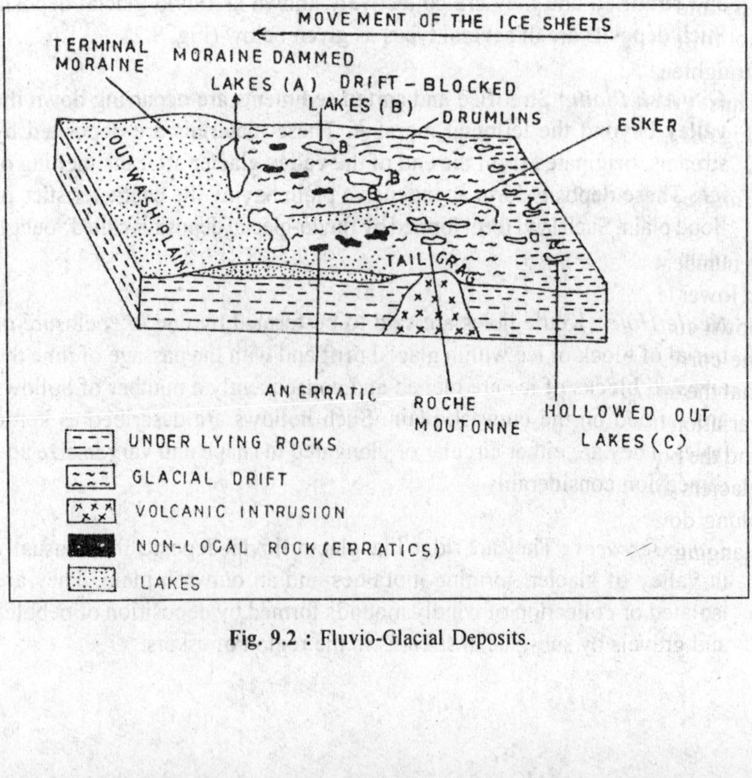

Fig. 9.2 : Fluvio-Glacial Deposits.

Esker : They are meandering network of ridges formed by glacio-fluvial deposits. They are sinuous and beaded extension of delta formed by sub-glacial streams other than terminal moraines. On the recession of ice the delta extended backward in the form of narrow band of deposits. A halt in the receasion of ice caused an increase in deltaic deposits which is shown by swelling in the otherwise narrow ridge. Several halts of ice are marked by series of swellings and the esker is said to be beaded.

Landforms Formed by Glacier Erosion

As discussed earlier that the plucked materials are abrading the valley floor as well as abutments of the glacier valleys. Consequently, a number of distinct topography are carved out by the glaciers. Some of the important landforms formed by the glacial erosion are given below.

U-Shaped Valley

Initially the glacier moves in the V-shaped river valley. Ice armed with rock debris, held firmly with the ice, abrades effectively its valley laterally and vertically. Moreover, the glacier moves very slowly and does not follow the sharp and acute bends found in the course of valley but it straightens up its course and cut or grooves the topographic projections. Thus V-shaped valley is deformed into steep walled U-shaped valley (Fig. 9.3).

Hanging Valley

A number of small tributary glaciers join the main large glacier lying relatively at lower level. The smaller tributary glaciers are relatively less effective at down cutting than the main glacier because of their lesser thickness and the corresponding reduced pressure. Ordinarily they join in a such way that the upper surfaces of glaciers are concordant or nearly so. Because of variation in the rate of erosion, valleys are deepended by different amounts and the bottom of glaciers may be at different level or elevation. When the glaciers are melted, the water is flowing down the tributary valleys and plung down in main valley as a waterfall. Such topography is called as Hanging valley (Fig. 9.4).

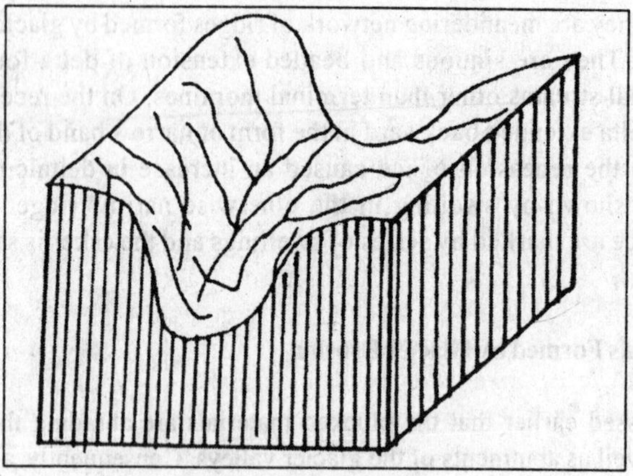

Fig. 9.3 : U-shaped Valley.

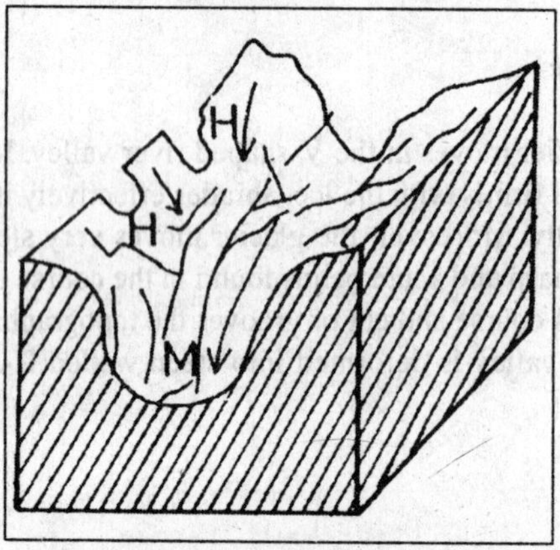

Fig. 9.4 : Hanging Valley.

Cirques

Cirques are amphitheatre or armed chair like depressions at the head of the glacier along the mountain wall. Valley glaciers excavate material in a number of ways. Rapid erosion occurs at the head of the glaciers. When the glacier moves down the slope a crevasses develop in the glacier which reach up to rock. This kind of opening, through which melted water reaches into cracks of head wall to work as frost wedging is called as 'bergchurund'. The shattered rocks are caught up by moving ice and through the plucking and abrasion process, the head wall is eroded back. The combined process of back ward erosion of head wall is known as Nivation. By this process the valley is eroded headward and an armed chair like depression is formed at the head of the glacier (Fig. 9.5). After the melting of ice, the depression is filled by water and converted into a lake. Such lake is known as Tarn.

Horn and Arete

When an adjacent cirque grows up along the opposite slopes of mountain, the wall between these two cirques, gradually becomes narrower and pointed due to headward erosion by them. Such sharp and irregular ridge is known as arete or when three or more cirques develop on various sides of the mountain. The various aretes form a pointed peak at the top of different cirques. This pointed peak shows pyramid shape, which is called as Horn or Mattern horn (Fig. 9.6).

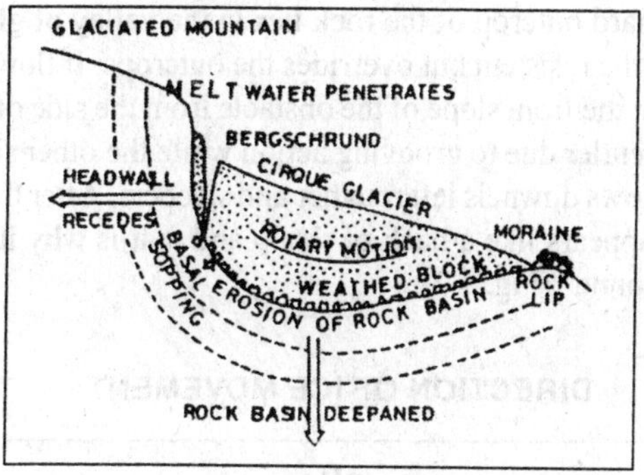

Fig. 9.5 : Formation of Cirque.

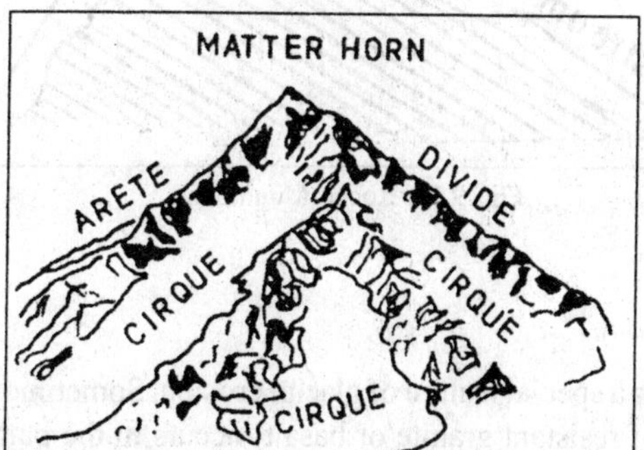

Fig. 9.6 (a) : Cirque Matter Horn A. R. etc.

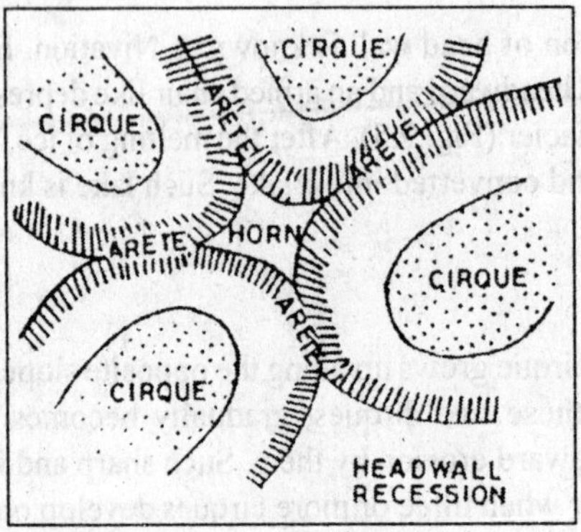

Fig. 9.6 (b) : Horn, Aretes and Cirques.

Roche Mountonnes

Sometimes, a hard outcrop of the rock lies in the valley of glacier, and it does not avoid the resistent but overrides the outcrops. It flows over them in a manner that the front slope of the obstacle from the side of the moving ice, becomes gentler due to grooving action while the other side leeward, where the ice flows down, is left rougher and steepers. After the melting of the glacier, it appears like a back of sheep and that is why it is called as "Roche Mountonne" (Fig. 9.7).

DIRECTION OF ICE MOVEMENT

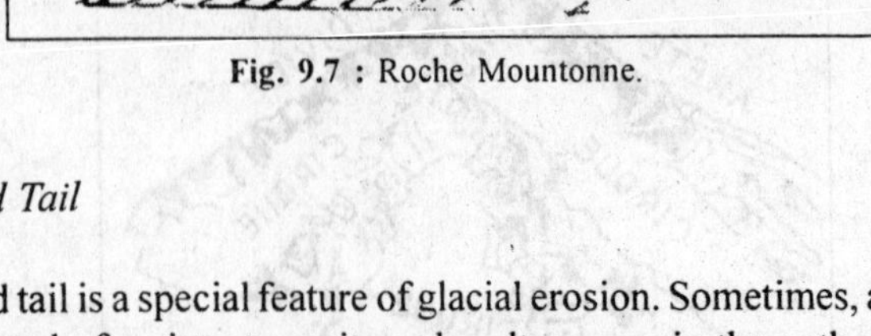

Fig. 9.7 : Roche Mountonne.

Crag and Tail

Crage and tail is a special feature of glacial erosion. Sometimes, a volcanic plug, formed of resistant granite or basalt, occurs in the path of moving glacier. This plug is surrounded by sedimentary soft rocks of especially

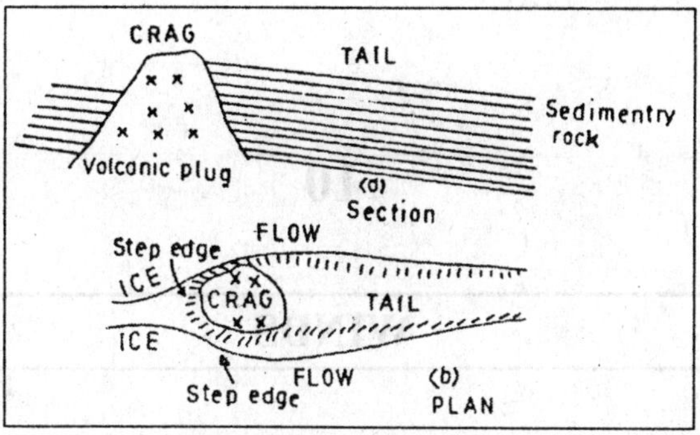

Fig. 9.8 : Crag and Tail.

shale or clay. The moving ice carves and scratches the sedimentary soft rock when it moves up the plug. After overriding the plug, the glacier moves down the sedimentary rock on leeward side. The sedimentary rock is protected and rather slowly eroded, so that leeward side remains gentler due to slow rate of erosion. Thus the ice front side shows steep and rough slope and leeward side is gentler unlike the roche mountonne (Fig. 9.8) such feature is called as crag and tail.

10

WINDS

Wind is an important agent of denudation in arid and semi-arid regions where scarcity of rainfall and very sparse vegetation occurs. The dry conditions discourage the binding of soil particles by water droplets and roots of vegetations. The absence of the vegetation may also expose the weathered rock waste mantle to the blowing winds. Thus the effectiveness of wind action in dry areas is very acute due to the absence of stabilizer such as vegetation which cover and protect solid from the erosion by friction and reduction of wind velocity near the surface. Similarly, the sand and silt when moist, can be eroded by the wind only with great difficulty but when they are dry, no restraints, exist, and surprisingly large amount of elastic sediments may be eroded rapidly. However, the loose particles produced by different process of weathering are blown off by the blowing wind from the desert to a great distance and are deposited there. Thus, like the river, winds also do the work erosion transportation and deposition.

Process of Erosion

The blowing winds do perform the erosional work in the desert region through three important processes, i.e. (i) deflation, (ii) abrasion, and (iii) attrition.

Deflation

The loose materials or mantle are picked up by blowing wind, and carried away from their original place. Consequently, the removal of dust particles exposes the hard unweathered underlying rocks which may be subjected to further erosion by winds. Thus the process of removal of dust particles by blowing winds along their course is called as Deflation.

Abrasion

The picked up sand materials by wind through deflats process help in the erosion made by the winds along their course of movements. When some country rocks lie in the path of blowing winds, the wind borne materials collide with the exposed rocks and ultimately both the attacking rock particles and attacked rocks are broken into finer sizes. Such type of wearing and tearing of both attacking and exposed rocks is known as Abrasion process.

Attrition

The wind borne materials have the tendency and opportunity to collide with each other in the way of their movement. Due to mutual collision of wind borne particles, further wearing and tearing of carried materials are taking place. Such process of breaking of rock particle is described as Attrition.

Process of Transportation

The loose particles, produced by either weathering or abrasion and attrition are transported by blowing winds from their original place to some other places. The process of carrying away of loose particles by winds takes in the following ways.

Saltation : When the sand particles are moved or transported due to forward movement of grains in a series of jumps. Such type of transportation process caused by blowing wind is known as Saltation.

Suspension : The finer size of dust particles are hanging within the turbulent winds and these particles do not reach on the earth surface and but along the velocity of blowing winds. Such type of carrying of dust particles by the wind is called as Suspension.

Erosional Works Done by Winds

The erosive force of winds is influenced by amount of the sand grains moved by wind either in the form of suspension or saltation, speed of the wind and length of time it continues to blow. These sand particles wear away the rocks by process of abrasion and polishing the rock surface just like the sand blasting machine. Thus through the deflation and abrasion process the wind forms different kinds of landforms especially in desert region. The important landforms formed by the aeolian erosion are discussed below :

Pedestal Rocks or Mushroom Rocks

The rock particles, which travel with the blowing winds, are commonly concentrated near the surface and the velocity of the wind is also greater at above the surface due to absence of friction. Thus the abrasion power of the wind is greater at a level slightly above the ground rather than the upward and downward from this level. When a column or block of rock occurs in the way of blowing wind, the upstanding rock is undercut and abrasion effect is the maximum near the base of the rock. Thus the base of the rock becomes narrower than the upper part of the rock. Consequently, pedestal rock is formed which has wider top supported by comparatively narrower base. This pedestal rock is also known as Mushroom Rock (Fig. 10.1).

Zeugen

Some desert area has the surface layer of hard rock underlined by softer rocks. The overlying hard rock layers is consisting of number of crackes and joints reaching to the underlying soft rocks. The wind enters into joints and erodes the softer rocks. Thus winds remove the underlying soft rock by the deflation and abrasion process through the rock opening in the overlying hard rocks. Thus the these gaps widen and number of furrows are formed separated by ridges of harder rocks. Such type of ridge and furrow landscape in desert region is known as Zeugen (Fig. 10.2).

Yardangs

When the desert area is formed of a number of alternate hard and soft rocks which are arranged vertically. The winds erodes the softer rocks easily by the process of deflation and abrasion rather than the harder rocks. Consequently, a number of furrows or valleys are formed at the place of softer rocks and the harder rocks appear as ridges separating the furrows from one another. Such type of ridges and furrows land scape is called as Yardangs (Fig. 10.3).

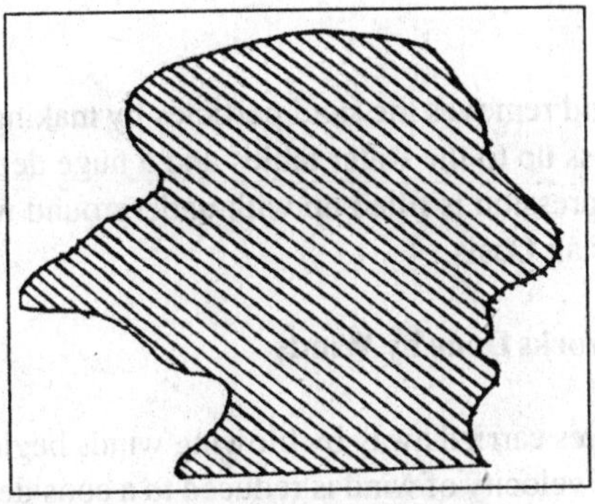

Fig. 10.1 : Mushroom Rock.

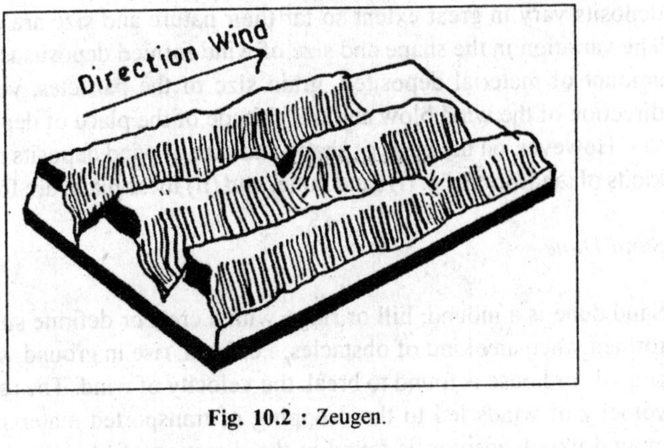

Fig. 10.2 : Zeugen.

Fig. 10.3 : Yardang.

Oasis

Sometimes, wind removes the sand particles by making erosion, through deflation process up to the water tables and a huge depression is formed later on this depression is filled up with underground water and converts into a swaup or an Oasis.

Depositional Works Done by Winds

The rock particles carried away by blowing winds begin to deposit at the place where the velocity of wind is reduced to a considerable extent. This deposition occurs either in the form of hillock or mound in thin horizontal layer, depending upon the sizes of the particles. Thus the wind formed deposits vary in great extent so far their nature and size are concerned. The variation in the shape and size of wind formed deposits is guided by amount of material deposited, grain size of the particles, velocity and direction of the wind blow and the altitude of the place of deposits.

However, on the basis of shape and size of wind deposits mainly two kinds of landforms, i.e. (i) sand dune, and (ii) loess plain, are formed.

Sand Dune

Sand dune is a mound, hill or ridge with a crest or definite summit. It is formed when any kind of obstacles, i.e. bush, rise in ground, skeleton of animals or house is found to break the velocity of wind. The reduction in velocity of winds led to the dropping of transported materials. Thus a mound like deposition is found in the direction of blowing winds. This mound shows gentler slope in windward side while steeper to Leeward side. This formation may also occur across the direction of the wind. Such kind of depositional features is known as sand dunes.

The Leeward slope is generally called slip face because along this face the sand particles slip down and rest on the base of the sand dune. They vary in their sizes from few feet to 250 feet in the height in the desert. Besides in desert, the sand dune are also found along the sea coast and beach. They usually migrate from one place to another place in windward direction in the absence of any vegetation to stabilize them at a particular place. During the migration of sand dunes, sometimes, some settlements, are burried under the sands. The sand dunes are formed at different angles to the direction of wind. Thus, again sand dunes may be classified into two groups owing to their location with reference to direction of flow of winds. Sand dune can be classified into two groups as given below :

Burchan : Burchan, a crescent shaped sand dune, is formed at across the direction of blowing of wind. Such typical type of sand dune is formed due to constant wind direction and limited supply of wind borne materials for the deposition. They are generally convex windward side and projecting two horns on the Leeward side. A burchan moves slowly forward as grains of sand are carried up the windward face and slip down the Leeward side. They range in height from few metres to over thirty metres and they may occur singly or in groups (Fig. 10.4).

Longitudinal or Seif Sand Dune : Such type of sand dunes run parallel to the direction of flow of the wind (Fig. 10.5). The winds blow straight through the corridors between the different parallel sand dunes. They are also called as self because they resemble the Sword in their shape. Sometimes, the self is formed when the cross wind blows and persist, for longer time and one of the horns of Burchan is elongated.

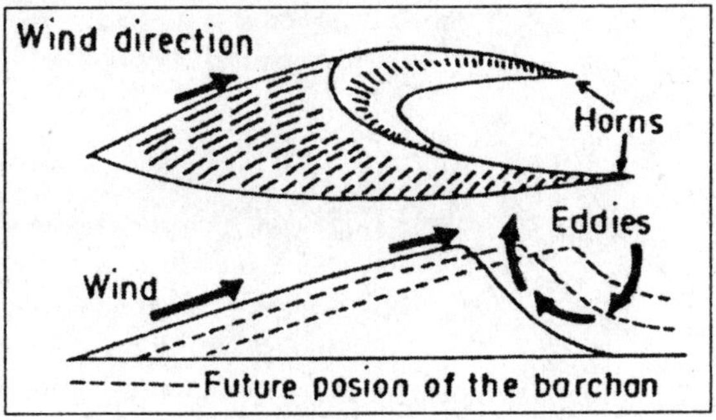

Fig. 10.4 : Burchan.

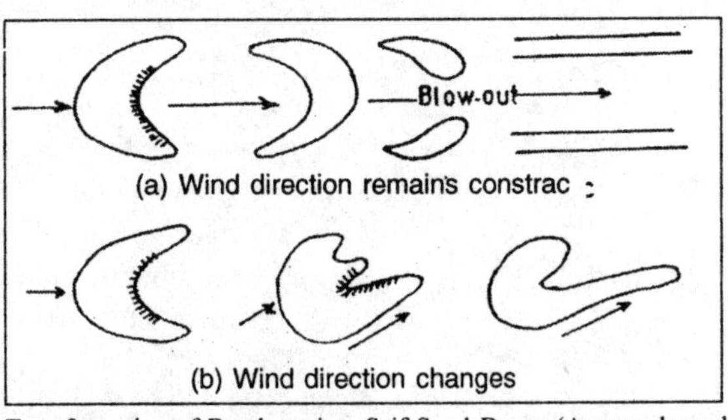

Fig. 10.5 : Transfomation of Burchans into Seif Sand Dunes (Arrows show direction of Wind.)

Loess Plain

The finest particles, which are carried away by wind in suspension from the desert to a considerable distance, are deposited in the desert or out of desert in a thin horizontal layer in the form of paper laminae under the favourable condition. Such aeolian deposit, or the plain formed by deposition of wind borne material is called as loess plain. Loess plain in North China is the typical example of this kind of landform of aeolion feature. Such feature is of loose materials and strata or laminas are almost absent. Sometimes the grasses or vegetation are buried and they are standing beneath the sand. After the decomposition a number of pores develop at the place of stems of burried plants. Sometimes, these pores may be filled up by rainwater forming underground water.

11

UNDERGROUND WATER

The water percolates down through the pores and openings of the rocks beneath the earth surface. The percolating water accumulates beneath the surface forming the underground water. It occupies a zone of saturation filled up by percolating water at some depth from the surface.

Conditions or Elements of Underground Water

There are mainly four characteristics which determine the amount of underground water in a particular area. They are porosity and permeability of rocks, climate and topography.

Porosity

The porosity of rock refers the amount of space or voids contained in it and it is stated as percentage of the total volume of material. The percentage of porosity varies with the differential nature of the rocks. The sedimentary rocks are rather more porous than the igneous or metamorphic rocks. The porosity may also vary within the same group of the rocks. For example, loose sand and gravels may have porosity to the extent of 50 per cent whereas the consolidated sedimentary rocks like limestone are definitely less porous. The igneous rocks rarely have the porosity more than 2 or 3 per cent.

Permeability

The term permeability indicates the capacity of the rock to drain the water through it. A rock of high permeability transmits water more rapidly and one with low permeability transmits it slowly. The permeability of rocks is controlled by mainly two factors, i.e. (i) Size of rock forming particles, and (ii) Interconnection among the pores or voids of the rocks.

The size of grains of rocks is very important elements in determining the permeability of rocks, because the size of internodular space is directly proportional to the grain size.

For example, coarse grained sedimentary rocks like gravel, which have larger internodular openings, are more permeable than the fine grained sedimentary rocks like silt and clay. If we double the grain size, the permeability will increase by about 7 times.

The interconnection between the pores and voids is another important element to determine the permeability of the rocks. The porous rocks are only permeable when the pores are inter connected with each other. That is why all porous rocks are not necessarily permeable rocks, but all the permeable rocks are porous rocks because the pores or openings are needed for movement of water.

Climatic Condition

Climatic condition also an important determinant of the amount of underground water. Cold and humid region, due to low evaporation and high precipitation, is having relatively more underground water, if other things being equal. Because more amount of water percolates down to form underground water. In contrast, dry and warm region may have lower volume of underground water because the more evaporation and low precipitation retard the percolation of water downwards.

Topography

Nature of terrain or landforms also decide the rate of water percolation. The hilly and slopy topography has relatively lower rate of percolation of water as the water from precipitation moves as run off very rapidly and it does not stand for a longer time for the percolation downwards. While the low lying or plain surface have accumulated water from precipitation for longer duration and consequently more and more water percolates down if other conditions are favourable.

Hydrological Cycle

The water, that reach on the earth surface by precipitation (snowfall and rainfall) is lost through different processes. Some part of it, evaporates and transpirate to the atmosphere. Some of it reaches to the ocean by rivers in the form of run off and remaining is soaked by soils and rocks beneath earth surface. The underground water, again, reaches the sea or ocean in the form of subsurface flow. Thus the water is moving continuously among hydrosphere, lithosphere, atmosphere and biosphere through the process of evaporation, transpiration, condensation and precipitation. This continuous and unending circulation of water among the different kinds of physical environment is known as Hydrological cycle (Fig. 11.1).

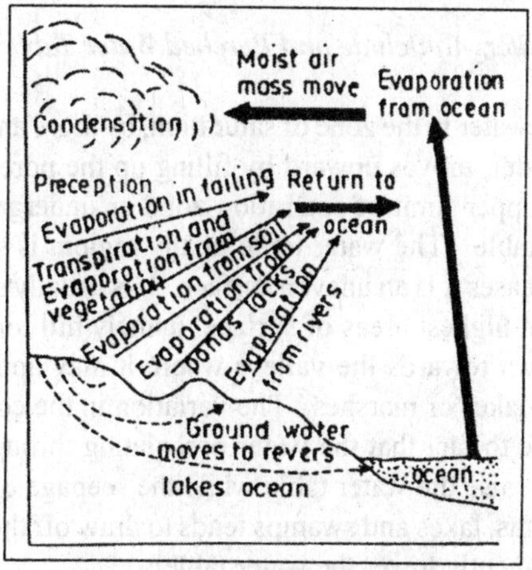

Fig. 11.1 : The Hydrological Cycle.

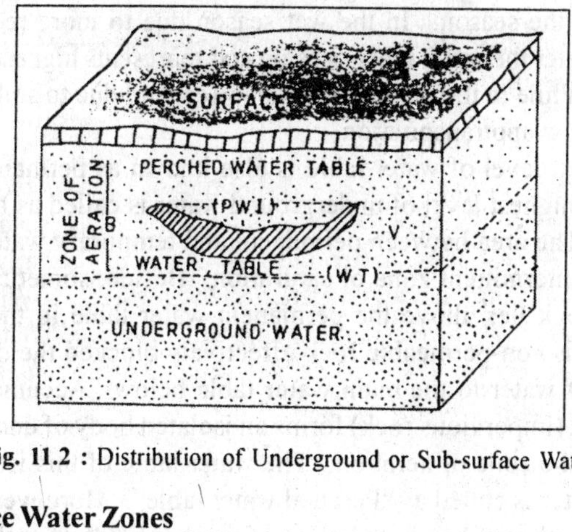

Fig. 11.2 : Distribution of Underground or Sub-surface Water.

Sub-surface Water Zones

As discussed earlier, the water percolates down through the pores and opening in the underlying soil rocks and its accumulates at certain depth where the rock becomes impervious. The region of sub-surface water may include two distinct zones : (i) Zone of aeration, and (ii) Zone of saturation (Fig. 11.2). The zone of saturation refers the region of underground where pores of rock are fully filled up with water and ground water is formed. Above the zone of saturation, the pores of the rock are usually partly or entirely filled by air. This is the area through which water moves downward from the ground in wet season. This zone is described as zone of aeration.

Water Table, Aquifer, Aquiclude and Perched Water Table

The accumulated water in the zone of saturation, by more and more supply of percolating water, moves upward by filling up the pores of overlying rock. Thus the upper limit of saturation zone or underground water is known as water table. The water table in any region, is seldom a level surface. In most cases it is an uneven surface. It is usually seen that water table is highest in highest areas of surface, namely hill tops and divides but descends down towards the valleys where it may appear at surface close to streams, lakes or marshes. The variation in the configuration of water table is due to fact that the water percolating through the zone of aeration tends to raise the water table while the seepage of underground water to the streams, lakes and swamps tends to draw off the underground water and consequently lower the water table.

Moreover, the water table or vertical thickness of saturation zone is dynamic with the season. In the wet season due to more rechange of percolating water the water table rises up and represents highest height in this season. While in the dry season it is lowest and due to soil moisture losses through evapotranspiration.

The lowest level of water table is also known as permanent water table and the highest level of underground water is called as temporary water table. The area between permanent and temporary water table is described as intermittent zone of saturation. Beside, sometimes a thin impervious rock lies above the permanent water table in the zone of aeration. This non-permeable rock effectively blocked the downward percolation of water to the main water table below. Accumulation of water above it (impervious rock) forms an isolated body of underground water within the zone of aeration. The water table of this isolated underground water is called as "Perched water table". Moreover, the area beneath the earth surface is consisting of various kinds of rocks of contrasting physical characteristics. The region, consisting of porous and permeable rocks, through which underground water may be stored and moved easily. It is referred as aquifer. For example, sand stone is both porous and permeable provides large ground water storage reservoir through which water may move easily. Contrary to this, the region of rocks of low permeability like shale which restricts the flow of underground water is known as Aquiclude.

Well

When a hole is dug on the surface through the zone of aeration upto or below the water table the water will seep into it. This hole is described as well. Any water drawn out will be replaced as long as water table remains above the bottom of the hole. Such hole is called as an ordinary well. There are two types of wells, i.e. (a) Temporary well (b) Permanent well. Temporary well is one in which water table goes down below the bottom of well in dry season and water cannot be obtained from it in this season. While the permanent well has its bottom always below the permanent water table so that water may be supplied continuously.

Artesian Well

It is a kind of natural objection or underground water to the surface water through a man made outlet under the pressure of column of underground water. The name artesian to this well is given after the name of province "Artois" in France where the first well of such type was sunk. Following conditions are essential for the development of such artesian wells (Fig. 11.3).

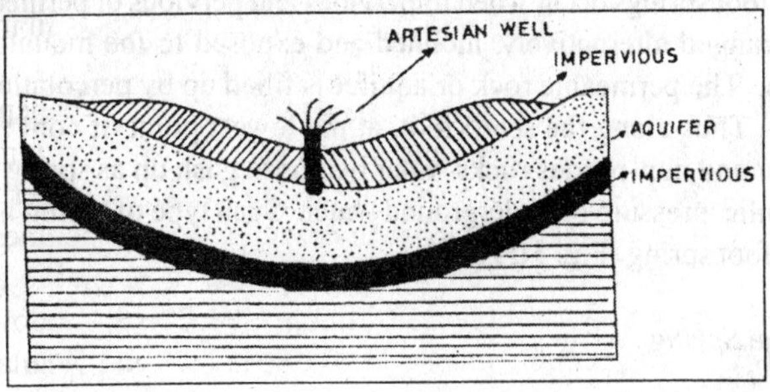

Fig. 11.3 : Artesian Well.

(i) The rocks should have cynclinal or tilted structure.
(ii) Porous and permeable rock must lie between two non-permeable rock's layers.
(iii) At least one extream or end of the porous and permeable rock

should be exposed to the surface so that rain water can percolate to fill the aquifer zone.

(iv) Sufficient amount of rainfall should be also in the region.

Moreover, Australia is famous for the artesian wells in the world. Great artesian basin of Australia in eastern interior part with an area of 1500,000 sq. km is well known for artesian wells. In India such wells are found in alluvial plain of Gujarat, Pondicherry and South Arcott district of Tamil Nadu province.

Springs

As we know that the water table is, always, not regular. Sometimes, when the beds of rock are inclined, the water table may intersect the surface especially in valley or low lying region and water begins to come up to the surface from the underground water. Thus the flow of water through the natural outlets from the underground water is described as spring. Artesian is different from the spring in the way that former is formed after digging the non-permeable rock overlying the aquifer. Springs may be classified into different groups according to the mode of formation and place of their occurrences.

Scarp Foot Springs

Scarp foot springs occur when impervious and pervious or permeable rocks are arranged alternatively, inclined and exposed to the mountain or hill slopes. The permeable rock or aquifer is filled up by percolation of rainwater. Thus along the scarp foot, at the lowest point of convergence of aquifer and non-porous rocks water begins to gush up as spring under the hydraulic pressure of underground water. Such type of spring is called as scarp foot spring (Fig. 11.4).

Fissure Spring

Groundwater frequently comes on the surface through a fissure in the bed rock, forming fissure spring (Fig. 11.5). The structure of such a spring may be identical with that of an artesian well. The water from fissure springs is usually wholesome, like artesian water. It is rather cold because it comes from greater depth where sun heat does not penetrate. The flow from such spring is usually constant.

Fault Springs

When the rock beds are dislocated or displaced due to formation of fault. Sometimes, the permeable rock is set against the non-permeable rock and faultline is exposed to the surface. Due to percolation of rain water, the permeable rock is filled with water and it moves out to the surface through the faultline. This kind of spring is described as Fault Spring (Fig. 11.6).

Dyke Spring

If there is a dyke having impervious material exposed at the surface and pervious strata lies on the both sides of it. The groundwater comes out on the surface through the space along either side of dyke in the form of spring, which is known as dyke spring.

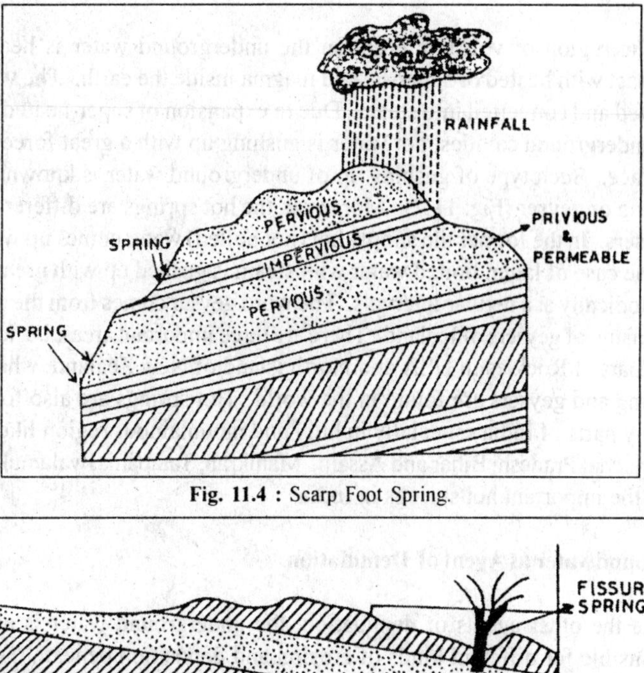

Fig. 11.4 : Scarp Foot Spring.

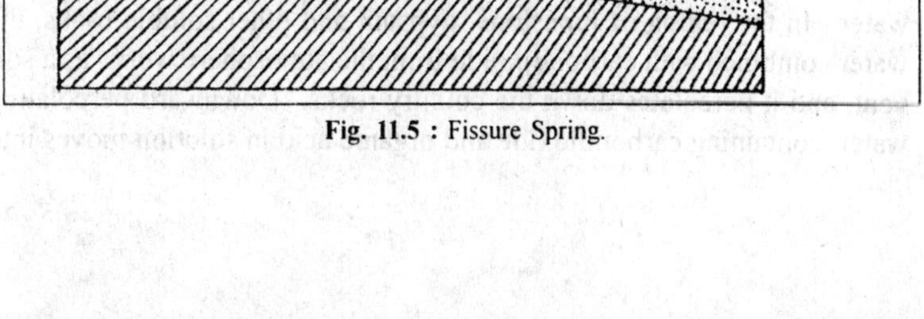

Fig. 11.5 : Fissure Spring.

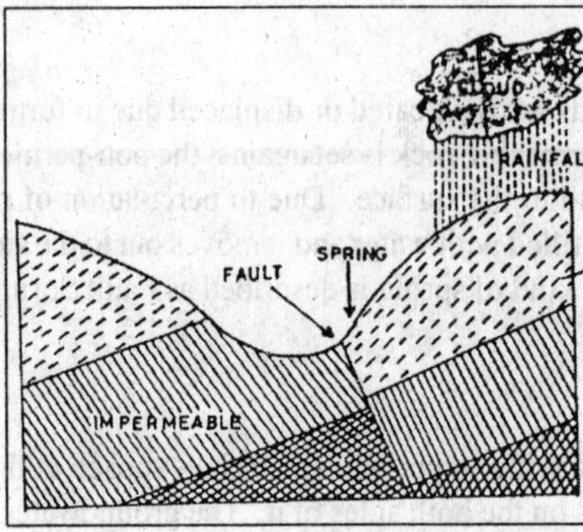

Fig. 11.6 : Formation of Fault Spring.

Hot Spring and Geyser

In the region of volcanic eruption the underground water is heated in contact with heated or super-heated magma inside the earth. The water is heated and converted into steam. Due to expansion of super-heated steam in underground cavities, hot water is gushing up with a great force to the surface. Such type of gushing up of underground water is known as hot spring or geyser (Fig. 11.7). Moreover, the hot springs are different from geysers. In the former the simply hot (not boiled) water comes up whereas in the case of latter, boiled water with steam is gushed up with great force periodically at a regular intervals. The word geyser comes from the geyser, the name of geysers of Iceland. There are three important areas, i.e. iceland, the part of Rockies in USA and North island of New Zealand, where hot spring and geysers are found in the world. Hot springs are also found in many parts of India especially in hilly and mountainous region like J&K, Himachal Pradesh, Bihar and Assam. Manikran, Tatapani, Jwalamukhi etc. are the important hot springs in India.

Groundwater as Agent of Denudation

Like the other agents of denudation, the underground water is also responsible for different kinds of erosion and deposition beneath the earth surface. Solution is the main process of erosion done by underground water. In the region of limestone, gypsum and other soluble rocks, the water combined with carbondixoide from the atmosphere works as a solvent, and it percolates down the country rocks. Downward percolating water, containing carbondioxide and organic acid in solution moves into

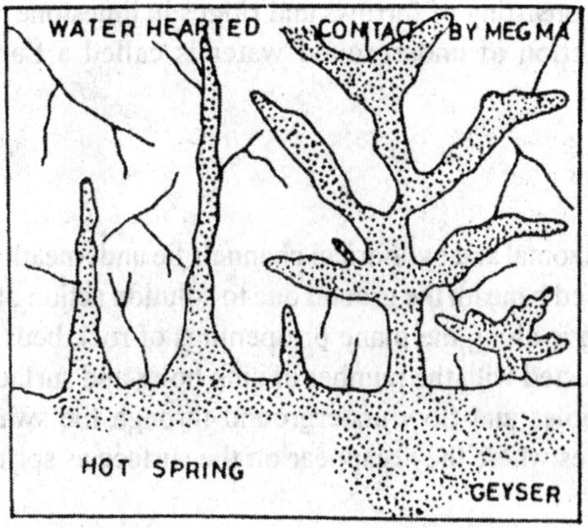

Fig. 11.7 : Hot Spring & Geyser

joints, along bedding plane and through more permeable beds of limestone. Due to solvent action of the water the rocks in the passage of water are dissolved and washed down. With the passage of time some openings became enlarge and various kinds of features are formed by solution of rock by underground water.

Landforms Formed by Solution Action or Erosion of Underground Water

Following landforms are formed by erosion of underground water.

Sink Hole

They are almost vertical funnel shaped depression in the region of limestone rock, where the rain water dissolves the rocks along the sides of joints in the surface rocks by solution action. Thus the joints become larger and deepens to few metres (Fig. 11.8). The sink holes have an average depth of three to nine metres and in area they may vary from one square metre to more. Limestone plateau of Kentucky of USA and Southern edge of Meghalaya's limestone region in India are well known for the distribution of sink holes.

Lapies or Clints

When the water combined with CO_2 moves down through the pores of limestone rock region, a number of long furrows with walls are formed.

Such feature consisting of furrows and ridges in limestone region formed by solution action of underground water is called a Lapies or Clints (Fig.11.9).

Swallow Holes

These are horizontal and cylindrical channels lie underneath the sinkholes. They are formed beneath the ground due to solution action of underground water as it move along the plane or openings of rock beds. Sometimes, they are connected with the number of sink holes and surface stream may sink in sink holes and flow underground through the swallow holes to other sink holes where they reappear on the surface as springs.

Doline

Dolines are huge size of sink holes. They are formed due to solution action of water as the sides of sink holes are laterally eroded and dissolved in solution and they increase in diameters.

Uvala

Uvala is a very extensive size of depression in limestone region and it is formed due to coalescence of a number of dolines together.

Caverns or Caves

Caverns are underground depression with intact roof formed by enlargement of underground opening due to solution action of water (Fig. 11.10). Concerns are usually above the water table and interlinked with each other by underground channel or passage. Sometimes, the roof of cavern collapse from two sides leaving central part standing as natural bridge. When the whole roof collapses, the huge and deep depression is formed, known as polje.

Depositional Feature of Underground Water

The soluble minerals which move with water to underground, are deposited beneath the earth surface especially in caverns. Following land futures are formed by the deposition of soluble materials of underground water.

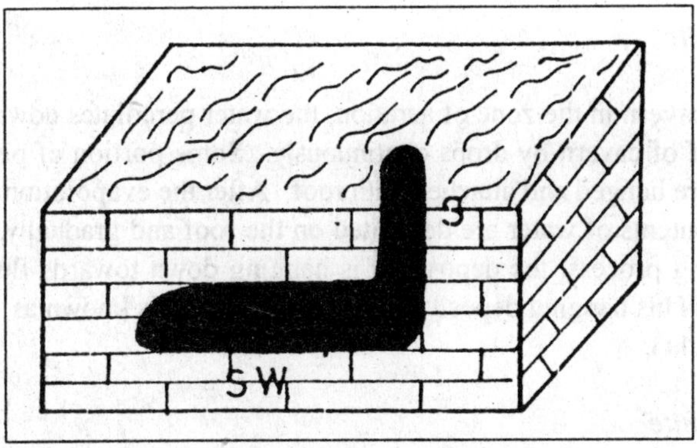

Fig. 11.8 : Sink Hole.
S — Sink Hole; SW — Swallow Hole.

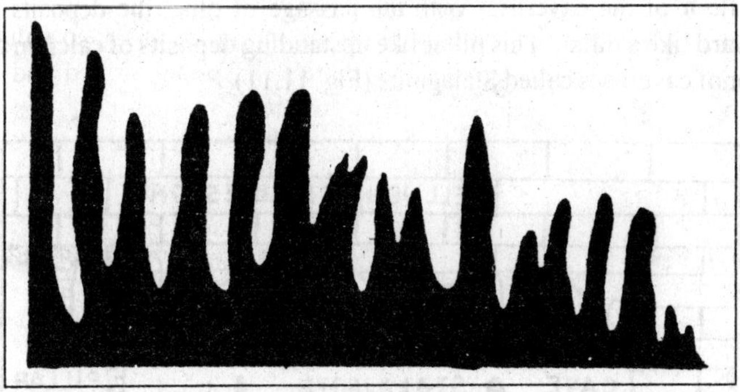

Fig. 11.9 : Lapies.

Fig. 11.10 : Cavern.

Stalactite

Being cavern in the zone of aeration, the water percolates down through the roof of cavern by drops continuously. Some portion of percolating drops are hanged and attached with roof. After the evaporation of water, lime contents of water are deposited on the roof and gradually, with the repeating process, the deposition is hanging down towards floor like a finger. This hanging deposit of lime in the cavern is known as Stalactite (Fig. 11.11).

Stalagmite

The percolating underground water also reaches on the floor or the cavern. After the evaporation of water the deposits of the calcium are left on the floor of the caverns. With the passage of time, the deposits move upward like a pillar. This pillar like upstanding deposits of calcium on the floor of caverns is called Stalagmite (Fig. 11.11).

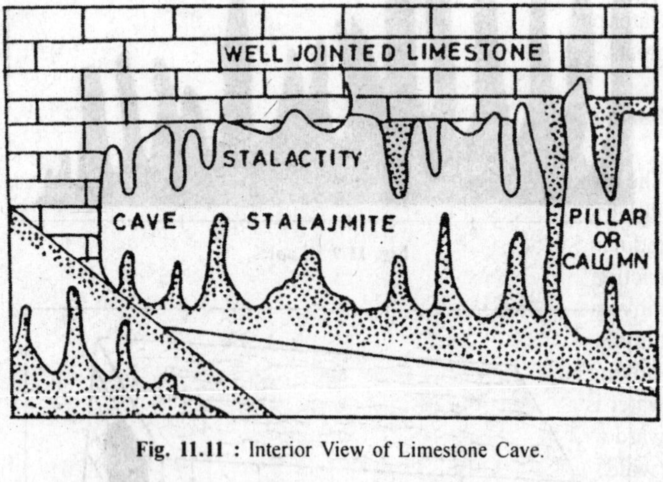

Fig. 11.11 : Interior View of Limestone Cave.

Moreover, thus limestone (Calcium) is deposited on both roof and floor of the caverns. With the passage of time, as the process goes on, these two deposits meet together forming pillar. This pillar is also known as column. Thus the limestone region is dissected by number of sink holes, swallow holes, Polje, blind valley and so on, which are formed by denudational work of underground water. Such type of topography in geomorphology is called as Karst Topography.

12

WAVE ACTION

The extensive bodies of water like lake, sea and oceans are not in sationary positions. Tide, ocean currents and waves are common phenomenon. These oceanic features are caused by number of factors. Tides are caused by gravitational force of sun, moon and the earth currents caused by variation in sea temperature and the waves are due to blowing winds over the water surface. The waves are regarded as oscillatory movement of the water particles in which the surface of water is deformed into troughs and crests.

Formation of Waves

The blowing winds over the water surface cause the friction and pressure. The friction exerted on water by the movement of overlying air is rather reducing with the height. The top layer of the air experiences the least friction and consequently moves with faster rate. Due to unequal rate of movement of air at different depths, the air of top layer moves ahead and tumbles forward. Due to absence of base of underlying layers of slow moving air, top layer falls down on the surface of water and consequently water is pressed down at its front. The down pressed water moves backward as sub-surface water and pressed up at the rear of the blowing air (wind) and water level is risen up. Hence the surface of water is disturbed in the wave form (Fig. 12.1).

However, the wave, which is formed in the sea, is consisting of many parts. The highest part of the wave is known as crest and the lowest part is called as trough. The horizontal distance between either two crests or two troughs is described as wavelength. The vertical distance between the crest and trough is wave height or amplitude (Fig. 12.2).

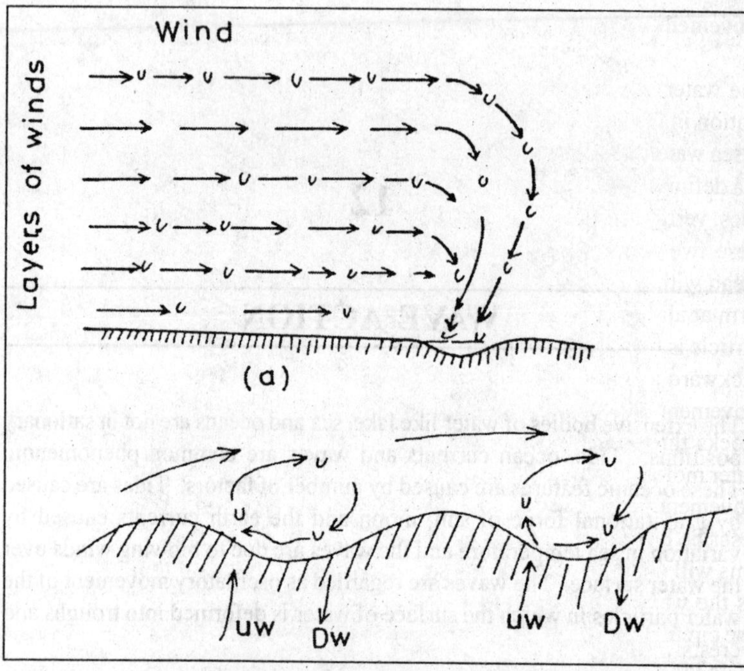

Fig. 12.1 : Formation of Sea Waves.
a — Water surface is deformed by pressure of toppled top layer of wind.
b — Trough and Crest are formed by downward (dw) and upward (uw) pressure of water.

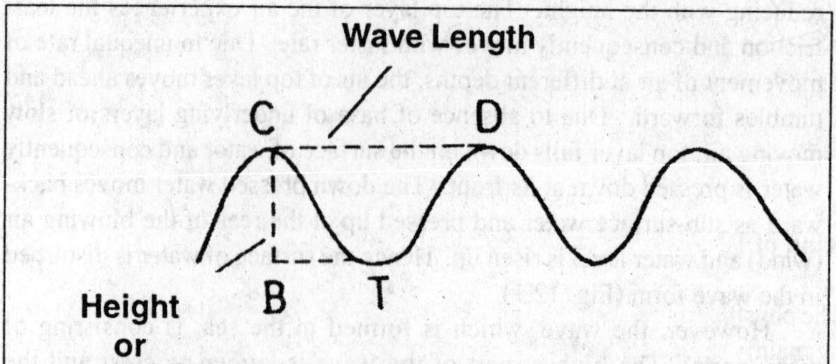

Fig. 12.2 : Wave.
C — Crest; T — Trough.

Movement of Water Particles in Sea Wave Motion

The water shows different trends and behaviour with different kinds of motion in the sea. Ocean currents, tides and waves are common features of sea water. The currents displace the huge amount of water horizontally in a definite direction from one region to another. While in the case of the tides, vertical motions, i.e. rise and fall of water, takes place. Contrary to these two forms, in the latter case (wave) water particles do not move ahead with wave but oscillate ups and down at the same place in circular form at all depth especially in deep water. At the crest the movement of particle is forward, at the middle of hinder slopes downward, at the trough backward and at the middle of front slope upward (Fig. 12.3). Thus the movement of waves in water is analogous to transmission of earthquake shocks through land where the shock travels instead of the land. The water molecules move downward in circular form as the wave passes. The movement takes place in such fashion that the particles near the surface describe larger circle than the successive circles beneath the one another. This will continue to a level when depth becomes equal to waves length. As the wave moves into shallow zone of water, the velocity decreases. The circular motion of particles is distorted into ecliptical form. Gradually the ecliptical form becomes horizontal. Consequently, the crest of wave becomes steeper and higher and the trough is flattened. Finally, the crest loops out and fall down in the form of wave break.

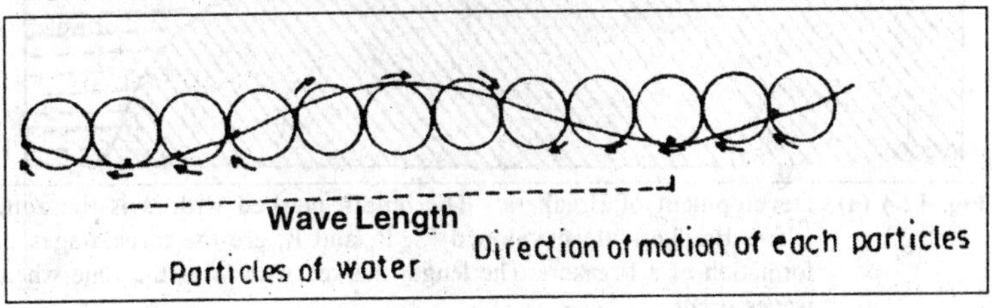

Fig. 12.3 : Wave Motion.

Kind of Waves

The physical characteristics like size and velocity of waves are determined by mainly three elements. They are velocity of wind, fetch and depth of sea water. Deep and extensive water with faster winds experience stronger and high waves and *vice versa*. Thus the waves are known by different names according to their location, shaps and strength along their courses of travel.

Sea

When the wave originates in the deeper part of sea water, it is very stronger and violent. Sea water appears very high. Such stronger wave is known as 'Seas'.

Breaker or Surf

When the waves (seas) move into shallow, water, the velocity and wave-length are decreased. Height of waves increases as the water particle's orbital movement becomes ecliptical and horizontal at the sea bottom due to decline in depth of water body. The troughs of the waves touch with bottom and move at rather slower rate as compared to the crest. The advancing crests of the wave gradually develop in concave forms and fall down and rush the coast. Thus the wave after breaking up in shallow zone of water is known as Breaker or Surf (Fig. 12.4a).

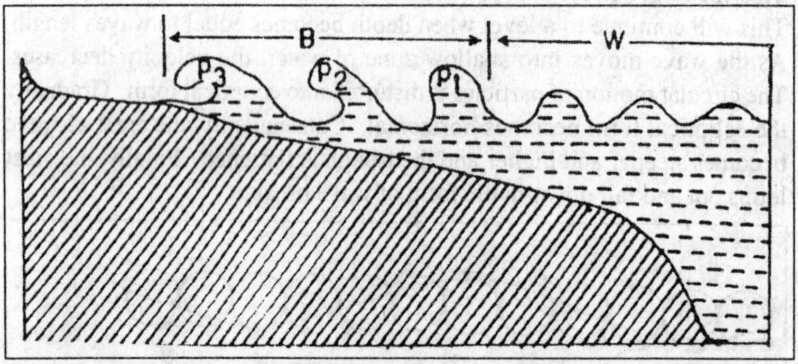

Fig. 12.4 (a) : Development of Breakers. The length marked with B is the zone where Breakers are developed P_1, P_2 and P_3 are the three stages of formation of a Breaker. The length marked with W is the zone where waves occur.

Swash and Backwash Current

After the breaking up of the waves near the shorezone in shallow water, a huge amount of sea water moves over the shorezones to strike the coastline. This upward (landward) flow of sea water is described as swash current (Fig. 12.4b). When the swash current drains back down from the shore zone to sea is known as backwash or undertow current. On the basis of the strength and function of the swash and backwash currents waves can be grouped into two types.

Fig. 12.4 (b) : Swash and Backwash Current.

(i) *Constructive Waves :* When the frequency of breaking up of the waves is lesser, the wave is able to continue its course without the interference of wave behind it and it is more active in erosion and deposition of material on the shorezone. This wave is called as constructive waves.

(ii) *Destructive Wave :* Sometimes, the waves are broken at a faster rates, the backwash of the wave mix into the swash of the wave behind and consequently the force of swash is neutralized by backwash. The back-wash becomes stronger and it removes pebbles and sand from the shore. They are knows as destructive waves.

Process of Erosion of Sea Coast by the Sea Waves

The erosional activity of the sea wave is confined to the coast environmental zone and nature of the erosion is linear. The rate of erosion is determined by nature of the rocks forming coast line, the strength of the waves, seaward slope and the depth of the water.

The strength of the waves is the product of a number of factors like velocity of winds, depth and extent (fetch) of the water bodies. Under the condition of turbulent wind on extensive and deeper ocean surface, the strength of waves will be greater. The erosional actions by sea waves are performed by following processes.

Hydraulic Action

The swash water currents are thrown up against the cliff and they enter into the crackes and joints of the rocks forming the coast. Thus the air in the cracks is compressed suddenly by intrusion of swash water. Contrary to this air expanded explosively when water retreated back from the openings. Due to repeating process of compression and expansion of air in the cracks or joints under the influence of entering and retreating swash water into cracks, the rocks are shattered and broken off. This process of disintegration of coastal rock is known as Hydraulic action.

Abrasion

The broken fragments of rocks produced by hydraulic action commonly travel with the advancing waves and strike against shore. The rocks of the shore are subjected to further erosion due to rubbing effect of impinging rock fragments. Thus the process of wearing and tearing of coastal rocks, due to grinding effects of invading rocks, is called as Abrasion.

Attrition

The broken fragments of rocks are moved up and down the shorezone by swash and backwash currents. During the movement, the boulders and pebbles are broken further into finer particles due to collision among themselves. Such process of wearing and tearing of rocks is described as Attrition.

Solution

The rocks of sea coast, consisting of some soluble constituents, are attacked by sea waves. The solvent action of water dissolves the rocks and remove them in the form of solution. Such process of disintegration of rocks of sea coast is called as Solution.

Erosional Features Formed by Sea Waves

Sea waves have linear erosional effects along the coast and various kinds of erosional coastal features are formed by them as discussed below —

Notch

Swash currents clash with the exposed coastal rocks and the rocks are weakened, loosened, and shattered off at variable height depending upon the level of wave water. Thus the rock are under cut and small hollow is constructed along the coast especially of softer rocks. The overlying resistant rock projects in the sea and hangs over the hollow. Such marine eroded hollow is called as Notch (Fig. 12.5).

Wave-cut Platform

The coastline of sea is eroded back by swash current moving up beach after breaking the wave in the shallow water. Due to recession back of cliff of the coast, a distinct zone between shore line and coastline is produced as a result of combined effect of swash and back wash currents. This wave worn platform is called as Wave-Cut Platform or terraces (Fig.12.6). The size and dimension may vary with regards to variation in rock hardness and structure along the coast, and the strength of the sea waves. A wide and extensive platform is built when the coastal zone consist of softer well jointed and thin layered rocks, and the swash currents are weaker than the back wash or undertow currents and *vice versa.*

Sea Cliff

The coast facing sea is eroded by wave erosion at sea level. The coast is undercut forming the notch. This notch gradually widened to undermine the hanging rock from above. When the hanging rock falls down due to repeating beating of sea waves against the coastline, the sea coast becomes very steeper like a scarp. Such scarp like sea coast is known as Sea Cliff (Fig. 12.7).

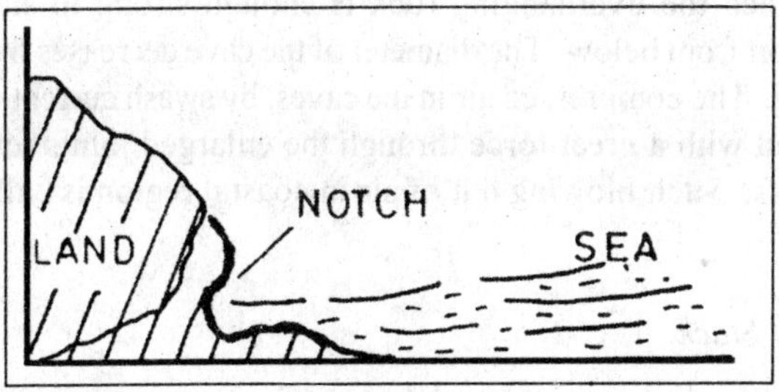

Fig. 12.5: Notch.

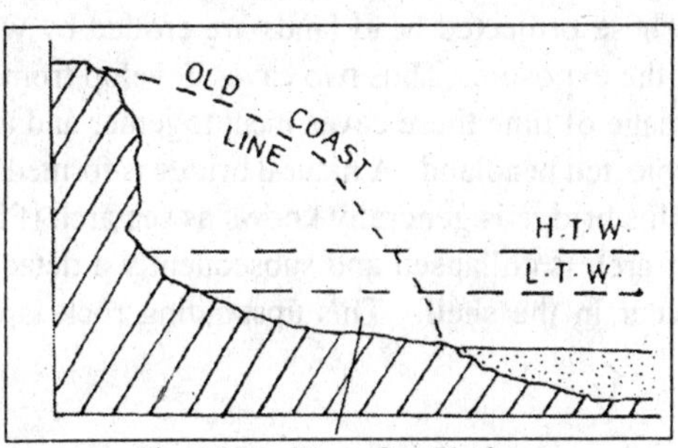

Fig. 12.6 : Wave Cut Platform.

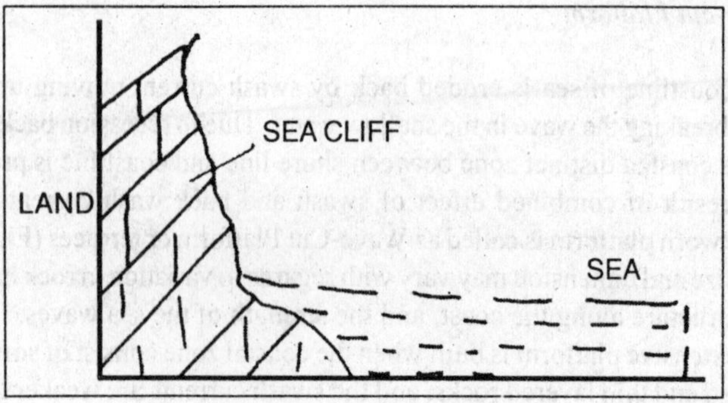

Fig. 12.7 : Sea Cliff.

Sea Cave

Sea cliff or the coast is formed of rocks of differential hardness and the rocks are associated with a number of joints, fractures and zone of weakness. Due to prolonged wave action a cylindrical or funnel shaped hole is formed at the base of sea cliff along the line of weakness. Such type of wave cut formation is described as Sea Cave (Fig. 12.9). It generally occurs when the overhanging rock is enough strong to stand without any support from below. The diameter of the cave decreases from entrance to the end. The compressed air in the caves, by swash current, sometimes, gushes out with a great force through the enlarged joints found in their inact roofs. Such blowing out of air in coastal region is called as Blow Hole.

Arch and Stack

Along the irregular coastline a number of resistant headlands are projecting in the sea. These projected head lands are eroded by waves from both directions of the exposure. Thus two caves develop from opposite sides. With the passage of time these caves meet together and a hole is formed across the projected headland. A natural bridge is formed connecting two sea caves. This bridge is generally knows as sea arch (Fig. 12.8). When the bridge or arch is collapsed and subsequently a detached upstanding rock may occur in the shelf. This upstanding rock is called as Stack (Fig.12.10).

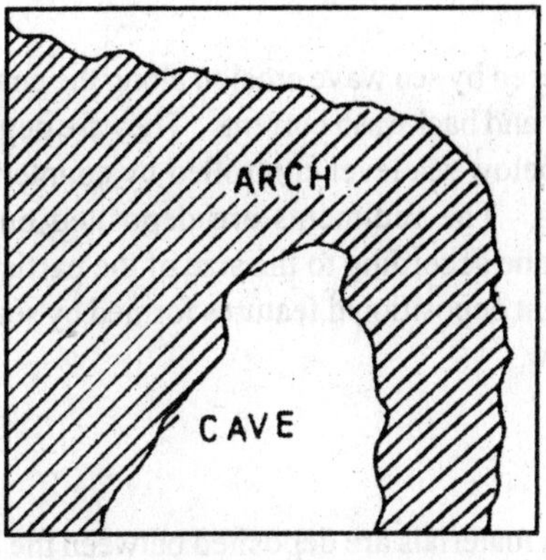

Fig. 12.8 : Formation of Sea Arch.

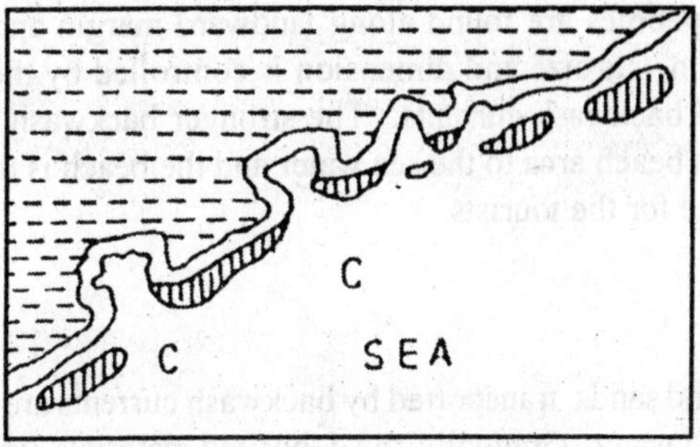

Fig. 12.9 : Cave and Bar.

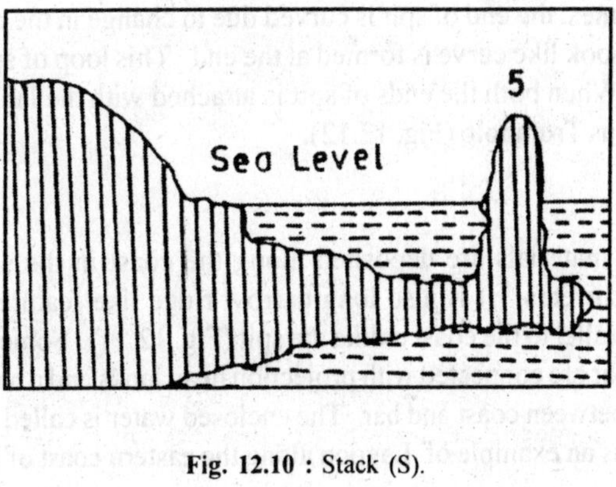

Fig. 12.10 : Stack (S).

Depositional Features Formed by Sea Waves

The material produced by sea wave erosion along the coast are transported to and fro by swash and backwash currents. They are deposited at different places above and below sea level controlled by counter balance between these two currents. The sediments are deposited on shore zone and continental shelf zone according to the size of the particles in descending order. The important depositional features formed by sea wave deposition are discussed below.

Beach

The sea wave worn materials are deposited between the high and low tide water. The deposited materials are sorted out and consisting of many gravels, pebbles and sands. It is temporary and seaward sloping features. Shingles and pebbles are found along landward margin and sand along seaward margin. Its size and dimension is controlled by the strength of the swash and backwash currents. The stronger backwash removes the sediment from beach area to the sea water and the beach is reduced. It is attractive place for the tourists.

Spits

The pebbles and sands, transported by backwash currents are deposited in the shallow sea water. Gradually, the deposited materials extend upward, sideward and seaward and develop in the form of a long low narrow ridge transverse to the coast. This ridge like formation is known as Spit. Spit is connected with mainland or island by atleast one end and it projects in the sea. Sometimes, the end of spit is curved due to change in the direction of wind and a hook like curve is formed at the end. This loop of spit is called as 'Hook'. When both the ends of spit is attached with the land blocks, it is described as Trombolo (Fig. 12.12).

Bar

Wave worn materials are deposited along the coast in the sea by long shore drift currents. It is also long narrow ridge like feature in the sea water but parallel to the coast unlike the spit (Fig. 12.11). Sometimes, both the ends of bar are connected with projection head lands and a narrow water is enclosed between coast and bar. The enclosed water is called as Lagoon, Chilka lake is an example of Lagoon along the eastern coast of India.

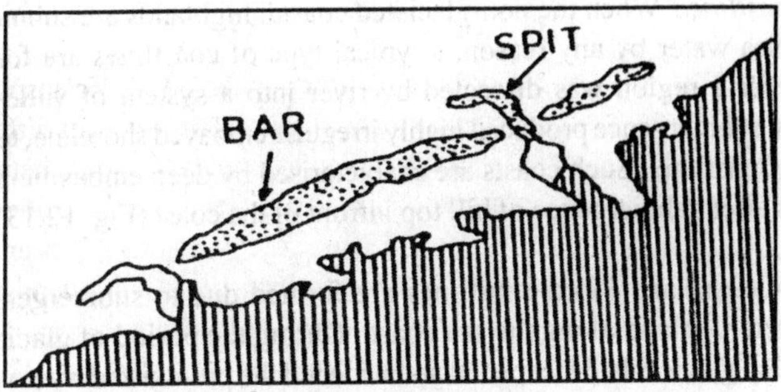

Fig. 12.11 : Bar and Spit.

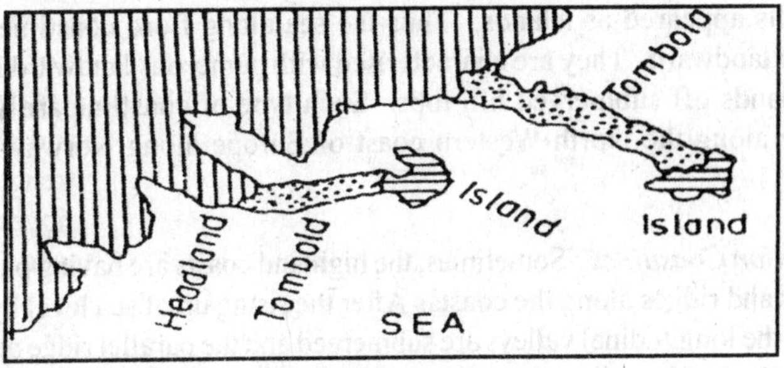

Fig. 12.12 : Trombolo.

Coastlines

The formation of coast and shorelines has not been unchanged. During the geological past, the coastal areas have undergone modification due to rise and fall of sea level and tectonic movement. On the basis of mode of formation of coastlines, they may be grouped into following groups, i.e. (i) Coastlines of submergence, (ii) Coastlines of emergence, (iii) Neutral shoreline, (iv) Faulted shorelines, and (v) Compound shorelines.

Coastlines of Submergence

When the coastlines have come into being due to either submergence of coastline area through the rise of sea level after melting of ice, or subsidence of coastal areas by tectonic movement. Such coasts may be formed by submergence of river eroded highland, glacial eroded highland or coastal plain areas under the sea water.

According to the Nature of topography that existed before the submergence, the coast lines, may be again grouped into following groups.

Ria Coastline : When the non-glaciated coastal highlands are submerged under sea water by any reason, a typical type of coastlines are formed. Wherever a region was dissected by river into a system of valley and divides, submergence produced highly irregular embayed shoreline, termed as Ria Coastline. Such coasts are characterised by deep embayment, and islands after submergence of hill top infront of the coast (Fig. 12.13).

Fiord Coastline : Fiord coastlines are formed due to submergence of glaciated highland along the sea coast. During the period of glaciation, the valley glaciers had steepened and widened the pre-existed river valley. After the melting of glaciers, the sea level has risen up and the coastal glaciated valleys had submerged under the water and the water divides and hills appeared as islands. Thus the sea along fiord coastlines are steeper landward. They are characterised with numerous landward inlets and islands off submerged hill tops. Such type of coastline are found mainly along the North-Western coast of Europe along Norway (Fig. 12.14).

Dalmation Coastlines : Sometimes, the highland coasts are having parallel valleys and ridges along the coasts. After the rising up of sea level by any reason the longitudinal valleys are submerged and the parallel ridge appear like the chain of parallel islands along the coastline. Such type of coastlines are designated as Dalmation or longitudinal coastline (Fig. 12.15). This type of coast occurs in Yugoslavia and along the part of Pacific coasts of North and South America.

Low Land Coastlines : Such type of coastlines are originated after the drowning of pre-existing lowland coast and sea water penetrates landward along the river valley to considerable extent. Marshes, swamps and mud flates are not uncommon at the low side. The flooded river valley is called as estuary. A series of bars enclosing lagoon run parallel to the coastlines. Such type of coast may be seen along the Baltic coast in Germany.

Coastlines of Emergence

Such kinds of coastlines are originated either due to unheaval of earth crust near the continental margins or due to falling of sea level during the period of glaciation. The land area, above the coastlines of emergence, was once the area of continental shelf. This kind of coast may be grouped mainly into following subgroups.

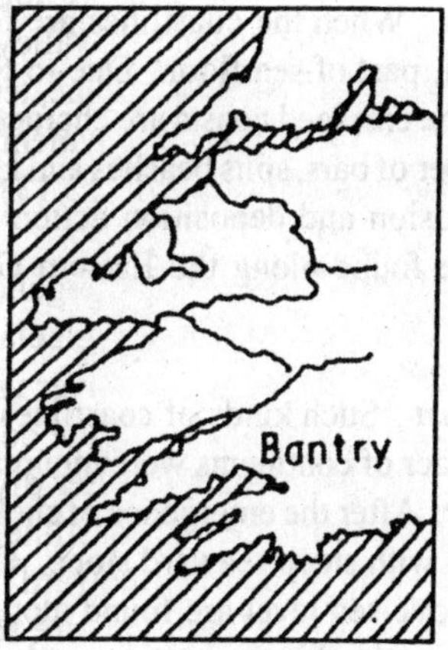

Fig. 12.13 : Ria Coast.

Fig. 12.14 : Fiord Coast.

Fig. 12.15 : Dalmation Coastline.

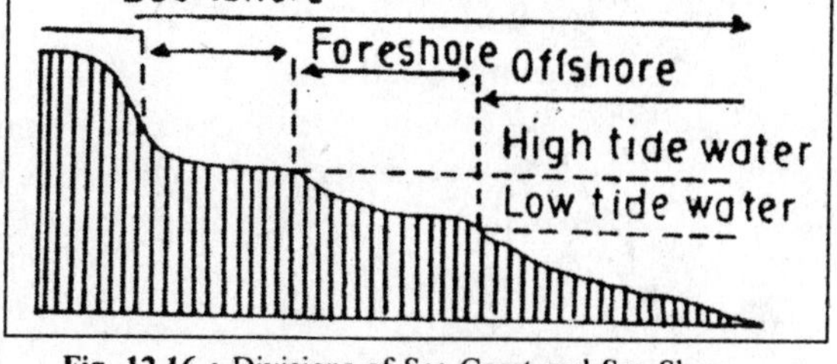

Fig. 12.16 : Divisions of Sea Coast and Sea Shore.

Coastlines of Emerged Coastal Plain : When the coastlines are formed after the uplifting of continental shelf, part of sea floor. Due to the flat and gentle sloping nature of shelves, the emerged coasts are characterised by existence of a coastal plain. A number of bars, spits beaches and lagoons are found along the coasts due to erosion and deposition action of sea waves and currents. Such coasts are found along the Eastern Ghat in India.

Coastlines of Emerged Highland Coast : Such kinds of coastlines occur usually in those regions where the border of continents were mountainous and continental shelves were narrower. After the emergence of shelf area, the coast lines appeared as a highland with steep seaward slope. Old sea cliffs, and beaches 7 to 30 metres above sea level are found along such type of coastlines. The coastline of Scotland is the best example of this category.

Neutral Shore Lines

Neutral shore lines are formed by deposition of eroded materials in the sea water. This kind of shorelines is known by different names according to the agents of erosion. For example, Alluvial Fan Neutral shoreline formed by deposits made by river, Outwash Plain Neurtal shoreline formed by deposition made of glacier etc.

Faulted Shore Lines

Such type of shoreline is originated by faulting of the earth crust along the margin of the continents. There is dropping down of segment of the coast on seaward side of the break and the corresponding landward side segment is standing up with escarpment. Western coast of India is the best example.

Compound Coast Line

Such coastlines are combined form of submergence, emergence as well as faulting coastal structures.

13

CYCLE OF EROSION

As discussed in the preceding chapters, the landforms on the earth surface have been evolved as the product of endogenic and exogenic forces of the earth movements. The diastrophism or internal forces led to the upheavals of low lands like ocean basin, or plain and subsequently, the external forces, i.e. forces of denudation become active with the objective of maintaining grading or levelling on the earth's surface. With the passage of time at slower or faster rates (depending upon the structure of the rocks) an uplifted land is transformed into a plain surface reaching at about sea level. This process produces, ultimately Peneplain. Thus the duration through which an elevated land is reduced to the sea level by river erosion is described as the cycle of the erosion. But this concept is rather more theoretical than practical. Endogenetic movement does not wait for the completion of the cycle but due to either falling of sea level or upheaval of river beds, rejuvenation of the river occurs. Restarting of erosion process again takes place.

The concept of cycle was first used by Hutton in 1785 in his work *Cyclic Nature of the Earth's History*. Subsequently, following Hutton concept and Charles Darwin's theory of evolution of organic species, an American geomorphologist W.M. Davis introduced the concept of geographical cycle of erosion in 1899. This concept of cycle of erosion became so pronounced that the American school of geomorphology was known as Davisian School of Geomorphology. His model of cycle was also applied in Karst geomorphology, aeolian geomorphology and coastal geomorphology. Later on he was criticised and during 1920s and 30s Walther Penck and his follower staged an opposition against to the Davis' cycle of erosion. He gave his own idea with reference to cycle of erosion. There are mainly two contrasting models of cycle of erosions in geomorphology.

(i) Davis, Model of Cycle of Erosion.

(ii) Penck's Model of Cycle of Erosion.

W.M. Davis' Concept of Cycle of Erosion

William Marris Davis was the first geomorphologist who evolved the concept of cycle of erosion in late 19th century. His first attempt was "complete cycle of river life (1889)" in which he sought progressive development of erosional steam valley and finally emphasised on the sequential development of landforms as a result of combined effect of endogenetic and exogenetic (Erosion) forces. In 1899 Davis propounded the concept of cycle of erosion taking guidance from the Darwin's model of evolution of organic life in which growth takes place in an irreversible sequence with the passage of time. He gave the dictum that "the landform is the product of structure, stage and process". This dictum is known as TRIO of Davis.

Assumption of the Concept

The Davisian model of cycle of erosion was built around four important assumptions.

1. Landforms have been evolved through interaction between endogenetic force (diastrophism) and exogenetic forces (weathering and erosion).
2. Transformation of landforms takes place in sequential order through the various time spans, which Davis described as stage, from uplifted land to the featureless plain called as peneplain.
3. The upliftment of the land is fast and short-time phenomena. This upliftment is the source of potential energy for the river erosion. Each stage of cycle of erosion was associated with declining potential energy as uplifted land are worn down.
4. Erosion does not take place until the upliftment is completely stopped. In other words, upliftment and erosion do not occur simultaneously.
5. Cycle of erosion continues until the river gets its graded level, wherein, down cutting of the river is ceased. He incorporated the concept of graded condition which is first achieved near the mouth of river spreading headwards through drainage lines towards divide.

Discussion

Under these assumptions he postulated the model with the statement that the cycle of erosion passes through three stages excluding the period of upliftment. These stages he described as youth, mature and old. The duration and length of time in which a particular set of assemblage of landforms occur depends upon the structure of the rocks and the agents of weathering and erosion (process).

Youth Stage

In Fig. 13.1, curve lines AC and BD show the line of water divide and river valley bottom respectively. In the initial stage (youth) following features are very common.

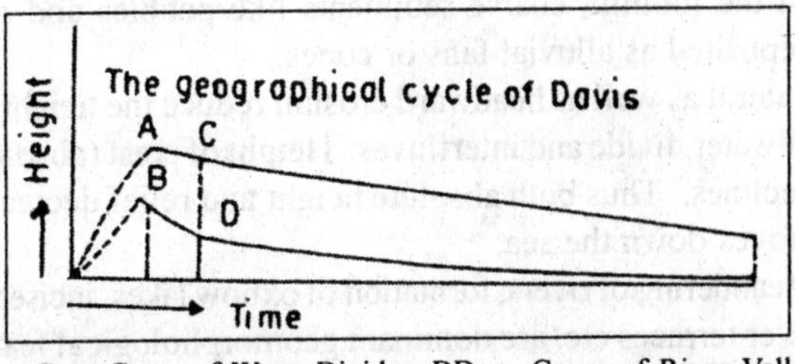

AC — Curve of Water Divide; BD — Curve of River Valley.

Fig. 13.1 : The Geographical Cycle of Erosion by Davis.

1. Water divide crest is not eroded. Narrow deep river valley develops due to steep slope. So the difference between crest of water divide and valley floor increases up. This difference of altitude between these two is known as relief. Relief as well as absolute height (Height of crest of divide from S.L.) are high.
2. Narrow and deep river valleys, known as gorge or canyon, are formed. Deepening of the valley is dominant activity.
3. Lateral erosion is minimum. Headward or back-erosion led the narrowing of the water divide.
4. Waterfalls and rapids are formed when some resistance rocks come in the way of river valley. Formation of pot holes is also very common.
5. Erosion and transportation are dominated while the deposition begins to take place in late part of youth stage.

Mature Stage

Mature stage of river cycle begins with the descending of river from uplifted region to the plain areas. The gradient of slope falls to a considerable extent. The velocity and transportation capacity and subsequently deepening of valley are reduced. Following characteristics are associated with this stage.

1. In the beginning of this stage, maximum level of relief is found. After this, deepening of the valley is reduced and lateral erosion takes place.
2. As the velocity and transportation capacity fall down, the deposition starts along the side and enroute on the bottom of river valley.
3. At the foothill, coarse sediments like pebbles and gravels are deposited as alluvial fans or cones.
4. Lateral as well as headward erosion reduce the height and width of water divide and interfluves. Height of crest (absolute height) declines. Thus both absolute height and relief decrease as river moves down the sea.
5. Meandering of rivers, formation of oxbow lakes, incised meander, river terraces etc. are dominant geomorphological features.
6. Erosion and deposition are balanced.

Old Stage

Old stage of cycle of the erosion, begins with very gentle gradient of river valley. The river is graded and widened at maximum level. Braiding of the river and floodplain and levees formation are also very typical feature of this stage. Uplifted land is transformed to a plain with residual mondonock which is called as Peneplain. Delta is formed at the mouth of river. The absolute height and relative relief are completely reduced. Both curve converge at sea level. Erosion is ceased and deposition dominant.

Evaluation of Davis' Cycle of Erosion

Davis' Cycle of erosion is very simple to understand theoretically. He is the first who propounded this concept. His normal cycle of erosion was also applied in arid, glacial and karst geomorphological regions. But his model is criticized basically on the pre-youth stage of the cycles. Because upliftment of lands takes at slow rate and continues for long period. Different orgenies of the world indicated that they were completed during millions of the years. It is also questioned about the beginning of weathering and erosion. Process of denudation does not wait for the completion of the upliftment, but both go on side by side. Geomorpho-logists also have opposite views on the concept of completion of the cycle assumed by Davis. Endogenic movement or climatic change may take place during the period of cycle at any time and restarting of the cycle may occur before the completion of cycle. This process is caused due to rejuvenation of the river. Completion of cycle rarely occurs in the world.

Penck's Cycle of Erosion

Walther Penck, a German geomorphologist, who tried to explain the process of development of landforms in fluvial geomorphological region. He presented this model in his work Morphological Analysis for the expression of landscape development. Penck's model is thought as an opposite efforts against to the Davis model. So he arbitrately avoided the cycle term of Davis and used Entewickelung (Development).

Principle of the Theory or Concept

Penck developed his model of cycle of erosion on the basis of certain following principles or assumptions.

(i) Landform is an expression of mutual interaction between endogenetic and exogenetic movement of the earth.
(ii) Most of the tectonic movements begin and end slowly. The common pattern of such movement is slow; initial uplift, an accelerated uplift, a deceleration in uplift and finally quiescence.
(iii) An initial quescence surface (Primarrumpf) follows slow long duration upheaval and process of gradation is set immediately after little upheaval. In other words, upheaval and erosion occur simultaneously without any wait for the completion of upliftment.

Discussion

W. Penck described his model with the explanation that primarrumpf (initial featureless surface) is gradually uplifted and exposed to the process of denudation. With the passage of time, the uplifted primarrumpf, with the help of river erosion and weathering, is converted into a feature less plain associated with some residual monodnock (Peneplain). The whole process, through which a peneplain is formed, has passed through mainly three phases. These phases are described by him as Entwickelung (development). Thus his cycle of erosion includes Aufsteigende wickelung (increasing development of landform), Gleichfarming Entewickelung (uniform development of landform) and Absteigende Entwickelung (decelerating development of landforms) (Fig. 13.2).

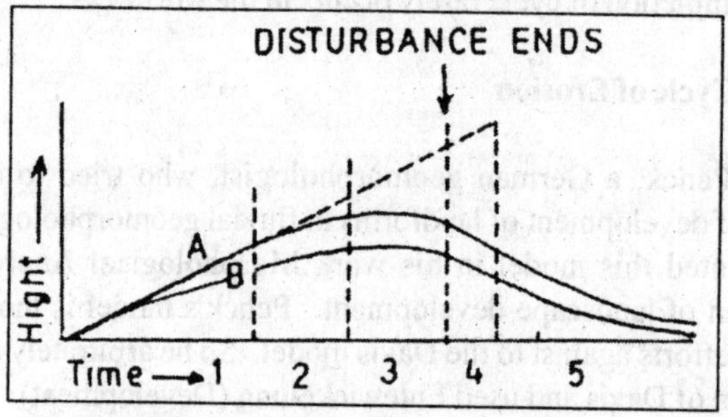

Fig. 13.2 : The Representation of Penck's Cycle of Erosion over a long period.

1. *Aufseigende Wickelung* (Waxing or Increasing Development of landform)

In this phase of cycle the primarrumpf remains featureless for some period when a slow long continuous upliftment occurs. As the rate of upliftment increases, the rivers begin to cut deep valleys through vertical incision and V-shaped valley is produced. Featureless primarrumpf is obliterated through the erosion. The crest of uplifted surface is not eroded now. Upliftment of both crest and valley bottom are continuously increasing up. The rate of upheaval is not balanced with incision of river valley. Consequently, the absolute height (height of crest of ridge) and relief (difference between altitude of water divide or crest of ridge and depth or river bottom) increase up in the stage of development of landform. The slope is convex because of extensive down cutting in valley bottom without any corresponding denudation on the upper slope (summit of the ridge).

Gleichfarmige Entwickelung (*Uniform Development of landforms*)

In this stage the curve of ridge summit and the bottom of river valley are parallel but they show different orientation or inclination in different parts of the stage. Development of landforms take place constantly at uniform rates, so it is described as the stage of uniform development of landform. It is divided into three phases depending on ratio between rate of upliftment and erosion taking place in this stage.

Phase-1 : In this phase of uniform development there is an acceleration of upliftment but not at same rate as in waxing or previous stage. Both upliftments of summit and river bottom are at uniform rate. Rate of summit erosion and deepening of valley is similar to each other. Thus the valley sides are straight and relief remains constant. Absolute height still increases up.

Phase-2 : In this phase, upliftment is still accelerated with reference to summit of divide and river valley bottom. Simultaneously, summit erosion and valley deepening are matched with the rate of upliftment. Further the rate of summit erosion and down cutting of valley is also at uniform rate. Consequently, the curves of summit and bottom or river valley are parallel as in Phase-I. Absolute height is maximum and relief is constant. The summit neither increases nor decreases as lowering of summit is offset by uplift.

Phase-3 : This phase is marked by stopping of upheaval and lowering of summit and deepening of valley. Relief remains constant as in 1 and 2 phase because rate of summit erosion and valley deepending is uniform. But the absolute height declines due to lowering of summit by erosion and weathering.

Abstegende Entwickelung (Wanning development of Landforms)

After the complete end of upliftment, the summit or the ridge, continuously, is declining due to denudation process. But due to lowering of slope (crest), deepening of river valley is reduced. Lateral erosion becomes dominant. Consequently, both absolute height and relative relief are declined and the upper and lower curve come closer to each other toward downstream. On the basis of slope of river valley and crest (summit curve), the stage is divided in two parts. The upper part called bosche steilwand (Gravity slope) which is steep and maintain its angle by parallel retreat although the height is declining. The lower part is called as Haldenhang or wash slope which is very gentle with declining gradient. A very sharp break point occurs between them. This sharp break is also known as knick point. The wash slope is extended up at the expense of gravity slope. Gradually the crest of ridge is lowered to minimum level with presence of some inselberge of resistant rocks. River is completely graded and absolute height and relief reach at its minimum level. The whole landscape is transformed into a denuded plain which he called as a Endrumpf (Peneplain).

Evaluation of Penck's Model

Penck's Cycle of erosion became able to remove the many difficulties of Davis model with reference to especially rate and period of upliftment. He assumed a primarrumpf with long slow, continued upheavals. Erosion and upliftment simultaneously are the most important rational and reasoned elements of geomorphic cycle. His model is more practical than theory unlike Davis Model. However, Penck's emphasis is only on the development of slope in the evolution of landforms. He did not care about the structure of the rocks but assumed a certain relation between upheaval and erosion irrespective of any geological and lithological character. It is more complex to understand the whole process of development of landscape. It is developed with a view of undermine the Davis' model of erosion which is considered more simple and theoretical.

Distinguish Between Davis and Penck's Model of Cycle of Erosion

Davis' Model	*Penck's Model*
1. Landform is the function of structure, stage and process.	1. Landform is the expression of phase and rate of uplitment and degradation.
2. His model is from point of view of geomorphologists.	2. His model is from point of view of geologists.
3. He believed in fast rate of upliftment in a short period.	3. Penck believed in slow, long and continued upliftment of the land.
4. Erosion of the land begins after the completion of the uplitment.	4. Upliftment of the region and erosion go together.

	Davis' Model		*Penck's Model*
5.	He believed in the development of landforms in sequential order in three stages, i.e. youth, maturity and old.	5.	He believed in the development of lands into three forms, i.e. waxing, uniform and wanning development of land scapes.
6.	Davis does not give more emphasis on the slopes in the description of landforms development.	6.	Penck gives more importance to the slopes, i.e. convex, uniform, concave in the development of landforms.
7.	Davis is an American geomorphologist.	7.	Penck is a German geologist.

14

SOILS

All the rocks, viz. igneous, sedimentary and metamorphic, are disintegrated and decomposed when they are exposed to the atmosphere. The disintegration or breaking up of rocks into different sizes of particles are continued for a longer time. Consequently, a layer of fine particles overlying the rocks is formed at least everywhere on the land surface with varying depth. Thus, the top layer of finer particles of weathered rocks of the crust is described as soils.

Soil is an important element of physical environment which controls the inhabitation and development of human civilization on the earth surface. It is a link between climate and vegetation owing to its physical as well as chemical characteristics. The soils are mixture of a number of component both organic and inorganic elements and they are dynamic entities in which physical, chemical and biological activities are taking place continually.

The study of soils is very worthwhile especially for the geographers and planners because it may facilitate in planning for agricultural as well as industrial development after taking into account the different characteristics like fertility, texture, structure and profile.

The deep and fertile soils attracted the people and many important ancient civilization like Indus, Babylonian and Mesapotamian, were found in the region of fertile alluvium soils of river valleys. While the region of infertile and thin soils especially in mountainous and desert region, very sparse population is found because the agricultural development could not take place owing to infertility of soils and inaccessibility of water to man for drinking and irrigation purpose.

Components or Elements of Soils

Soils are the mixture of a number of elements found in physical environments. These elements are derived from different realms of environment like lithosphere, atmosphere, hydrosphere and biosphere.

They include the matter of all state like solid, liquid and gaseous. These matters are of both organic and inorganic nature. Thus the elements of the soils may be grouped as follows:

(i) Inorganic Materials

Soils are consisting of a number of minerals derived from the parent rocks as they are the aggregate of two or more minerals. Iron, silica, aluminum etc. are important minerals found in the soils.

(ii) Water

Soils are having a number of pores between the constituent particles. These pores are filled by water generally in humid region and remain dry in dry region. The presence of water in the pores is described as soil moisture.

(iii) Air or Gases

Soils are also equipped with various kinds of gases which are found in the pores when they are dry especially in dry season or in dry region. CO_2 Nitrogen, Oxygen, Hydrogen etc. are the important gases occurring in the soils.

Organic Materials

In the organic materials all the living organism, i.e. animal and plant, and decomposed matters from their dead remains are included. The decomposed form of remanent of biotic organism is described as humus. Humus is added to the soils at the last stage of development of soil profile. Different kinds of bacterias, fungy, insects and rodent animals are commonly residing in the soils.

Characteristics or Properties of Soils

Soils have two kinds of properties. They are described as physical and chemical characteristics.

The physical properties of soils include texture structure and colour of the soils. They vary from soil to soil. Thus the main physical properties are discussed below.

Texture

Texture of soils refers to the size and shape of particles which constitute the soils. The particles vary from gravel and sand to fine clay according to the following International particle size classification system.

Size of Particles	*Diameter in mm*
Gravel	2 and more
Sand	2 to 0.02
Silt	0.02 to 0.002
Clay	less than 0.002

Thus on the basis of these particle size a number of textural classes, i.e. sandy, clay and silt have been recognised. When these textures are mixed in any soil, loam texture is obtained and sandy loam, silty loam, clay, loam and so on are further recognised keeping in view the proportion of different sizes of the particles in the mixtures (Fig. 14.1).

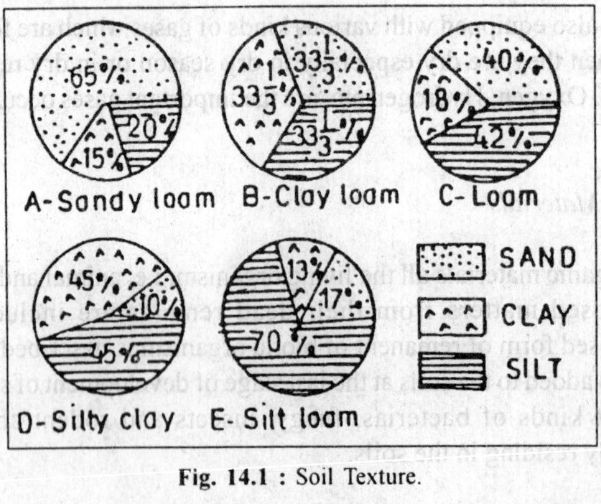

Fig. 14.1 : Soil Texture.

The texture of soils largely determine the water retention capacity, aeration, workability, movement and availability of water and the content of plant nutrients. For example, the sandy soils are very permeable and well drained and low water retentive and need more frequent irrigation for successful plant growth than the fine textured soils like clay which are more water retentive and subjected to waterlogging resulting poor aeration and workability. While the loam, i.e. mixture of sand and clay are very good soils for crop cultivation as they have the advantage of both sand and clay.

Structure

Soil structure means the arrangement of soil particles. It is one of the most important factor which determines the aeration, permeability and workability of the soils. There are mainly four types of soil structures, i.e. (i) Platy, (ii) Blocklike, (iii) Prism like, and (iv) Spheroidal. Soil structure is influenced by soil management, absorbed cations, microogranism and variation in soil moisture.

Colour

The colour of soils gives an idea about the soil conditions and other physiochemical properties. There are mainly two sources through which the colour of soils is derived. First, parent material, consisting of various kinds of minerals, influence the soil colours. For example red, yellow or brown colour are mostly related to the differences in degree of oxidation, hydration and diffusion of iron oxides in the soils. The second is the organic matters, i.e. decomposed part of plants gives usually black or dark soil colour. For example blacky colour of soils of temperate grasslands of world is only due to rich humus in them. The colour of the soils is determined with the aid of "Munsel Colour Chart", an international accepted system of colour designation.

Soil Profile

A vertical cross-section of soil through its different layers right from the parent rocks is known as Soil Profile. The soil profile have three important layers, one lying upon the other. These layers are known as horizon and designated as A, B and C. These horizons vary in thickness which are distinguished from the morphological characteristics which including colour, texture and structure etc. (Fig. 14.2). Horizon A is a upper layer of the soil consisting of both inorganic matters as well as organic like humus. Humus is found in large quantity and this horizon is also known as top soil. In the dry region, where water moves up from sub-soil layer to top soil layer due to excess evapotranspiration than the precipitation. "A" horizon is also called as Zone of Illuviation because some soluble minerals associated with water of sub-soil are accumulated here. While

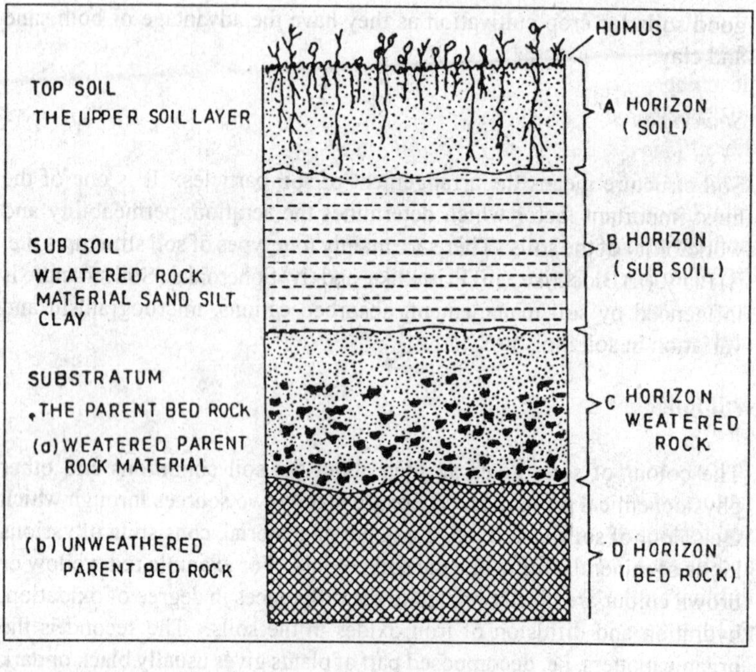

Fig. 14.2 (a) : Soil Horizons and Bedrock.

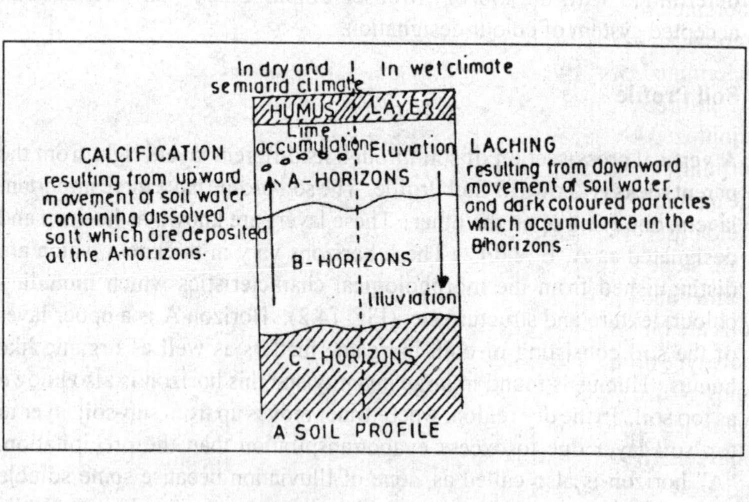

Fig. 14.2 (b) : A Soil Profile.

this horizon in humid region is called as layer of elluviation because the soluble minerals wash down from this layer to underneath layer due to leaching process.

"B" horizon is next to the A horizon and sand, clay and silts are found here. This is called as sub-soil and rock materials are found in weathered state. It is also called as layer of illuviation (zone of accumulation) and elluviation (zone of wash) in humid and dry region respectively. The content of humus decreases in this layer quite appreciably.

"C" Horizon is last layer of soil profile consisting of weathered matters of rocks. The particles are found in relatively large size. The roots of plants, and trees do not reach to this layer.

Soil Forming Process and Factors

A Russian pedologist Dukuchaiev firmly established that there are five factors which control the operation of soil formation. They are called as soil formers and act together to develop a soil. Some of them like climate and biotic factors are active and remaining are passive agents. Five soil forming factors are as follows : (i) Parent Rocks or Materials, (ii) Landforms or Topography, (iii) Climate, (iv) Biological Activity, and (v) Time.

Parent Rocks

Initially soils are formed by weathering of different kinds of rocks and the inorganic materials, *viz.* mineral contents are added from the underlying parent rocks. Colour and texture of soils are, to some extent, also determined by the rocks from where they have been originated.

Moreover, the effect of parent rocks on the nature of the soils is minimized by the influence of climate and time. Some of original rock forming minerals undergo the changes chemically with the time, hence the matured soils show the properties fully different from the parent rocks. Field observation has indicated that in hilly and dry region where relief is high, slope is steep, soils are very similar to their parent materials but in warmer and humid plain region the similarities between parent rocks and soils are blurred. Sometimes, same type of parent rocks produces two or more type of soils under the influence of varying local climate and different kinds of rock produce single type of soil in uniform climatic condition. For example, granite rocks have produced laterite soils in moist part of monsoonal region and non-laterite soils to its drier margin. Likewise granite and sand stone of constrasting nature have produced only sandy soils in Rajasthan desert owing to uniform climatic condition.

Climate

Climate is an important soil forming active agent. Climatic elements involved in soil development are moisture, temperature and wind. Moisture is the water of soils which operate the chemical and biological activities in them. Thus the moisture condition influence the soil formation into mainly two ways. In the region of heavy rainfall, the water moves down through the pores of soils from top surface to underlying horizons through leaching process. The soluble minerals are leached down by percolating water. The processes of downward migration of soil minerals is called elluviation and deposition of leached material in underlying horizon is called as illuviation. Thus soils in these region like equatorial rainforest, are infertile and deficient in calcium, sodium, magnesium and portassium as they are leached down to sub-soil region leaving behind only iron in A horizon leading red and yellow colours.

Contrary to it, in arid, semi-arid and sub-humid region where evapotranspiration exceeds the precipitation of the soil moisture and underground water move up to the surface by capillary action and the soluble materials like salts in water is evaporated. Calcium carbonate forms hard pan in the soils. While in the area of moderate rainfall like mid-latitude steppe, calcium carbonate is found in low proportion.

Similarly, the temperature also controls the soil forming process in two ways, i.e. (i) bacterial activity, and (ii) chemical activity. In warm region bacterial activities are vary intensive and they consume the humus of the soils, consequently, the soils become infertile devoid of humans. For example, the laterite soils of tropical rainforest region are lacking humus despite of excessive amount of leaves and stems of plant there : In cold continental climate bacterial activities are reduced and sufficient amount of humus is preserved in soils because of low consumption of humus. That is why black soils (Chernozem) in middle latitude are rich in humus.

However, the chemical activity is more prominent in tropical soils. The derived materials are thoroughly changed and the soil character becomes dissimilar to the parent rocks. While in the cold region, mechanical activity is most dominant in soil formation consequently the polar soils have very much similarity with the parent rocks.

Biological Factors

Biological factors are also described as active soil formers. They include both plant and animal kingdoms. The plant kingdom, i.e. macro and microflora influence the soil formation in three ways.

(i) They add the humus to the soils when their dead remains are decomposed and burnt. The humus gives a brown or dark colour to the soil;
(ii) The macroflora like tree absorbs the bases from the lower horizon of soils in leaching dominant area. Through the leaching process the bases like calcium, magnesium and potassium move down to the lower layers of the soils. Absorbed bases enter into the stems and leaves of trees. When dead stems and leaves decompose, these bases are added to the soils along the humus and consequently the loss of fertility of soils, through leaching process, is balanced; and
(iii) During the process of formation of humus (humification) organic acids are formed. These acids help in decomposing inorganic minerals of soils from parent materials.

Moreover, the animal kingdom also influence the soil formation process into three ways:

(i) Influence of animals on soil is mechanical and the burrowing animals like rat, fox, prairie dogs, ground squirrels and insects like earth worms and ants break and rearrange, the soils and mix up the soils of upper layer and lower layers in each other resulting in alteration in texture and structure;
(ii) Bacterial activity also determine the humus content in upper soil. In tropical region due to intensive bacterial activity, all humus are consumed by bacterias and a humus deficient soil is formed. While in the cold region where bacterial activity is lower more humus is found in the soils; and
(iii) Some bacterias in the soil take the nitrogen from the atmosphere and convert it into chemcial forms which are used by plants through the nitrogen cycle fixation process.

Landforms or Topography

Topography refers the configuration of ground surface. It influences the soil formation in two ways.

(i) The soils on mountain slope are generally thinner and immatured because the weathered materials can remain no longer due to erosion from the run-off. The flat surface of upland and low land have thick soils and matured profile as the soil formation process continues for longer time due to standing of weathered materials for a longer time in place of soil formation; and

(ii) Direction or slope of land surface exposed to sunlight also influence the characteristics of the soils. The southern sunlight facing slope has different soil from the north facing dark slope.

Time

Time refers the duration through which soil forming processes have been continued. The length of time determines the maturity of soil profile, i.e. development of horizons. The soils, on which soil forming process have acted for long time, will have well developed horizons. Whereas in the soils, that are evolving from recently deposited alluvium of glacial till, the characteristic horizons are absent or poorly developed. The time required for matured soils is not same for all, but varies from one to another depending upon the variation of soil forming elements. The soils formed from porous shale rocks require lesser time than those formed by impervious basalt rocks. The soil formation process is faster in tropical region than the polar region.

Classification of Soils

The Russian pedologist have developed zonal classification of soils in which he recognised strong relationship between climate, vegetation and soil regimes throughout the world. Hence the soils are classified as zonal and intrazonal groups. Zonal category refers the strong influence of climate on the soil forming process and intrazonal groups reflect domination of local factors in soil formation. While the azonal indicates the soils of immatured profile. Hence, the soils can be classified mainly into two broad groups and they are again subdivided as follows (Fig. 14.3):

Pedalfer

Pedalfers are a group of soils which are usually found in moist climatic regions of the world. There are mainly three important regions, i.e. tropical rainforest region, coniferoces forest and mid-latitude deciduous forest

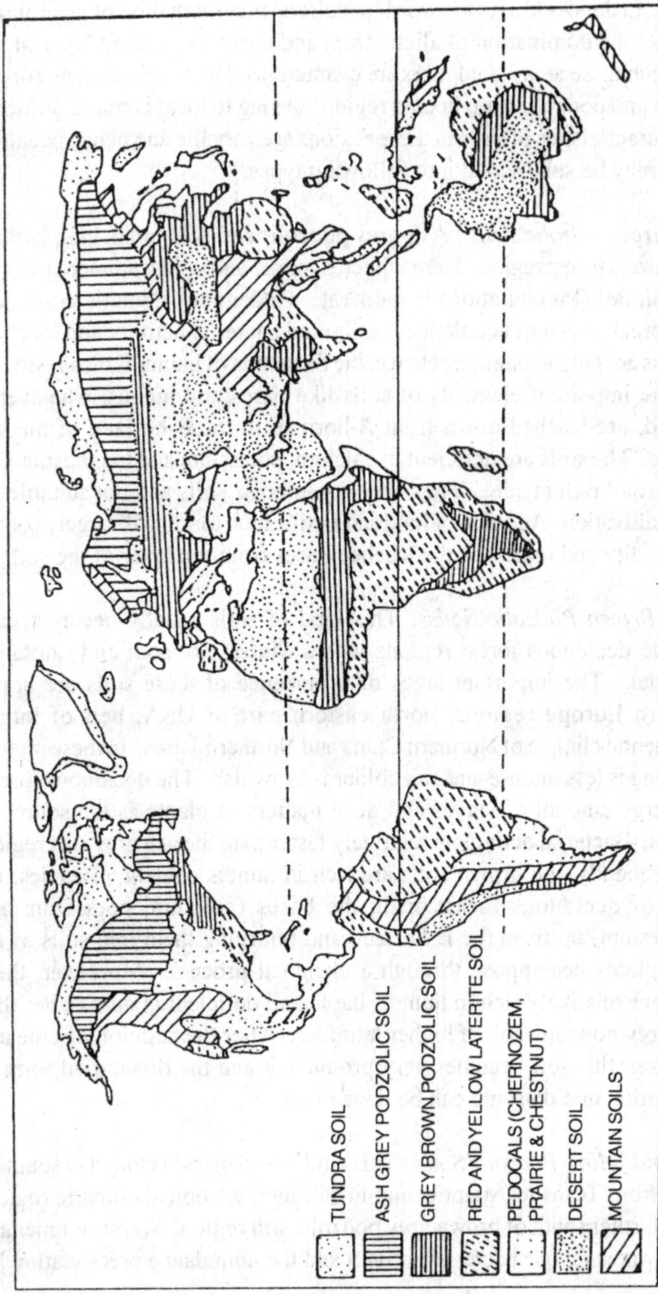

Fig. 14.3 : Soils of the World.

region, in the world. As the word 'pedalfers' is combination of ped+al+fer, indicate the domination of alluminium and ferrous in the top layer of soil profile in these areas. Pedalfers are characterized by laterization in tropical region and podzolization in cold region. Owing to local climatic influence the characteristics of soils in these regions are variable and hence pedalfers again may be subdivided into following types.

Ash-Grey Podzolic Soils : Ash-grey podzolic soil is found in high latitude coniferous forest region. Here winter is longer and cold, summer is shorter and rainfall (Precipitation) is moderate. Under such climatic condition, slow break down of needle leaves, limited bacterial activity and leaching process are not uncommon. Hence, the formation of humus is also restricted and the important elements of soils like bases and humus, whatever is formed, are leached down from A-horizon to the B-horizon of the soil profile. The soils are deficient in calcium, potassium and magnesium and humus and rich in acid. They are very infertile soils and not suitable for crop cultivation. Addition of lime and fertilization to soil will largely correct soil acidity and replenish the leached bases from top layer of the soil.

Grey Brown Podzolic Soils : This kind of soils usually occur in mid-latitude deciduous forest regions where winters are cool and rainfall is seasonal. The important areas of occurrence of these soils are north-western Europe regions, north eastern part of USA, belt of humid continental climate of Northern China and Northern Japan. In these regions leaching is less intense and soil colour is brownish. The deciduous forests add large amount of leaves and dead matters of plants as the source of humus. Bacterial activity is relatively faster than the ash grey soil region. Consequently, the brown soils are rich in humus content. Besides, the trees of deciduous forest bring the bases (calcium, potassium and magnesium) up from the B-horizon and returning them into soils as the dead plants decompose through a chemical process. Moreover, these soils are relatively rich in humus, bases and deficient in acid rather than Ash grey podzolic soil of higher latitudes. After the addition of lime and fertilizer, this soil becomes very productive and the diversified form of cultivation and dairying can be developed.

Red and Yellow Podzolic Soils : This kind of soil group is found in southern USA from Texas to Atlantic, and in sub-humid tropical climatic regions lying farther south of brown grey podzolic soil region. Warm summer and mild winter favour bacterial activity and the abundance precipitation led to the leaching process dominant in soil forming process. Thus the soils are deficient in humus, in bases and rich in acids, and hydrogen and iron oxide, as the laterite soils in tropical rainforest region. The abundance of iron oxides has given the red colours to this soil.

Red-Yellow Laterite Soils : Soils of humid tropical and equatorial zones are described as laterite soils owing to the dominant of laterization in soil forming process. On account of higher temperature and humidity, mechanical and chemical weathering of parent rocks are completed. Leaching of soluble minerals is also dominating and it leaves the concentration of aluminum and iron oxide in A-horizon of the soils, giving red and yellow colour to them. The leaching process washes away the silica and bases from A-horizon to B-horizon and these soils are deficient in bases, and rich in acids. Humus is also poor due to intense bacterial activity. Because the bacterias consume the humus at faster rate than that of formation. They are very infertile soils and with the application of fertilizers, different kinds of coarse crops may be cultivated.

Pedocals

Pedocals are those group of soils which are rich in calcium as evident from combination of pedocal (Pedo+Cal), where 'Cal' stand for calcium. Such type of soils are mainly found in arid, semi-arid and sub-humid regions where evapotranspiration exceeds the precipitation and water is in deficit. Hence, evaporation of soil moisture from the lower horizons through capillary action is common here. Unlike the leaching process, soluble salts and bases reach to the A horizon from B-horizon and this process of upward movement of calcium is called calcification. On the basis of local variation in the characteristics of pedocals soils under different local climatic conditions, different sub-groups may be recognised as follows.

Chernozem Soils : Chernozems or black soils, are very important type of pedocal soils. They are concentrated to steppes of Ukraine area, Black sea, central USA, Central Africa, South America and Australia. They rich in humus, calcium and unleached unlike the podzolic soils. The black colour of the soil is due to greater concentration of humus. The origin and development of these soils is due to climatic factor which has contributed to incomplete leaching and evaporation of soil moisture owing to water deficiency. Chernozems are very fertile soils and need little fertilizing when used year after year. They also require little irrigation due to their higher water retention capacity. The areas of chernozem soils are famous for the harvesting of bumper grain crops like wheat and oat in the world.

Prairie Soils (Brunizem Soils) : The prairie soils are occurring between the zone of chernozems and grey brown podzolic soils in USA. This soil group is, apparently, similar to the chernozems but differs in that they are lacking in calcium carbonate. These soils are transitional type between pedalfer and pedocals. They are extremely productive combined with the fertility of chernozems and moist climatic condition.

Chestnut Soils or Brown Soils : Chestnut soil, occur on the drier side of chernozem region. They are found in semi-arid, middle latitude, steppe land in North America and Asia. They are similar to chernozems in the profile but are deficient in humus. The shortage of humus has led to the brown colour of soils. They are fertile under the condition of adequate rainfall and irrigation.

Desert Soils : The soils which are found in both temperate and tropical deserts of the world. They are of different characteristics in temperate and tropical climatic condition.

In temperate desert region, colour of the soils is grey due to little humus because of sparse growth of vegetation. Horizons are not clearly differentiated, though present. At the depth of less than one foot a lime crust or hard pane is found due to deposit of calcium carbonate. This deposition near the surface is formed during prolonged dry period when the underground water rises upward by capillary action and water is evaporated leaving behind the deposits of calcium at shallow depth near the surface.

Moreover, the tropical deserts are having red colour. They are deficient in humus due to lack of vegetation. The presence of oxides or iron gives the red colour to this group of soils. Profiles is very poorly developed and textures are coarsy. The deposits of calcium carbonate is also common. They are very unsuitable for cultivation but crops can be grown in the area where texture is fine and water for irrigation is available.

Other Soils

Alluvium Soils : Alluvium soils are formed by deposition of sediments brought by various kinds of agents of denudation. Such soils occur mainly in the area of river basins, lake and coastal lowland regions allover the world. They are considered very fertile soils because every year new materials are added to them by sedimentation. These soils are different from local parent rocks on which they are bedded because they owe their origin to other parent rocks located at far distance from the place of present soils. They are very thick and matured soils. Different plant nutrients are abundant and so they are very suitable for agriculture. Consequently, all ancient civilizations flourished in the area of alluvium soils in the world.

Mountain Soils : This type of soil is located along the mountain slopes. It occurs in all types of environments. Immature profile, thin thickness and the infertility are common features of this soils. It is subjected to severe erosion especially after deforestation. It is widely spread in Central Asia, Himalayan region, South East Asian regions, Alps region, Rockies and Andes and so on. Altitudinal soil zones are also found due to variation in inclimatic and vegetative characters in the mountainous regions. Laterization is common at lower altitude near foothills and podzolization at higher elevation in tropical region. While in the cold region podzolization is common feature at every elevation level in mountain soils. This soil is not suitable for agriculture. It can be cultivated only after the terracing with additional dose of fertilizers.

Bog or Peat Soils

This kind of soils developed in area of marsh, permafrost and delta regions. In the polar and sub-polar regions marsh and swamp area develop after melting of ice (perma frost) during short summer season. Due to cold marshy conditions, the vegetable matters do not decompose but produce a liquid of blue colour. Thus the soils of immature profile and undemoposed vegetation matters are described as bog soils.

15

CLIMATOLOGY

Climatology consists of two words — climate and logy which mean the study of Climate. Climatology can be described as the subject of description of climate and atmosphere. With the passage of time, it has undergone many changes and its subject matter, gradually, widened since 6th century B.C. Climatology is concerned with atmospheric conditions, climate and weather on the earth surface. It also deals with the analysis and effect of various climatic elements on physical and cultural environments of the earth surface. It studies the same process of the atmosphere as meteorology in addition to seeks short-and long-term impact of atmospheric circulation and the solar radiation in different geographical regions. The climate and weather vary on the space, hence the climatology also deals with spatial variation of climates and their effect on the different realms of natural and cultural environments, on and above the earth's surface.

Elements of Climate and Weather

Climate is the aggregate condition of mean of different elements of the weather for longer periods, i.e. more than 35 years. Temperature, pressure, wind, precipitation and cloudiness are called as elements of the climate and weather. These elements are discussed below.

Temperature

Temperature is the driving force of the atmosphere. Various types of physical, chemical and biological process in our environment occur, directly or indirectly, due to temperature. Sun is the main source of the temperature. It emits energy which reachs to the atmosphere and on the earth surface in the form of insolation. Amount of insolation varies from place to place on the earth surface owing to variation in the length of days or sun shine, angle of sun rays, latitudes and transparency of the atmosphere. Vertical rays, passing through transparent atmosphere, increase the insolation as compared to slanting rays through opaque atmosphere. Amount of insolation decreases from equator to poles. It, also, rises in summer and reduces in winter season due to variation in hours of sun shine.

Due to variation in the amount of insolation, temperature also varies with space, time and altitude. It is because that insolation is absorbed by the atmosphere and the earth and converted into form of heat energy. This heat energy heats up the atmosphere and the earth. The degree of hotness of these bodies, is called as temperature. Temperature is measured by an instrument, called as thermometer. The unit of measurement is degree Celsius (°C) or Fahrenheit (°F). In Fahrenheit system freezing point is indicated by 32°F and boiling point by 212°F respectively on the thermometer. While in Centigrade (Celsius) system, these two points (freezing and boiling) are shown with 0°C and 100°C respectively. Both units may be converted easily with the help of following formula :

Fahrenheit = (1.8 x °C*) + 32°F

Or

Centigrade = **(°F – 32) ÷ 1.8

where * means the given temperature of a place in Centigrade.
** means the given temperature of a place in Fahrenheit.

Moreover, the temperature is shown on a map by a line connecting the places of equal temperature. This line of showing temperature is described as Isotherm line. It is studied in the form of maximum and minimum temperature, diurnal an annual range of temperature and so on.

Maximum and minimum temperature are recorded daily by maximum and minimum Thermometer. The difference between maximum and minimum is described as daily or diurnal range of temperature. The annual range of temperature refers to the difference of mean temperature of January and July months. Ranges of temperature varies from coast to interior location and from equator to poles. Various kinds of atmospheric phenomena like pressure, winds, humidity and turbulence are directly controlled by temperature. Agriculture, cultural activities, transportation, health etc. are also influenced by it.

Winds

Wind refer to the horizontal movement of air in a definite direction. The movement and direction of winds are determined by the distribution of pressure on the earth surface. They are the main source of evaporation and the precipitation. Wind direction and velocity are recorded by wind vane and anemometer respectively. The velocity of wind is estimated following Beaufort wind scale which was designed by Admiral Beaufort in 1805 for estimating wind speed.

There are various kinds of winds which are recognised by the direction of movement and the velocity. Planetary, periodic and local winds are very commonly designated in meteorology. Easterlies (Trade winds), westerlies and polar easterlies are categorised as permanent or plantetary winds. Local winds include shinook, mistral, sirocco, bora, loo etc. which affect human life in different ways locally. Monsoon is a very important seasonal winds which change their direction according to rhythm of seasons. South, South-East and East Asia are characterized as monsoon regions due to influence of these winds on the climate of the areas. Temperature as well as water are transported and mixed up in different contrasting regions by winds. Airmasses, fronts turbulence are the important elements of weather and climate.

Pressure

Atmospheric pressure is an important element of climate and weather. It is the weight of overlying or hanging layer of the air on a given area of earth's surface. It is measured by an instrument called as Barometer. Its unit of measurement is millibars or inches or cms. The pressure is variable responding temperature, subsidence of air, gravitational force, coriolis force and altitude from the earth's surface. The average pressure at sea level is 1013.25 mb or 76 inches. There are seven belts of pressure in the world. Pressure is, usually, inversely related with the temperature. Diarnal as well seasonal variation occur in the distribution of pressure. Coastal regions, mountain and valley region as well as the desert show the change of pressure in day and night which has resulted in local winds. Monsoon is also the effect of changing pressure on the land and water surface seasonally.

Pressure is very important guide of weather change. Lowering of pressure is the harbinger of arrival of storms. The rising of pressure predicts the occurrence of rain weather. High pressure discourages the precipitation and the low pressure increases the probability of precipitation if air is moist. Pattern and trend of pressure distribution are the main key of weather forecaste. Pressure is indicated by an imaginary line connecting the places of equal pressure reduced to sea level on the map. This line is described as isobar.

Humidity

Humidity is also very important component of atmosphere and an element of weather and climate. Humidity is described as the presence of water vapours in the given volume of the air. There are various ways, i.e. specific humidity (g/kg), absolute humidity (g/m^3) and relative humidity (%) to represent the humidity in the air. It is measured by hygrometer. It is variable in both time and space. Air over lying water bodies, forest area, and humid region shows high amount of humidity. Contrary to this, desert and dry region exhibit low humidity. Rainy season has rather more humidity than dry and hot season. Temperature and velocity of winds are the important determinant of humidity.

Out of all, relative humidity determines the various weather phenomena. Condensation of water vapours takes place when the relative humidity reaches to 100 per cent. This stage of air is called as saturation of the air and the temperature of saturated air is described as dew point. Formation of clouds, fogs, mist and precipitation etc. are determined by the presence of humidity at given temperature in the air. Cooling effect of body and dampness of weather are also the manifestation of the humidity. It also provides energy to the different storms like cyclones, thunderstorm and tornado by releasing latent heat after the condensation of water vapours. It is because that the water vapours have extra heat hidden in them.

Cloudiness

Cloud is a form of condensation of water vapours of an ascending moist air at higher elevation from the ground. As air rises up, it expands and cools down by rate of adiabatic change of temperature. Adiabatic change of temperature refers the fall and rise of temperature of an air by expansion and contraction respectively without any addition or subtraction of temperature from the surrounding environment. Air is expanded when rising up and contracted when moving down. Clouds, however, are combined form of water droplets and ice crystals hanging in the upper part of troposphere. Clouds are classified into various kinds, on the basis of their shapes, sizes and altitude of their location.

Clouds also control the amount of insolation reaching on ground. Temperature is rather high on the cloudy night as terrestrial radiation is blocked to reach the atmosphere from the earth. Formation of fogs and dews is also discouraged by cloud covers. Cloud shadow provide relief to the people from sunburn in hot summer days. Diurnal range of temperature is extremely greater in tropical deserts of clear sky as compared to Equatorial region of overcast or cloudy sky.

Weather and Man

Climate and weather affect human life in various ways. Man's activities of different nature are being shaped by prevailing climates and dynamic weather conditions in numerous geographical areas. His works, food habits, social rituals, religious traditions, political organisations, physiological and psychological characters, the cognitive behaviour and the economic organisations etc. are the reflection of composite effect of atmospheric conditions on the man in a given time and space. The relationship between weather and man may be understood by following discussion.

Weather and Agriculture

Climate showed very close relation with diversification of agriculture in the world. Temperature and moisture have established different patterns of agricultural activities. Temperature determines the growing or germination of plants. Tropical and sub-tropical regions are characterized by long growing period as compared to polar and sub-polar areas. Multiple cropping patterns are common in former region. Throughout the year, temperature is sufficient for the growth of vegetation. Monoculture and limited varieties of crops are found in latter case due to short growing season especially in summer.

Water of moisture is the next important determinant of agriculture. Plants get the moisture through the roots and transpirate through leaves by the process of photosynthesis. Water is life giving resource both for animals and plants. High temperature is useless in the absence of adequate amount of moisture in the soils. Desert region, where evapotranspiration exceeds the precipitation, is characterised by poor developed agriculture. Few crops of less demanding water may be cultivated. Nomadic herding is common activity. Contrary to this the monsoon region, equipped with both high temperature and adequate moisture, are the bowls of grains and cradles of large number of crops. Intensive multiple cropping system is practised throughout the region. Variation in amount of moisture have led to the specialization of crop combination region. Area with more than 100 cm annual rainfall is dominated by paddy as first crops while the region below 100 cm rainfall per annum is wheat dominant area. Eastern and Coastal India, Myanmar, Bangladesh, South East Asian countries and Far East region are growing paddy twice or thrice in a year. Temperate region especially grasslands are the bowl of wheat.

Beside, other weather element like direction of winds, distrurbances, diurnal range of weather etc. affect the various kinds of agricultural operations in any area. Sowing as well as harvesting operations are delayed or advanced due to varying nature of onset of monsoon especially in South, South East and Far East Asian countries. For example, winter crops like wheat, gram, barley and peas are sown by late if retreat of monsoon delayed. Similarly, various crops demand different rate of temperature and humidity at different stages of growth of various crops all over the world. Diversification of agriculture like horticulture, aquaculture, animal rearing or herding etc. have been the outcome of variation of agro-climatic conditions.

Climate also affect the quality of soils which regulate the cropping pattern. Soil moisture, texture and humus are the result of climatic phenomena. Thus the temperature, moisture and soil condition are important determinates of agriculture, consequently, the scientists have recognised different agroclimatic region for rational and scientific utilization of land for agriculture.

However, in modern era the control of weather and climate has been minimized by scientific development of various innovations. Vegetables can be grown even in Siberian region of negative temperature in glass house. Banana may be grown in polar region by providing artificial agro-climatic condition but they are not economically viable. Through assured irrigation desert may be cultivated under different crops. For example Thar desert, central Asian region have been converted into greenery by canal irrigation. Development of short duration crops enabled the farmers in temperate region to increase the yield per hectare. Despite of such technological change, agriculture still, in tropical southern hemisphere is the gamble of weather change.

Weather and Culture

Evolution of human civilization has been very much interrelated with the climatic conditions of different geographical regions. Ancient civilization grew up usually in the region of adequate temperature and humidity. Mode and means of life-styles of human beings have been the outcome of climate of their habitats. Their food habits, dress, social rituals and religious traditions vary from one region to another due to variation of weather and climatic condition in both time and space. People in tropics, generally, are vegetarian due to long growing season which encouraged the diversities of vegetable and their growth. Animals play very little role in agriculture and in the food habits of people. Different religions like Hinduism, Jainism, Buddhism came into being with prohibition of killing of animals and birds. While the people in temperate regions are not much vegetarians. Low temperature urges the people to take more energy with a view to protect them from the severity of cold. Animals are fatty. Vegetable growth is limited due to short growing season. All the temperate grasslands of world follow the mixed farming. Livestock rearing play an important contribution in agriculture. Fish is also greatly caught up and demanded only due to cold climatic conditions.

Clothing style is also the reflection of man weather relationship. Heterogenous nature of dresses is due to variation in weather condition in time and space. The people of middle latitudes and higher latitudes are well dressed throughout the year. They wear animal first during winter to adopt with temperature below freezing points. Woollen cloths are greatly demanded. Woollen textiles have been very much developed. Coat and tie are the typical dress of Europeans. Contrary to this, the people in tropics are half naked and bare footed. They wear light dress especially in equatorial climatic region due to hot condition throughout the year. Many tribes and primitive men wear only leaves and barks of tree to cover their sexual organs. Footwears are rarely used by these groups of population. Seasonal change also affect the dress style in sub-tropical and monsoon countries where rhythm of summer, rainy and winter seasons prevails. They wear light dress, in summer and woollen heavy dress in winter season. Dry tropical region like deserts encouraged the long loose light cloth which provide ventilation for evaporation of perspiration to get cool to the body. The winds remove the moist warm layered of body and encourage more evaporation. They wear turbans to protect head from sun burns. Hindu women do not wear full dress at the time of cooking in kitchen, priests wear only *dhoti*, which cover the body below waist during the worship of idols. Jains do not use foot wears. Saints wear narrow cloth to protect and hide only their sexual parts, in India. These dress habits are the outcome of climatic condition of the region under discussion.

Moreover, the nature of habitat and design of house of man are well related with the type of climate and weather. Closed house with low ventilation is common in the area of cold breezes especially in Tundra and Taiga type climatic region. Igloo is a typical form of closed house of Eskimoes. Courtyards are rarely constructed in temperate regions. While the houses in tropical and sub-tropical regions are very much ventilated. A house is consisting of verandah, courtyard and open space with a boundary wall. Such design of house is due to hot weather condition. People sleep under open sky in the night which gives them relaxation from scorching heat. In winter season again people get sun shine in open area of house. Projected and steep roofs are common in area of heavy rains of equatorial and monsoon regions. Contrary to this, flat floors are typical example of arid or semi-arid environment. Tents are used by nomadic herder in tropical desert like Sahara and Arabia.

Weather and Transportation

Weather and climate have affected the distribution of network of various kinds of transportation in the world. Airways, waterways, railways and roadways are the outcome of combined effect of climate and other physical and socio-economic attributes.

Civil aviation is very closely related with the atmospheric conditions of both surface and aloft of the earth. Flights of planes are restricted to the tropopause. Earlier, the aeroplanes could not fly in all kinds of weather. So the demand for scheduled flight based on the knowledge of three dimensional picture of temperature, pressure and humidity was greatly increased. A detailed knowledge or forecasts about temperature pressure, winds along the route of the planes are needed for a pilot. He usually avoids the route of turbulence. Now, through technological advancement in aircraft design, equipped safety devices and radiosode network, the dependency of pilot on meteorologists has reduced. But still the pilot, of small craft, depends upon the weather forecasts. Pilots are also very much interested about the surface weather conditions in both source and destination stations. At the time of take off and landing fogy condition is very dangerous. Wind speed, direction, type of precipitation, visibility etc. are important weather conditions which control the success of landing of the aircraft in any aerodrome. Aircraft industries have developed radar and other instruments attached with planes by which aircrafts may be landed under very unfavourable conditions. Flight of planes is also affected by clouds. It is found that the supercooled water droplets freeze on the airplane surface especially on the landing edge of wing. The change in the shape of wing reduces the effectiveness of planes and a number of accidents may be resulted from icing of aircraft.

Beside, the speed of wind at different height levels from the ground also affect the air flight. Forecasts about the turbulence and violent winds are very essential for secure flight in both upper and lower parts of the atmosphere. A new problem for air travel has developed with the advent of high speed jet aircraft. In certain parts of stratosphere, pilots find themselves in the areas where turbulence can disturb the comfort of the passengers and at time even endanger the plane itself.

Like the aircraft, navigators and sailors are very much related with changing weather conditions. Direction of winds were the main source of energy for shipment of goods during ancient periods. The planetary trade winds are designated only due to their roles in sailing the ships in the ocean. Weather forecasts about the atmospheric condition like temperature, pressure, direction and velocity of winds and various kinds of violent storms are very essential for silors or navigators. They must be well warned in advance by meteorological department from safety point of view.

Frozen condition of water also discourages the navigation during winter reasons in cold and temperate regions. Navigable rivers, oceans and seas are closed for sailors during this period. Many rivers in Russia, Canada, China, Korea become unnavigable during winter due to freezing of water while in spring and summer, they again open for navigation.

Surface transportation like roads and railways are also affected by the changing weather and climate. Hot and desert conditions retard the development of road and railways network due to excess heating. Soils are very loose. Construction of roads and laying of the tracks of rails are very difficult. Railway tracks are heated and expanded. Consequently, they may be disturbed and distorted. Fast blowing winds displace the sand dunes in desert areas. This mobile sand dune sometimes, rests over and cover settlement, railway line and road, while the moderate climate condition encourage the high concentration of surface transport network. Rainy season, usually, disturbs the transportation systems. Movement of vehicles and man is reduced. Flood water destroys the road and destabilizes the railway tracks. Construction of many bridges are also required. Similarly, the foggy weather also disturbs the surface transportation like the airways. Both rail and bus services may be delayed and the fixed schedules are disturbed.

Weather and Health

Climate and health are two inseparable elements of human life. Climate has led the health care style of man. During the history of human beings, men have adopted their life in different climates of earth's surface. Adapted to the climates, people are still affected by weather changes taking place day to day in various ways. Generally the incidence of cold, cough and throat infection is due to frequent change of weather. Body can not adjust immediately with weather change. Wind is the another weather element that causes the comfort or discomfort to human beings. Strong winds with chilled low temperature make man uneasy. He feels cold because the cold winds replace the warm air's layer about the body with fresh cold air. Contrary to this, hot and dry winds like *loo*, and Shinook heat and burn the body during hot and dry season in tropical and sub-tropical regions. Incidence of dehydration and cholera is very common problem due to high rate of perspiration from the body in hot and dry condition. Damp and humid conditions with high relative humidity cause the genesis of fever and sleeping disease. During the excess rainfall and flooded conditions, diarrhoea, malaria and cholera become very common diseases.

Weather and Hydrology

Hydrology is the study of distribution of water in lithosphere, hydrosphere, atmosphere and biosphere. Distribution of water at every level affect the human life in various ways. It is known as life giving resource. The presence of water, in quantity and quality at every point of occurrence, is very much related with the type of climate and weather. Hydrological cycle is the combined function of man and climate. Atmosphere is related with water hydrology at the point of transfer of water into different states, i.e. solid, liquid and gaseous. The affect of atmospheric phenomena on the water distribution on and beneath the earth's surface is dealt by hydro-metrology. Evaporation, precipitation and transpiration are important climatic processes which affect the quantum of water from different sources, i.e. surface, sub-surface, glaciers and water vapours. Evaporation shows very great relation with temperature and humidity of the air.

More than 80 per cent water vapours in atmosphere comes from ocean surface of tropical and sub-tropical regions. Evaporation is common in summer season in cold region. Dry and tropical desert climate, sub-tropical anti-cyclonic region and rainshadow zones on leeward side of high lands are characterized with the scarcity of both surface and underground water. While the equatorial region, mountainous regions along windward side in monsoon region exhibit high amount of precipitation surface and underground water. Fall in temperature reduces the liquid water and stores it in solid state. Cold regions of high latitude and high altitude are remained frozen or ice capped. Similarly, the rainy season accelerate water availability and dry season causes the scarcity of water for human beings.

Soil moisture is also a component of hydrological cycle. It has intrinsic relation with weather condition. Dry weather reduced the soil moisture as the evapotranspiration exceeds the precipitation. Rainy weather adds the moisture unlike dry weather discharge the moisture in soils. Amount of drinking water is also controlled with weather condition. Usually dry and semi-dry regions suffer from deficiency of drinking water when long dry spells occur.

Weather change also controls the discharge of rivers, seasonally. The discharge of frozen rivers like Obe, Yeneise, Lena of Russia, Amur river of Manchuria increases during spring and summer due to rise of temperature. Many rivers in monsoon regions flow in dry season after melting of glaciers in mountains during summer and flooded up during rainy season. Flooding of rivers in East, South East and South Asia cause a great damage of life and property every year during rainy season.

Weather and other Human Activities

Weather changes affect rather more business and activities than discussed earlier. Construction of building is hindered by wet, stormy, snowy or hot weather conditions. Construction engineers have developed many methods by which they can continue building, although at a slower pace, during unfavourable weather. Similarly, washing cloth activity by washermen is also delayed by bad weather like rainy or cloudy conditions. Housewives feel difficulties during rainy weather in their kitchen at the time of cooking food due to wet wood fuels. Schools and colleges are affected by weather changes. Extreme weather like too rain, too hot and too cold led the educational institutions to be closed. Efficiency of workers in offices factories, farms etc. are also reduced in bad weather.

Sport events are also controlled by weather changes. Outdoor games are subjected to whims of weather. Cricket, football, hockey, baseball can not be played in rainy weather. Many scheduled games are postponed due to bad weather. Yachting can be played only in dry or rainless weather.

Weather change also influences transmission lines of electricity. A major problem in power transmission is the occurrence of an ice storm. The coating of ice, can, seriously, overload the powerlines and break up them. Consequently, the supporting pylons collapse. Transformers are burnt unless protected from lightings. The power demand also increases up during cold condition. Modern air conditioning need extra electricity during rainy summer days. Sudden cloudy sky also increases the use of electricity by housewives as lighting the rooms to remove darkness in day.

Moreover, tourist industry is also very much affected by weather. Maximum number of tourists travel during spring season especially in Europe. Rainy and foggy weathers are unfavourable conditions for tourism development. Mobility is hindered by disturbances of transportation systems and natural beauty of landscape is also weathered during bad weather conditions.

16

ATMOSPHERE

The earth is surrounded by an ocean of air. This ocean of air is hanging under the pull of earth's gravitational force. It has a thickness of more than thousands of kilometres above from the earth surface. This thick extensive envelop of gas around the earth is called as atmosphere. Due to the presence of the atmosphere the earth is an unique planet in which life is found.

Significance of the Atmosphere

Atmosphere is an important component of physical environment and it affects our life and activities in different ways as discussed below.

1. It provides the life of both animals and plant species on the earth surface. Because it contains oxygen for man and animals and CO_2 for the plants. Animals and man inhales oxygen while the plant use CO_2 in photosynthesis process for the preparation of their food.
2. It controls the flow of energy from sun to the earth and from the earth to the space and regulates the temperature on the earth surface. It is transparent for the shortwave radiation especially for visible light and opaque for long wave terrestrial radiation from the earth surface. Thus it keeps the earth average temperature about 15°C. In the absence of the atmosphere there would have been extremes of temperature to the tune of 260°C between day and night. Under this situation the life would have been impossible on the earth surface.
3. It protects us from the ultraviolet rays which are very injurious to the plants and animals on the earth surface. The ozone gas (O_3) in the atmosphere absorbs these rays and save the earth surface from the danger of them.

4. It is the storehouse of water vapours which produce condensation and precipitation on the land and water surface of the earth.
5. It is also useful for the development of communication system through electronic media. Radio-waves transmitted from earth surface, return back from the ionosphere in the atmosphere.

Composition of the Atmosphere

The earth's atmosphere is consisting of different kinds of materials like solid, liquid and gases. These elements in the atmosphere are having their definite contributions in the composition of it and play an important role in controlling the human activities on the earth planet. The description of different constituents of atmosphere is given below.

Dry Air

Atmosphere is made of different kinds of gases. Nitrogen, Oxygen and CO_2 are the important gases which are found in their uniform or stable proportion within the lower layer of the atmosphere closed to the earth. The proportion of different gases to the volume of total dry air is given in Table 16.1. Important gases of dry air are discussed below.

Table 16.1 : Composition of Dry Air by Volume

Gas	Percentage	Gas	Percentage
Nitrogen (N_2)	78.08	Neon(Ne)	Trace
Oxygen (O_2)	20.95	Helium (He)	Trace
Argon (A)	0.93	Methane (CH_4)	Trace
Carbon dioxide	0.003	Hydrogen (H_2)	Trace
(CO_2)		Ozone (O_3)	Trace

Nitrogen : It is very important constituent of the atompshere. It accounts for about 78 per cent of total volume of dry air. N_2 gas works as a diluent and neutralizes the combustion phenomena in the atmosphere by diluting the oxygen. Besides, it is also used by plants and trees to make protein which is very useful for their growth.

Oxygen : O_2 is the second important constituent of the atmosphere. It shares about 21 per cent of the volume of dry air. The presence of oxygen in the atmosphere is very useful. Man and animals inhale O_2 and exhale the CO_2. The survival of life on the earth surface is impossible in the absence of oxygen. It also helps in burning of the substances. In combustion process, O_2 is also used.

Carbon dioxide : The share of Carbon dioxide (CO_2) in the composition of the atmosphere is very insignificant. But role played by it on the earth as well as in the atmosphere is very much appreciable. The advantage of CO_2 are as follows —

(i) It is a selective absorber and reflector of short wave radiation. For visible sun light the CO_2 is transparent and opaque for the terrestorial radiation from the earthsurface. Consequently heat is trapped between atmosphere and the earth surface and ultimately temperature on the earth is regulated.

(ii) CO_2 is also consumed by plants because they utilize it in the formation of food through the process of photosynthesis.

However, the CO_2 is added to the atmosphere by combustion of fossil fuels like coal, petroleum and oils. Automobiles industries and deforestation have been the important agents of add CO_2 in the atmosphere. CO_2 is increasing at tremendous rate in the atmosphere. More than 10 times it has increased over a century from 1890 to 1990. Rising of its content in the atmosphere will cause global warming through greenhouse effect.

Ozone : Ozone (O_3) is very minor constituent of dry air in the atmosphere. It accounts for only 0.00006 per cent of volume of dry air. It lies at the elevation between 19 to 35 km above the earth surface. This zone of the atmosphere is known as ozonosphere. The presence of ozone gas is very useful for life on the earth. It absorbs the ultraviolet sun rays and protects us from sun burn. In the absence of this valuable gas, the ultraviolet radiations will reach the earth surface and renders it unfit for living world.

Moreover, study reveals that the ozone layer is depleting due to release of nitrogen oxide from supersonic jet planes and chlorofloro carbon (CFC) from refrigerator industries. This depletion in the amount of ozone gas has resulted in blood cancer, cataract etc. diseases in developed countries like USA and Canada.

Water Vapour

Water vapour is also an important constitutent of the atmosphere. It is variable element and closed to the earth surface. About 90 per cent vapours are confined upto 5 km height in the atmosphere. The content of water vapour varies from as little as 0.02 per cent of air in desert environments to over 4 per cent in the humid tropics. The vapour in the atmosphere is added by evaporation from water bodies like oceans, seas, lakes, rivers etc., by transpiration from the plants and precipitation from the atmosphere. The share of vapour, though, is very little to volume of air, but climatically it is playing very significant role. It affects climate and weather in following respects.

(i) It is the source of all types of condensation like snow, fogs, cloud, mist etc. and precipitation on the earth surface. It directly determines the type (rain, Hail, sleet) and amount of precipitation;

(ii) It is regulator of the heat on the earth surface. Vapour is almost transparent for shortwave solar radiation and opaque for longwave terrestrial radiation. Thus the selective sun rays reach on the earth surface without much obstacles and the terrestrial radiation is not allowed to go back to the space. Hence, the earth surface and the lower atmosphere are heated up by greenhouse effect;

(iii) It is also the source of energy which helps in heating the atmosphere to some extent. After the condensation of vapour, latent heat is released in the atmosphere. This latent heat is also steering the turbulence in the atmosphere; and

(iv) It also affects the rate of cooling of the human body. Amount of vapour in the atmosphere is inversely related with the rate of cooling effect of body.

Dust Particles

Dust particles are also very useful constituents of the atmosphere. Like the vapour, they are also variable in both horizontal and vertical distribution. Dust particles are in larger numbers in dry desert environment and lower in humid environment. These particles are added in the atmosphere by cyclonic storms, volcanic activity, breaking of salt, along the sea coast, ploughing of land and so on. Their presence in the atmosphere is very essential and affects the weather condition in the following ways.

(i) Dust particles scatter the radiation from the sun in the atmosphere. The varied colours i.e. red and orange at sunset and sun rise is only due to scattering of selective sunlights. The blue colour of sky is attributed to the scattering of blue light in the atmosphere. The unusual amount of dust particles especially over town and urban centres reduce the insolation on the earth surface due to the scattering of sunlight in the atmosphere; and

(i) They help in the process of condensation of water vapour in atmosphere. Dust particles like salt particle act as an hydroscopic nuclei around which the condensation of water vapour takes place. In the absence these particles, there would have been not formed precipitation for and snow on the earth surface.

Structures of the Atmosphere

The atmosphere has great thickness around the earth surface. Its thickness or height above the earth surface is not still well known. However, on the basis of informations from sounding, baloon, radar, satellite, rockets etc., the altitude of atmosphere is estimated upto 10,000 km from sea level. The air is compressible and is more dense near the surface of the earth, so that 50 per cent of atmosphere lies below 5-6 km and 90 per cent is confined to the height of 29 kms.

The atmosphere may be divided into different roughly concentrating layers or strata. The layers may be demarcated on the basis of temperature, pressure and ionization and chemical properties of the air. Therefore following layers of atmosphere based on variation in temperature and pressure with altitude have been recognised (Fig. 16.1).

Troposphere

Troposphere is derived from Greek words Tropein (turning or rotating) and spheurous (sphere). It is lower most layer of the atmosphere. It is closest to the earth surface. The height of this layer is variable from equator to poles and from season to season. At the equator, it has maximum (16 km) height and at poles it is lowest about 6 kms. The maximum height at the equator is due to convectional movement of the

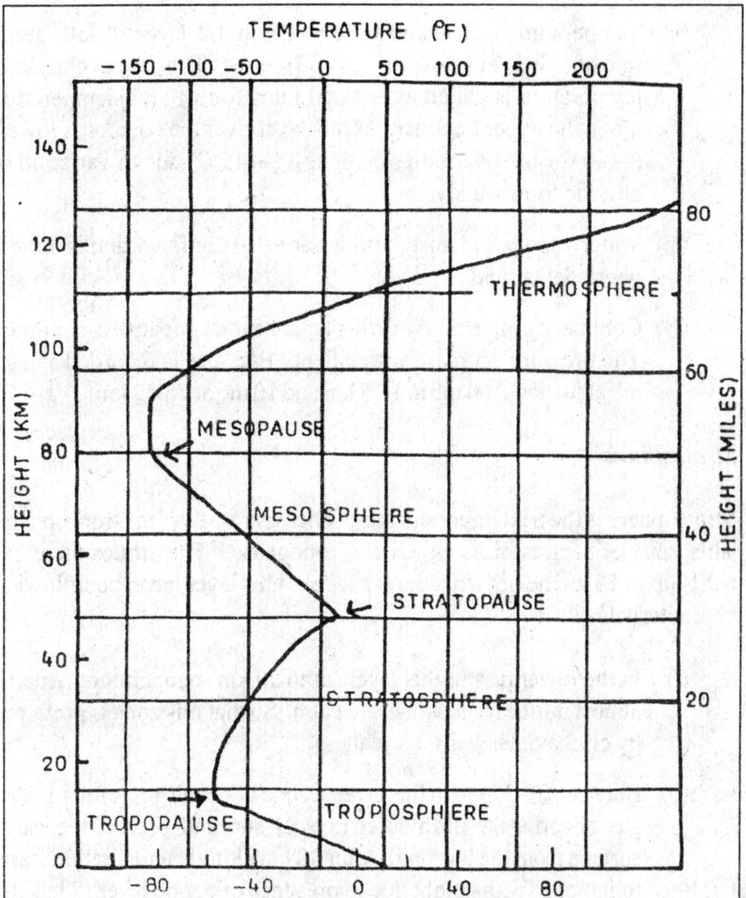

Fig. 16.1 : Structure of Atmosphere on Temperature Basis.

air currents. Similarly, its altitude in summer season is higher than in the winter season. This layer of atmosphere have the following characteristics.

(i) All kinds of weather phenomena like winds, clouds, fog, dew, snow, precipitation, cyclones, anti-cyclones etc., are confined to this layer.

(ii) Temperature declines with the height in this layer. It falls at the rate of 6.5°C per km or 1°C for 165 metres. This rate of change in temperature is called as Normal Laps Rate. Thus temperature along the upper boundary of this layer over the equator is lowest (– 80°C) and over poles is highest (– 45°C) due to variation of altitude from sea level;

(iii) Water vapour and dust particles are also confined to this atmospheric layer; and

(iv) Commercial aircraft fly in this layer near tropopause is common. The pressure decrease with height. Pressure is recorded as 500 mb at 5-6 km, 100 mb at 16.5 km and 10 mb at 28-52 km.

Stratosphere

Stratosphere is the next important atmospheric strata over the troposphere. This zone of separation is called as Tropopause. The stratosphere extends up to 50 km height from the sea level. This layer have the following characteristics.

(i) In the lower part of this layer (upto 20 km) temperature remains almost uniform and winds are calm. So that this zone is preferred by civil aviater to fly Jet plane;

(ii) Between 20-35 km in this layer, a zone of ozone gas is found. The gas absorbs the ultraviolet rays of sun and protect the earth surface from the burning. After 30 km altitude temperature starts to increase with height due to presence of ozone layer. Thus the temperature at upper limit of the layer reaches upto 0°C;

(iii) Weather phenomena like cloud formation storms and eddies, precipitation etc. are almost absent. Only horizontal movement of air can be observed in this zone; and

(iv) Rarely high clouds are found between 20-30 km height. It is called as mother of pearls or Nacareous cloud.

Mesosphere

Mesosphere is the third strata of atmosphere lies above the stratosphere. The former layer is separated from latter by a narrow zone called tropopause. The mesosphere extends between 50-80 km above sea level. The main characteristic of this layer is that the temperature decreases, with altitude. It reaches upto – 80°C at the upper limit (80km) of mesosphere. In this layer, there is appearance of clouds in high latitude region in summer season. The clouds glow in the night because it is consisting of cosmic dust with highly nickle content. It is also known as noctilucent".

Ionosphere

This layer of the atmosphere extends between 80 km to 640 km above sea level. It is ionized and electrically charged layer. It is very important atmospheric layer which reflects the radiowaves of various frequencies on the surface of the earth, so that broadcasting communication system is persisted. It is consisting of different following sub-layers with distinct features.

(i) D-layer : It lies between 60 to 99 km altitude. The main feature of this layer is to reflect the low frequency radiowaves to the earth and to absorb the medium and high frequency waves. This layer disappears in the night because it is associated with the sunshine.

(ii) E-layer : It stretches between 99 to 130 km height above the D-layer. It also appears with the sunrise because it is formed due to reaction between ultraviolet photon and nitrogen and nitrogen molecules. Medium and high frequency waves are reflected by this layer. This layer is also known as Kennelly Heaviside layer.

(iii) Sporadic E-layer : This layer occurs occasionally and formed under special condition. Very high frequency waves are reflected.

(iv) E_2 Layer : It extends upto 150 km. It is originated by interaction between ultraviolet photon and oxygen molecules, hence it also vanishes after the sunset.

(v) F-Layer : This layer overlies the E-layers and is known as Appleton layer. Unlike the above mentioned D and E-layers. F-layer exists in night. This layer is very much significant for long distance broadcasting communication as it reflects medium and high frequency waves in both days and nights. It is also divided into F_1 and F_2 layers. F_1 extends between 145-240 kms and F_2 between 240-965 kms.

Exosphere

It is the top most layer of the atmosphere overlying the ionosphere. It extends upto the space. The temperature increases with the height and air is very rarified in this layers. Hydrogen and helium are predominant gases and temperature is very high more than 1000°C. But the efficiency of this temperature is not so much as temperature on the earth surface has. If an astronaut putout his hand from satellite orbiting the earth in exosphere, he would not feel hot.

Stratification of Atmosphere on the Basis of Composition

As discussed earlier that our atmosphere is composed a number of gases and other elements like water vapour and dust particles. The proportion of different gases in the composition of atmosphere at certain heights is uniform. But above this limit the proportion of gases varies at different levels of altitude. The atmosphere may be divided mainly into two layers (i) Homosphere, (ii) Hetrosphere.

Homosphere : It is an atmospheric layer in which the proportion of gases at different height remains uniform. It extends up to 88 km from the earth surface, where a uniform mixing of gases are found. The uniformity of mixture is the result of turbulent mixing in this region of atmosphere. It is most important zone of the atmosphere because atmospheric elements like CO_2, O_2, water vapours etc., required for survival of man, animals and plants, are found in due proportions. It is formed of three important atmospheric layers like troposphere, stratosphere and mesosphere. These three layers are separated from each other by thin layers, i.e. tropopause and stratopause respectively.

Hetrosphere : This is upper most layer of atmosphere on the basis of chemical composition. It starts from the Mesopause and ends upto space. The main characteristics of this zone is that the gases are having heterogeneous proportion at different levels of height. The gases are no longer uniform mix because turbulent mixing is negligible. The molecules of atoms of elements found here, tend to separate and arrange themselves in the layers of distinctive composition. Nitrogen, oxygen, helium and hydrogen, the important light gases, are found in this layer. These gases are located at different level from heavier to lighter one into different layers, i.e. Molecular Nitrogen (90-200 km), Atomic Oxygen (200-400 km), Helium (1100-3500 km), and hydrogen (35 km upto space). The hetrosphere is also known as thermosphere because temperature increases with height throughout the layer.

17

INSOLATION AND TEMPERATURE

Sun is the main source of energy for all the planets of the solar system. The earth, as being a planet of solar system, receives the energy from the sun too. Sun rediates energy in huge amount per second from its surface due to atomic fusion process. The temperature at the sun's surface is about 6000°C. The earth is about 150,000,000 km away from the sun. Due to such long distance only minute fraction of solar energy reaches to the earth surface. The energy which strikes with the earth is 1/20,000,000 part of energy emitted by the sun's surface. This insignificance amount of solar energy is responsible for different kinds of physical, chemical and biological phenomena taking place on the earth surface. The solar traditions travel as a speed of 300,000 kms per second from the sun's surface and they reach the earth surface in 8 minutes and 20 seconds. Thus the incoming solar radiations received by earth's surface is described as insolation. The amount of insolation on the earth surface is rarely uniform. But the average amount of solar radiation or energy received by one square cm area in one minute is 1.94 calories. This amount of solar energy received by one unit area at the top of atmosphere is called as Solar constant.

Factors Affecting the Distribution of Insolation on the Earth Surface

The amount of insolation on the earth surface is not evenly distributed with time and space. The unevenness in insolation distribution is attributed to the following factors.

A. Major Factors

(i) Angle of sun rays.
(ii) Length of the day or duration of sun shine.
(iii) Transparancy of the atmosphere.

Angle of Sun Rays: The axis of the earth is inclined by 66½° angle to the plane of the earth's orbit. The sun's rays strike on the different latitudes at different angles. The angle of sun's rays controls the amount of insolation on different places into two ways (Table 17.1).

Table 17.1 : Distribution of Angle of Sun Rays on the Earth Surface on Summer and Winter Solestice

Latitude	Angle between the rays of the sun at noon day and the surface of the earth at summer solestice	Angle between rays of the sun at noon day and surface of the earth at the winter solestice	Difference
90°N	33½°	0°+	23½°
70°N	43½°	0°+	43½°
66½°N	47°	0°*	47°
50°N	63½°	16½°	47°
23½°N	90°	43°	47°
20°N	86½°	46½°	40°
10°N	76½°	56½°	20°
0°	66½°	66½°	0°
20°S	46½°	86½°	40°
23½°	43°	90°	47°
50°S	16½°	63½°	47°
66½°S	0°*	47°	47°
80°S	0°+	33½°	33½°

* Rays are just horizontal.

\+ Rays are below the horizon.

(i) When the sun rays strike with the earth surface at right angle, they cover small unit area. Thus the minimum possible area is heated and consequently amount of received insolation increases. On the other hand, when the sun rays strike at acute angle, they cover relatively larger area and the intensity of heating is rather lesser. Consequently, the amount of received insolation decreases (Fig.17.1).

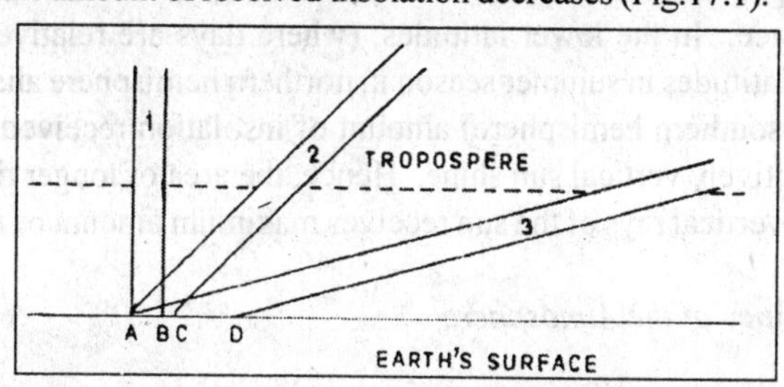

Fig. 17.1 : Inclination of Sun's Rays and Efficiency of Insolation.

(ii) The vertical sun rays pass through the shorter passage of the atmosphere as compared to the slanting or oblique rays which pass though longer passage of atmosphere. Hence, the amount of loss of insolation by turbidity (Reflection, scattering and absorption) of atmosphere is lesser. Thus net amount of loss of insolation received is greater in those area where sun shines vertically and lower in the area where sun rays strike obliquely [Fig. 17.2 (a) and 17.2 (b)].

Length of the Days

The duration of sunshine or length of the days is not uniform on every latitude in a given time. On the equator days and nights are equal throughout the year because the circle of illumination divides the equator into the equal parts in all seasons. The length of days increases from equator to poles during the summer solestice (21st June) when sun shines vertically over tropic of cancer. 24 hours duration of days occur on Arctic circle and six months day on the north pole. Thus the length of days is longer on all latitudes in Northern hemisphere during the summer season while the nights are longer in the southern hemisphere during the same season. During the winter season situation is reversed in both hemispheres. Days are longer and nights are shorter in southern hemisphere and *vice versa* in northern hemisphere. 24 hours day occurs on Antarctic circle and six month day on the south pole.

Thus the duration of sunshine determine the amount of insolation if all the other conditions are constant. Under this rule the higher latitude should have higher amount of insolation in summer season in northern hemisphere and in winter season in southern hemisphere. But in reality case is reversed. The slanting rays, which fall in higher latitudes and snow capped surface reduce the net amount of insolation received by earth surface. In the lower latitudes, (where days are relatively shorter than high latitudes in summer season in northern hemisphere and in winter season in southern hemisphere) amount of insolation received is greater due to relatively vertical sun shine. Hence, the area of longer duration of days with vertical rays of the sun receives maximum amount of insolation.

Transparancy of the Atmosphere

Transparency of atmosphere means the capacity of atmosphere to transmit the light from the sun to the earth and *vice versa*. The degree of

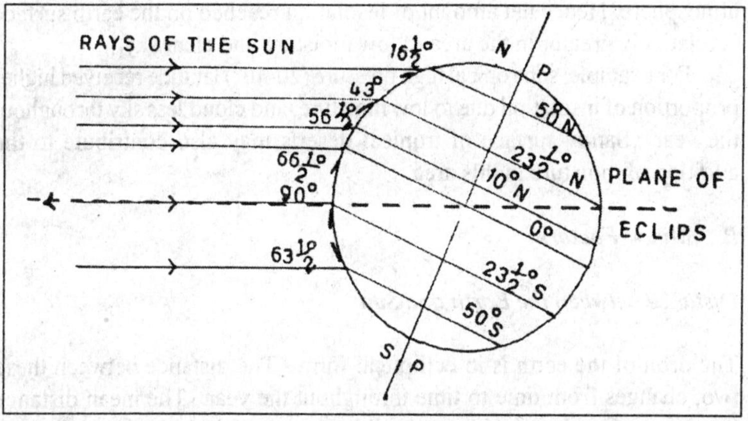

Fig. 17.2 (a) : Incidence of Sun's Rays at various Latitudes at the Winter Solastice.

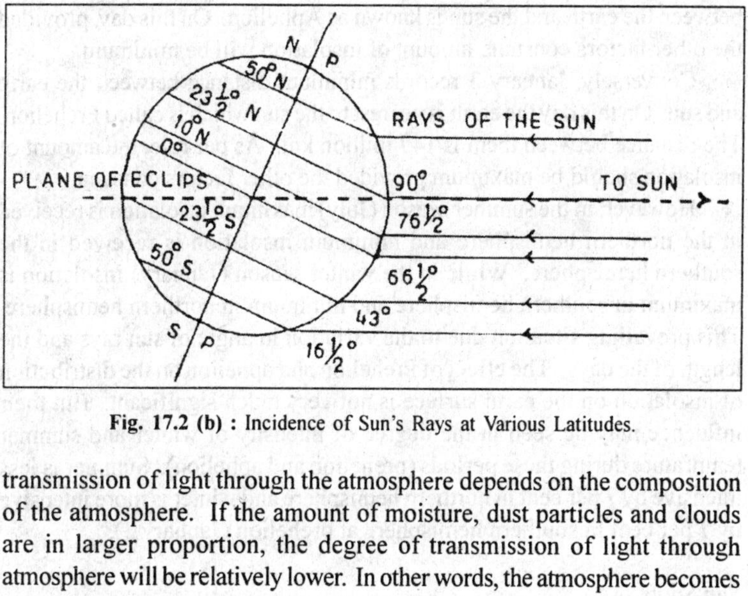

Fig. 17.2 (b) : Incidence of Sun's Rays at Various Latitudes.

transmission of light through the atmosphere depends on the composition of the atmosphere. If the amount of moisture, dust particles and clouds are in larger proportion, the degree of transmission of light through atmosphere will be relatively lower. In other words, the atmosphere becomes opaque for various wavelenghs of solar radiation and terrestrial radiations. More amount of solar radiation and terrestrial radiations. More amount of solar radiation is lost through the process of reflection, scattering as well as absorption of sunlight in the atmosphere. Hence low amount of insolation is received on the earth surface.

Contrary to this, when atmosphere is free from the moisture, dust particles and clouds, the loss of solar radiation through the atmosphere is lower due to low rate of reflection, scattering and absorption of light in the atmosphere. Hence net amount of insolation reached on the earth surface is relatively greater in the area of low moisture and clear sky.

For example, subtropical high pressure (20-40°) latitude received higher proportion of insolation due to low moisture, and cloud less sky throughout the year. Sandy surface of tropical deserts may also contribute to the addition of moisture in this area.

B. Minor Factors

Distance between the Earth and Sun

The orbit of the earth is in eclliptical form. The distance between these two, changes from time to time throughout the year. The mean distance between earth and sun is about 149 million km. On July 4, maximum distance (152 million kms) is found between them. The day of farthest distance between the earth and the sun is known as Aphelion. On this day, provided the other factors constant, amount of insolation will be minimum.

Conversely, January 3 records minimum distance between the earth and sun. On this day the earth is nearest to the sun which is called Prehelion. The distance between them is 147 million km. As per rule, the amount of insolation should be maximum provided the other factors constant.

However, in the summer season (July) maximum insolation is received in the northern hemisphere and minimum insolation is received in the southern hemisphere. While in the winter season (January) Insolation is maximum in southern hemisphere and minimum in northern hemisphere. This prevailing situation due to the variation in angle of sun rays and the length of the days. The effect of Prehelion and aphelion on the distribution of insolation on the earth surface is not very much significant. But their influence may be seen in the degree of intensity of winter and summer temprature during these periods (prehelion and aphelion). Summer is less intensive by 7 per cent in northern hemisphere and winter is more intensive by 7 per cent in southern hemisphere at prehelion (January 3).

Sun Spots

The number of sun spots on the surface of the sun also influence the amount of insolation reaches on the earth surface. Their number changes every year. They are caused by internal disturbance and explosion, in the sun. Amount of insolation is directly proportional to the number of sun spots. Larger the number of sun spots, higher the amount of insolation received on the earth surface and *vice versa.*

Process of Heating and Cooling of the Atmosphere

The atmosphere is heated directly not by the sun but by the earth surface. The incoming short wave radiation is absorbed by the earth and then it is rediated by the earth into form of long wave radiation or terrestrial radiation. The terrestrial radiations are directly responsible for the heating of the atmosphere. There are following processes by which atmosphere is heated from the earth surface temperature.

Radiation

Radiation is very efficient method of heating of the atmosphere. In this process, medium is not needed for the transmission of heat energy from one body to another. It can transmit through vacuum or emptiness of air. The upper part of the atmosphere is heated mainly by radiation process. The radiation process follow few principles as given below.

(i) Every object, either hot or cool, radiate energy;
(ii) Wavelength of energy emitted by an object is inversely related with the temperature of the bodies. Higher the temperature, shorter the wavelength and *vice versa*;
(iii) Amount of radiation emitted by a body is directly proportional to the temperature of body; and
(iv) Good absorber of energy is also good emitter of the radiation.

Conduction

Conduction is a process of transmission of heat in which energy moves from particle to particle. This process is continued unless the both ends of a body are heated equally. When two bodies of different temperatures are brought in contact, energy flows from the body of higher temperature to the body of lower temperature. The lower layer of air closed to earth surface is heated by this process. Upper portion of the atmosphere is not affected by this process of heating.

Convection

Convection is also very important process of heating of upper air. It is a method in which air is heated due to molecular movement of air. Excessively heated on the earth surface. Heated air becomes lighter and move upward and the cold upper air moves downward to take place of displaced hot air. Hence the upper part of the atmosphere is heated by convection process.

Latent Heat of Condensation

Water vapour is an important constituent of the atmosphere. The vapour contains latent heat when it is converted from liquid water to gaseous form. The water vapours condense into different forms like dew, fogs clouds etc. When temperature reaches at dew points. The condensation of water vapour causes the release of latent heat which is added to atmosphere. About 19 units of total terrestrial energy is added to the atmosphere by condensation of latent heat.

Distribution of Insolation

The amount of insolation on the earth surface is the product of various factors like angle of sun rays, length of days, transparency of the atmosphere etc. These variables are directly or indirectly related with the distribution of the latitudes.

Hence, the amount of insolation decreases from equator to poles. It is maximum at equator and minimum at poles. Poles receives above 40 per cent of amount of insolation received at the equator. There are three latitudinal zones of insolation. They are known as tropical, temperate and polar zones.

Tropical zone extends between tropic of cancer (23½°N) and tropic of capricorn (23½°S). The region receives maximum insolation due to the fact that at every point on it, the sun shines vertically twice a year. Two maximum and two minimum amounts of insolaton are received here due to apparent movement of sun in Northern and Southern hemisphere according to the season.

Temperate zone lies between 23½° and 66½° in both hemispheres. Maximum insolation is received once a year by every point of this zone on summer solestice (21st June) in northern hemisphere and on winter solestice (23rd September) in southern hemisphere. While the minimum insolation occures on winter solestice in Northern hemisphere and on summer solestice in southern hemisphere. Seasonal difference in amount of insolation is greater. No day is free from the sunshine in this zone.

Moreover, polar zone extends between Arctic circle (66½°N) and north pole in N.H. and between antarctic circle (66½°S) and south poles in S.H. This zone also have one maxima and one minima in a year. Maximum insolation is received in N.H. on 21st June and 23rd September in S.H.

While the day of winter solestice in N.H. and day of summer solestice in S.H. receive minimum insolation. For few days, insolation is absent due to not direct shining of the sun in both hemispheres. This zone has minimum amount of insolation, i.e. 40 per cent of insolation at equator.

Heat Budget

Heat budget may be defined as balance between incoming short wave radiation on the earth and out going radiation from the earth surface. It may be classified as (i) positive and (ii) negative heat budget. When amount of incoming short wave radiation exceeds the outgoing terrestrial radiation is described as positive budget. Negative budget occurs when terrestrial tradition exceeds the amount of short wave radition. On the earth surface present heat budget may be expressed and understood through the following system and diagrams.

Let the amount of insolation strikes the top of atmosphere is 100 units. Out of which, 27 units are reflected by clouds, 6 units are scattered and diffused by dust particles in the atmosphere and 2 units are reflected back by earth surface as short wave to the space. Thus 35 units of short wave radiation do not reach the earth surface and it is described as albedo of the earth. Besides 14 units are absorbed by atmosphere. Only 51 units of short wave radiations are received and absorbed by the earth surface.

After the absorption of solar radiation, the earth emits the energies in the form of long wave radiation. Long wave radiation is also known as terrestrial radiation. Out of 51 units (absorbed by earth) 17 units are, directly returned to the space, 19 units are used by water vapour or evaporation and 9 units are spent by convection and turbulence. Besides 6 units of terrestrial radiation is directly absorbed by atmosphere. Thus 48 units of terrestrial radiation are absorbed by atmosphere by one or other ways. Moreover, the radiations are absorbed by atmosphere (14 short wave radiations + 48 long wave radiations) are finally returned to the space. A balance is set in heat budget (Fig. 17.3).

Latitudinal Heat Balance

Like the amount of insolation received by earth, the amount of terrestrial radiation also varies from equator to poles. Below 40° latitudes more

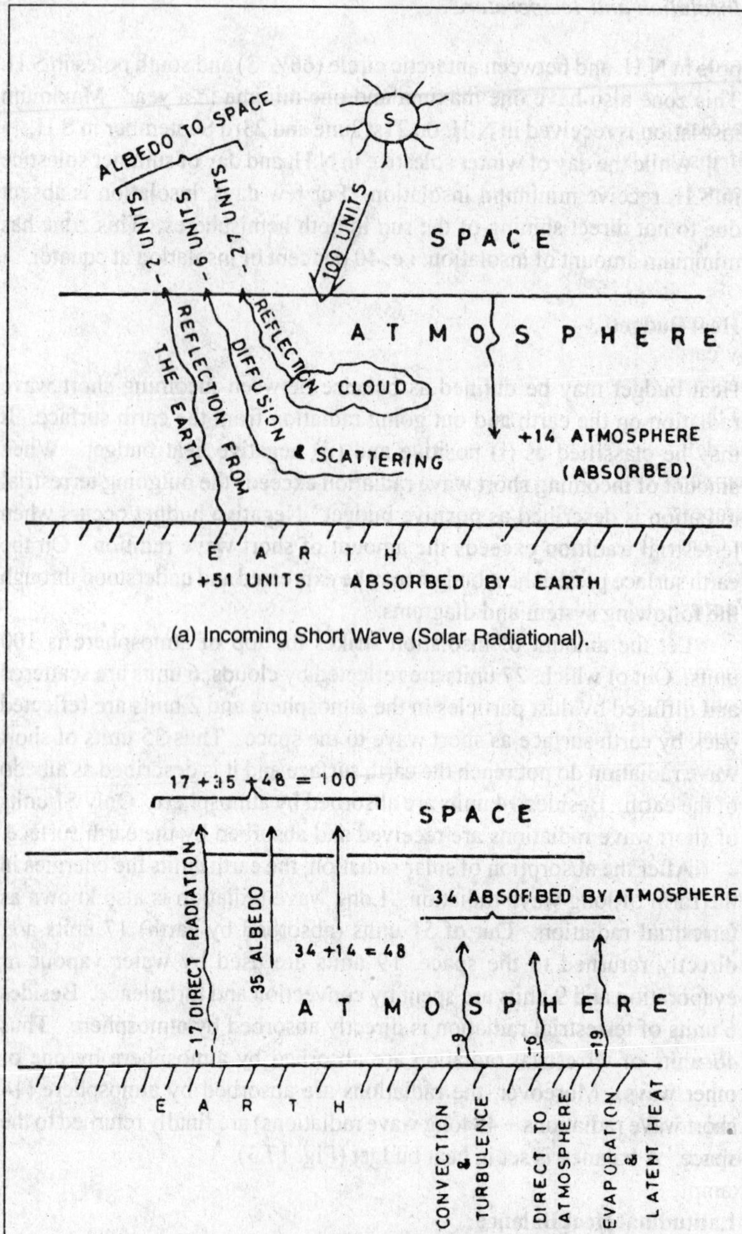

(a) Incoming Short Wave (Solar Radiational).

(b) Out-going long wave (Terrestrial Radiation) from the earth.

Fig. 17.3 : Heat Budget.

insolation is received than the lost to space by the earth.

Whereas the region above 40° latitude receives lower amount of insolation than the outgoing terrestrial radiation. Hence surplus and deficit of insolation occur in the region of low latitudes and higher latitudes respectively.

TEMPERATURE

The earth receives the insolation in short wave radiation. It is absorbed by earth and converted into form of energy. The energy consequently heat up the earth surface as well as the atmosphere. The degree of hotness of the earth and atmosphere is described as temperature. Thus heat may be defined as amount of the energy which heats up the body while the temperature may be described as degree of hotness of a body. The temperature is measured by thermometer and recorded either in Fahrenheit or Centigrade.

Factors Determine the Temperature

The mount of temperature on the earth surface is not uniform. It varies from one place to another place both in time and space. The spatial variation of temperature is the combined effect of a number of factors discussed below —

(i) Latitude.
(ii) Altitude.
(iii) Water and land surface.
(iv) Distance from the oceans and seas.
(v) Prevailing winds.
(vi) Ocean currents.

Latitudes : are very important factor to determine the distribution of temperature on the earth surface because the amount of insolation varies from equator to poles. In general insolation decreases from equator to poles, that is why temperature also decreases from equator to poles. For example tropical region experiences higher temperature than polar region.

Altitude : or height also determine, the temperature variation on the earth surface. Temperature decreases with the altitude at the rate of 1°C for 165 metres height. Hence, the place located at different heights from sea level show different temperatures even along the same latitude. For example, two cities of Quito and Guyaqail, both in Equador, are near the equator and are relatively closer to each other. But annual mean temperature at Guyaqail is 25.5°C as compared to Quito's mean temperature of 13.3°C due to difference in the elevation of these two cities. The former is only 12 metres above sea level whereas latter is 2800 metres above from sea level in Andes mountain.

Water and land surface are contrasting temperature due to their different nature of heating and cooling. Land surface is heated and cooled down very rapidly rather than the water surface. The land surface shows higher temperature in the summer and lower temperature in winter whereas water surface like ocean or sea has moderate temperature in both seasons.

Distance from the coast is also responsible for the variation of the temperature. The places located near the coast are having moderate temperature due to oceanic influence while the places located in interior parts (away from the coast) have relatively higher temperature in summer and lower temperature in winter due to continental effects. The annual as well as daily range of temperature is lower in the coastal region and higher in the region away from the coast.

The direction of the winds influences the temperature on the earth surface. The wind coming from colder region reduces the temperature of warmer region where it is blowing. The warm winds increase the temperature of colder region where it is blowing. That is why temperature along the coastal region is moderate due to inflow of sea breezes. Chinook, a hot wind along the eastern slope of Rokies, increases the temperature of Grassland in U.S.A. to considerable extent. Summer monsoon, a warm and moist air, declines the temperature in month of May and June in Indian subcontinent. In winter season cold wind from central Asia blows in India as cold wave and it reduces the temperature to considerable extent in North India.

Ocean currents are also influencing the temperature of coastal region. Warm current, increases the temperature in colder region and cold current decreases the temperature of warm region. For example, warm Gulf stream current has raised up temperature of coast of Europe in winter season above freezing point while the Conary cold current has reduced the temperature of Sahara coastal region.

Distribution of Temperature

The temperature distribution on the earth is studied into two ways, i.e.

(i) Horizontal distribution,
(ii) Vertical distribution.

Horizontal Distribution of Temperature

The horizontal distribution of temperature indicates the study of temperature across the latitude or the changing pattern of temperature from equator to poles. Latitudinal distribution of temperature is the best expression of absolute amount of temperature because the amount of insolation along the latitude is uniform. The horizontal distribution of temperature is shown by isotherms. Isotherm is an imaginary line which is drawn connecting the place of equal temperature reduced to sea level. These lines usually run east and west parallel to the latitudes. They are deviating or fluctuating north and south of latitudes due to other local temperature determining factors. Isotherns are having some spatial distance which indicates the steepness of the change of temperature gradient. Widely spaced isotherms indicates small rates of temperature change per unit distance and *vice versa.* Besides, isotherms take sudden bends where land water contrast is maximum. The study of isotherm maps of month of January and July give an idea about the distribution of temperature of summer and winter season in the worlds.

January — In the month of January the sun shines vertically near the tropic of capricorn and winter is experienced in Northern Hemisphere and summer in Southern Hemisphere. Following features are common in term of temperature distribution in the world in the month of January (Fig. 17.4).

(i) Maximum temperature is recorded in the landmass in southern hemisphere. 30°C is found over North West Argentina, East Central Africa, Borneo and Central Australia;
(ii) Minimum temperature is recorded on landmass (North East Asia) in Northern hemisphere;
(iii) Isotherms bend poleward when crossing over the land and towards equator when crossing the ocean surface in southern hemisphere due to higher temperature on the continents rather than on the ocean surface;

Fig. 17.4 : Distribution of Temperature (January) in °C

(iv) Isotherms bend equatorward when crossing over the land and poleward when crossing the ocean surface in northern hemisphere, due to higher temperature on the water surface and lower temperature on the land surface; and

(v) The isotherms are regular and widely spaced in southern hemisphere due to absence of landmass and the presence of oceans or water masses. While Isotherms are irregular in northern hemisphere due to domination of landmasses.

July— In the month of July sun shines vertically near the tropic of cancer, hence summer season is experienced in Northern Hemisphere and winter season in the Southern Hemisphere. Following features are common with respect to distribution of temperature in the month of July (Fig. 17.5).

(i) Maximum temperature is recorded on landmass in Northern Hemisphere, 30°C temperature occurs between 10-40°N latitude especially in South Eastern U.S.A., the Sahara, Arabia, Iraq, Iran, Afghanistan, larger part of China and small part of India.

(ii) The maximum temperature is also recorded in the Northern hemisphere. Below zero degree celcius is recorded in central part of Greenland.

(iii) Isotherms bend equatorward, in the Northern hemisphere, when crossing the water surface and polewards when crossing the landmasses.

(iv) Isotherms bend equator ward, in the Southern hemisphere, when crossing the land surface and polewards when crossing the water surface.

Thermal Anomaly

By rule, the temperature of different places should be uniform along the given latitude because each point on the latitude receives equal amount of insolation. But in practice temperature varies from place to place even along the same latitude due to different local factors like altitude, land and water contrast, prevailing winds etc. The difference between the mean temperature of the parallel is described as temperature or thermal anomaly. There are two kinds of temperature anomalies.

(i) *Negative Anomaly :* When the temperature of any place is lesser than expected temperature of the latitude, it is described as

Fig. 17.5 : Distribution of Temperature (July) in °C.

negative anomalies. For the year negative anomaly occurs over the land between 40-90° latitudes and between 0-40° latitudes on the ocean.

(ii) *Positive Anomaly :* Positive anomaly occurs when the mean temperature of any places is higher than expected mean temperature of latitude. It is found between 0-40 latitude over the land and between 40-90° latitudes on the ocean surface for the year.

Vertical Distribution of Temperature

There are two kinds of trends about the vertical distribution of temperature between the earth and troposphere. They are : (i) Normal laps of temperature, and (ii) Inversion of temperature.

Normal Laps Rate of Temperature

As we move from the surface of the earth, temperature falls with increasing height at a rate of 1°C for 165 metres or 6.5°C for 1000 metres. This rate of change of temperature with altitude is described as normal Laps rate. (Fig. 17.6).

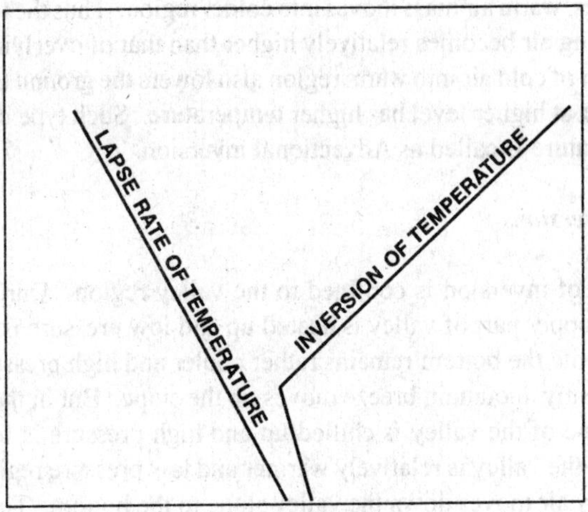

Fig. 17.6 : Laps Rate and Inversion of Temperature.

Inversion of Temperature

Sometimes due to certain reasons the temperature increases with increasing height instead of decreasing temperature. This kind of phenomenon is described as inversion of temperature in meterological text. On the basis of causes involved for such kinds of vertical distribution of temperature, inversion of temperature may be grouped into following types.

(i) Surface or Radiational Inversion.
(ii) Advectional Inversion.
(iii) Valley Inversion.

Radiational Inversion of Temperature

During long cold night, clear sky and calm wind condition, the rate of terrestrial radiation is rather faster and consequently, the earth and the lower layer (closed to earth) of air is chilled and temperature falls to considerable extent. While the temperature of the overlying air is relatively higher and inversion of temperature occurs. Such kinds of inversion known as radiated or surface inversion.

Advectional Inversions

Sometimes, warm air mass moves into colder region. Thus the temperature of overlying air becomes relatively higher than that of overlying air. The movement of cold air into warm region also lowers the ground temperature and the air at higher level has higher temperature. Such type of inversion of temperature is called as Advectional inversion.

Valley Inversions

This type of inversion is confined to the valley region. During the day time, the upper part of valley is heated up and low pressure is developed there. While the bottom remains rather cooler and high pressure occurs. Consequently, mountain breeze moves up the slope. But in the night, the upper slope of the valley is chilled up and high pressure is found. The bottom of the valley is relatively warmer and low pressure region. Hence cold heavy air moves down the valley slope to the bottom. The warm air is displaced from the bottom to upper slope of the valley. Thus the upper slope shows higher temperature than the bottom and the inversion of temperature develops. Such inversion is described as valley inversion of temperature. Due to this condition, tourist resorts, apple and mulbury gardens are usually found along the upper slope of the valley region.

18

ATMOSPHERIC PRESSURE

Our atmosphere is hanging on our heads under the gravitational pull of the earth. It is an indispensable part of the earth. On the every bit of earth surface, both water and land, the overlying air exerts a pressure equal to its weight. The amount of weight of a column of overlying air on a given unit area of the earth surface is usually called as Pressure. Every moment, a man carries more than thousands kg of air on his head but he does not feel it due to natural balance between pressure inside his body and the pressure outside surrounding him. At the time of mountaineering, nose bleeding, nausea and faintness often occur due to low pressure and small content of oxygen at higher altitude.

Significance of Atmospheric Pressure

Unlike the temperature and precipitation, the atmospheric pressure is thought as insignificant element of weather and climate. The direct effect of change of pressure on climatic condition of a region is not felt. But indirectly the study of atmospheric pressure is very useful for students of every descipline as it causes a number of changes in climate and weather condition in any region at any time. Such consequential weather changes have, indirectly or directly, great effects on the human life. However, the pressure may deserve for a systematic and detailed study due to following facts.

(i) The change of atmospheric pressure tells us about atmospheric disturbance likely to occur on the earth surface. Fall in pressure is the indication of occurrence of storms and cyclones while the rise in pressure reveals fair weather and occurrence of anticyclone. Thus the meteorological observatory may save the lives and property of man by forecasting about the weather condition on the basis of change in pressure of a particular day;

(ii) The change in pressure determines the direction and velocity of winds in a given time and region. Winds of high velocity may damage crops and building structures. Study of velocity of winds is of great significance in agriculture because different crops may require a specific velocity of winds on many occasions of agricultural operations. For example, the high velocity of winds at the time of flowering and ripening of wheat is very injurious;

(iii) The direction of winds, as a consequent of change of pressure, affects the temperature and precipitation of the area where they blow in. Dry and cold air from polar continental area, reduces the temperature and precipitation in tropical continental region. Warm and moist air in tropical region decreases the temperature and causes the precipitation in monsoon region in summer season; and

(iv) The change in pressure, with altitude also affect the physiological phenomena of the man. Nose bleeding and breathing difficulties may occur at the time of mountaineering or mountain climbing due to decrease of pressure and reduction of oxygen at higher altitude.

Measurement of Pressure

As discussed earlier, atmospheric pressure is the weight of a column of air over a given areas. It is recorded by two important instruments, i.e. (i) mercury barometer, and (ii) aneroid barometer. Mercury barometer is a simple vertical glass tube contains mercury and opens into reservoir of mercury. The length of tube is 90 cm. Normally the mercury is filled up to 76 cm and above the mercury a vacuue is found. While the aneroid barometer is without liquid. It consists of a small metal box with corrugated sides, from which most of the air is evacuated. Such a box react very readily to change of pressure either increases or decreases respectively. These movements of boxes are recorded by a pointer attached with scale which reads directly the value of atmospheric pressure.

The atmospheric pressure may be expressed as a force of overlying air on a unit area of the earth surface. Various units of measurement are used by meteorologists. Most common are cm, gram, inch, and milibar. One gram weight of air is approximately equivalent to one milibar per square cm area. A pressure of 1000 milibar is equal to 1.053 kg per square cms. It will be approximately equal to the weight of 75 cms or 29.53 inches high column of mercury in barometer. Hence the atmospheric pressure at sea level on 45° latitude is 15°C temperature is 1013.25 mb or 76 cm or 29.92 inches.

Variation of Atmospheric Pressure

Like the temperature, atmospheric pressure is also not static with time and place. It varies between 982 and 1033 mb all over world. Both diurnal as well as seasonal changes are common. The atmospheric pressure changes are unlike the temperature in recording daily maxima and minima. The former records two maxima and two minima in 24 hours. Such changes are taking place into three forms.

(1) In the higher or middle latitude fairly steadily rise and fall in pressure is recorded. It may last for few hours or for as long as several days. These changes are associated with the weather condition. Fluctuation in pressure changes is very frequent but amplitude is very small.

(2) In tropocal region especially at equator, a rhythm of change of pressure is observed. The pressure increases to maximum at 10 a.m. and decreases to minimum at 4 p.m. It increase again to second maximum at 10 p.m. and falls away to second minimm at 4 a.m. The amplitude of change is well marked.

(3) The upward moving of parcel of air also shows sudden pressure change. They are often associated with violent phenomena squalls of the winds, heavy showers of rains, thunder and lightning.

Moreover, the atmospheric pressure also fluctuate seasonally with the apparent movement of thermal equator. The permanent latitudinal pressure belts shift north and south in month of July and January respectively as the sun oscillates. Continuous pressure belts are broken in cells over land and water. Generally high pressure occurs over the continents in winter and over the ocean in summer due to differntial heating of land and water surface.

Determinants of Atmospheric Pressure

The distribution of atmospheric pressure is very uneven both in time and space. Such variation is attributed to the following causes.

Temperature

Air is compressible substance. When it is heated, it expands and when it is cooled, it contracts. Thus expansion and contraction in volume of a given air due to changing temperature led to change in air pressure

Temperature is inversely related with atmospheric pressure. Higher the temperature, lower the pressure and *vice versa*. For example, land surface like central Asia, in winter season experience high pressure and in summer lower pressure. Equatorial low pressure and polar high pressure belts are the outcome of high and low temperature respectively.

Rotation of the Earth

Temperature is not fully responsible for distribution of pressure. Deflective force of the earth due to its rotation is very much effective in determination of pressure belts in world. Effect of earth's rotation is felt at every points on the surface as well as in atmosphere. The effect is increasing from equator to poles. Along the Arctic circle (66½°N) and Antarctic circle (66½°S) low pressure occur, through the temperature is very low. This exception is due to fact that air of this region is displaced away north and south by deflective force of the earth and thus these regions become zones of scarce air. This effect is also observed in upper atmosphere where polewards winds from equator are deflected to west-east direction near tropic of cancer and capricorn. Consequently, piling up of air takes place above these latitudes near tropopause and subsidence of air occurs. Due to this effect high pressure belts occur in subtropical (23½° to 35°N & S) despite of high temperature (Fig. 18.1).

Gravitational Pull of the Earth

The weight of a column of overlying air is, to some extent, determined by gravitational force of the earth. This attraction is not uniform at every point of the earth's surface. It is inversely proportion to distance from the centre of the earth. The atmospheric pressure is directly proportion to the attraction of earth.

Distribution of the Pressure

There are two types of pressure, (1) high and (2) low pressure. These pressures are studied into two ways, i.e. Horizontal and Vertical.

Horizontal Distribution of Pressure

The atmospheric pressure on the map is shown by isobars which is an imaginary line drawn connecting the places of equal atmospheric pressure

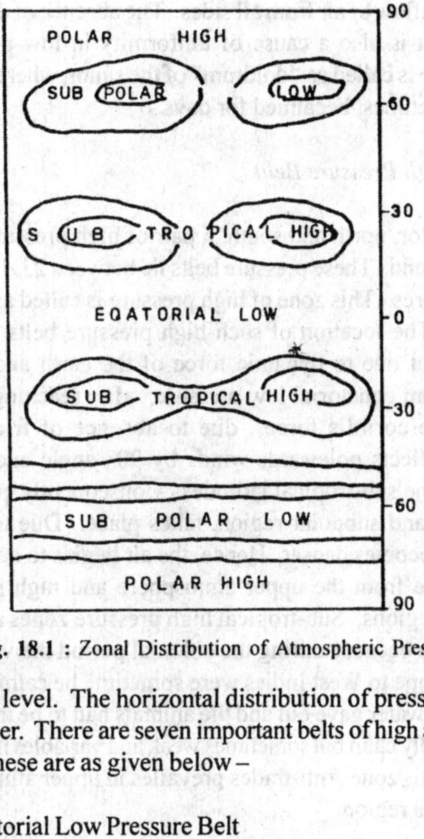

Fig. 18.1 : Zonal Distribution of Atmospheric Pressure.

reduced to sea level. The horizontal distribution of pressure is exhibiting a zonal character. There are seven important belts of high and low pressure (Fig. 18.1). These are as given below –

(i) Equatorial Low Pressure Belt
(ii) Sub-Tropical High Pressure (Southern)
(iii) Sub-Polar Low Pressure (Northern)
(iv) Polar High Pressure Belt (Northern)
(v) Sub-Tropical High Pressure (Northern)
(vi) Sub-Polar Low Pressure (Southern)
(vii) Polar High Pressure (Southern)

Equatorial Low Pressure Belt

Equatorial low pressure lies around the equator between 10°N and S, latitudes, High temperature is the main cause of low pressure. The air is heated and becomes lighter. Consequently, ascending movement of air takes place and the region experiences low pressure. The low pressure, here, is uniform in time and place because it is localized system and filled up by the radial inflow of air from all sides. The absence of deflection of air by coriolis effect is also a cause of uniformity in low pressure in this region. This zone is called as "doldrum" of the sailor, where sailing of ship might have, sometimes, becalmed for days.

Sub-Tropical High Pressure Belts

Away from equator, north and south, a pair of high pressure, one in each hemisphere, is found. These pressure belts lie between 23½ to 35°latitudes in both hemispheres. This zone of high pressure is called as Sub-Tropical high pressure. The location of such high pressure belts are not due to thermal cause but due to dynamic force of the earth and the air. The ascending air from equatorial low pressure, after reaching aloft, moves polewards. The coriolis force, due to absence of friction in upper atmosphere, deflects polewards winds by 90° angle and they become westerlies over the sub-tropical latitudes. Consequently, piling up of air, from equatorial and subpolar region, takes place. Due to cooling with altitude the air becomes denser. Hence, the air begins to descend down in sub-tropical zone from the upper atmosphere and high pressure of air occurs in these regions. Sub-tropical high pressure zones are also known as horse latitudes because during the colonial period the vessels carrying horses from Europe to West Indies were sometime be calmed so long that horses' feed and water gave out and the animals had to be thrown from the ships. Air is usually calm but sometimes weak and variables if any horizontal wind occurs in this zone Anti-trades prevailes in upper atmosphere above this high pressure region.

Sub-Polar Low Pressure Belts

Sub-polar low pressure lies between 45° to 66½° North and South latitudes in both hemispheres. The origin of these low pressure regions are not thermal but dynamic. The rotation of the earth causes the deflection and dispersion out of air from these regions and consequently the sub-polar belts experience scanty air and light air despite of very low temperature.

Polar High Pressure Belts

Both north and south poles hold high pressure of the air. The explanation of these high pressure belts', origin is not well known. Polarward from Arctic and Antarctic circles aerological data are scanty, so that the pattern of pressure distribution is not well known here. The earth's deflection force is maximum at the poles, so that the air should be very thin. But due to too low temperature the air becomes very denser and consequently high pressure occurs in Polar areas unlike the sub-polar regions. It is generally assumed that fairly shallow surface highs, of thermal origin, occupy the innerpolar areas.

However, all aforesaid pressure belts are modified and shifted north and south according to seasonal migration of thermal equator. They are occurring in cellular forms and their continuity as regular belts are broken over land and water due to unequal heating and cooling of lands and water surface in summer and winter seasons. High pressures are broken over land and low pressure over the oceans in summer and winter seasons respectively. Such modifications are very common in northern hemisphere due to premodification of land and less common in southern hemisphere due to prevalence of water surface. Sub-polar low pressure belt is well developed in southern hemisphere in both seasons.

Seasonal Distribution of Pressure

Atmospheric pressures are very much affected by change in the seasons. The pressures are shown by isobar lines on the maps for month of July and January. An outlook of these maps exhibits the following facts about the pattern of distribution of pressure of world in different seasons, i.e. summer and winter.

Atmospheric pressure in January

In month of January following pattern of distribution of Pressures are exhibited (Fig. 18.2).

(i) There is belt of continuous low pressure extending the whole length of equatorial regions, with local minima over the continents south of equator (South Africa, South America and Australia);
(ii) Sub-tropical high pressure belt of southern hemisphere is broken over the continents and remain confined to the oceans. Its development is maximum along the eastern coast of the ocean where cold current flow. The northern hemisphere exhibits well developed subtropical high pressure over the continents;
(iii) Sub-polar lows are well marked in continuous belt in about 60°S due to absence of landmass whereas the northern hemisphere

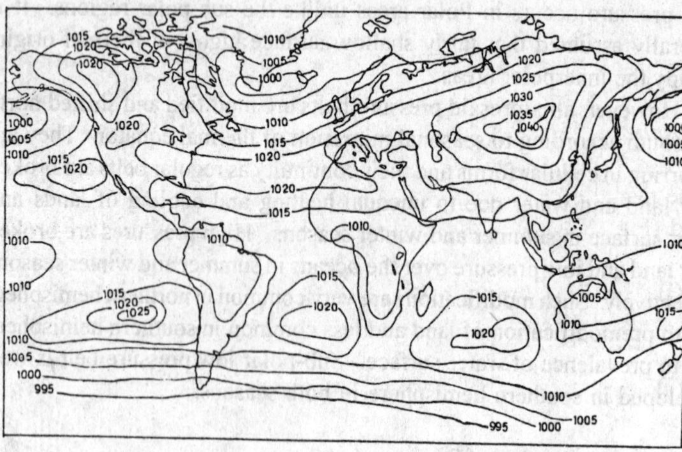

Fig. 18.2 : Distribution of Pressure (July) in Millibars.

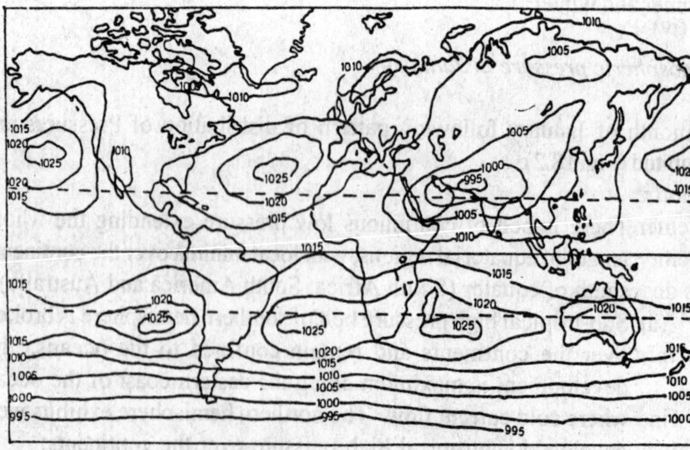

Fig. 18.3 : Distribution of Pressure (January) in Millibars.

has two broken cells of lows over the ocean. One area known as Aleution low which is centred near the Aleution islands in latitude 50°N and the second area, known as Icelandic low, lies south-west of Iceland in 60°N; and

(iv) The atmospheric pressure, polewards from these sub-polar lows, rises again especially towards the Antarctica continent and Arctic ocean. In Arctic ocean high pressure is well marked as Siberian and American anti-cyclones.

Atmospheric Pressure Condition in July

In the month of July following pattern of distribution of pressures are found. (Fig. 18.3).

(i) The low pressure is still well marked continuously about equatorial region with minimum which lies north of equator near Jacobabad in Pakistan about latitude 30°N;

(ii) Sub-tropical high pressure has disappeared from the continent, and confined to north Atlantic and Pacific oceans in northern hemisphere while in southern hemisphere this high pressure belt is almost continuous along 25°S latitude;

(iii) The sub-polar lows in southern hemisphere are very well developed and extends continuously in belts all about the ocean and they are faintly developed in Northern hemisphere. Aleution low has disappeared and Icelandic low still persists but with very low intensity; and

(iv) Polewards from the sub-polar lows, pressure again rises in January.

WINDS

Air on the earth surface seldom remains static but moves from one place to another under different climatic conditions. There are two kinds of movement of air, i.e. (i) horizontal, and (ii) vertical. The horizontal movement of the air is called as wind whereas the vertical movement of air is described as air current.

Determinants of Winds

The winds are outcome of combination of a number of factors. Their direction and velocity are determined by the following factors.

Pressure Gradient

The rate and direction of change of atmospheric pressure per unit horizontal distance in declining order is described as pressure gradient or barometric slope. This gradient is indicated by drawing the isobar lines and it lies at the right angle to the isobars. The space between the isobars indicate the steepness of pressure gradients. Closer the isobar lines, steeper the gradients and *vice versa.* The pressure gradient affects the blowing winds into following ways.

(a) The direction of winds is from high pressure to low pressure following barometric slope like water moves down hill under the gravitation force. The direction of winds is usually at right angle to the isobar lines, if other factors remain constant; and

(b) Velocity of winds is also determined by pressure gradients. Steeper the gradient, higher the velocity and *vice versa.*

Coriolis Force

Due to rotation of the earth the moving air on the earth is deflected from its direction of flow. This deflection force is called as Coriolis force. This force deflects the winds to its right side in northern hemisphere and to its left side in southern hemisphere. It also deflects the air in upper atmosphere because the atmosphere is attached with the earth by gravitational force.

Friction on the Earth

The distribution of various kinds of landforms, vegetations and constructed structure cause the hinderance in the flow of winds in lower part of the atmosphere. This friction affects decreases from earth's surface to aloft and from land to water. The rate of deflection is rather retarded by the earth's friction. Hence on the land surface, generally, the winds are blowing obliquely. While the winds, in upper atmosphere and over the water surfaces are parallel to the isobar lines.

Kinds of Winds

The pattern of atmospheric pressure varies from place to place, time to time and latitude to latitude. At some places, they occur in cell or oval forms, and at some places, in permanent continuous zones. At some places, they are changing from time to time alternating with high and low pressure due to unequal heating and cooling of the land and water surface in the same latitude. Hence, the winds may be classified into various forms as given below :

1. Planetary winds.
2. Seasonal winds.
3. Local winds.
4. Variable winds.

Planetary Winds

The planetary winds are blowing in the same direction throughout the year following the permanent global pressure belts (Fig. 18.4). The region

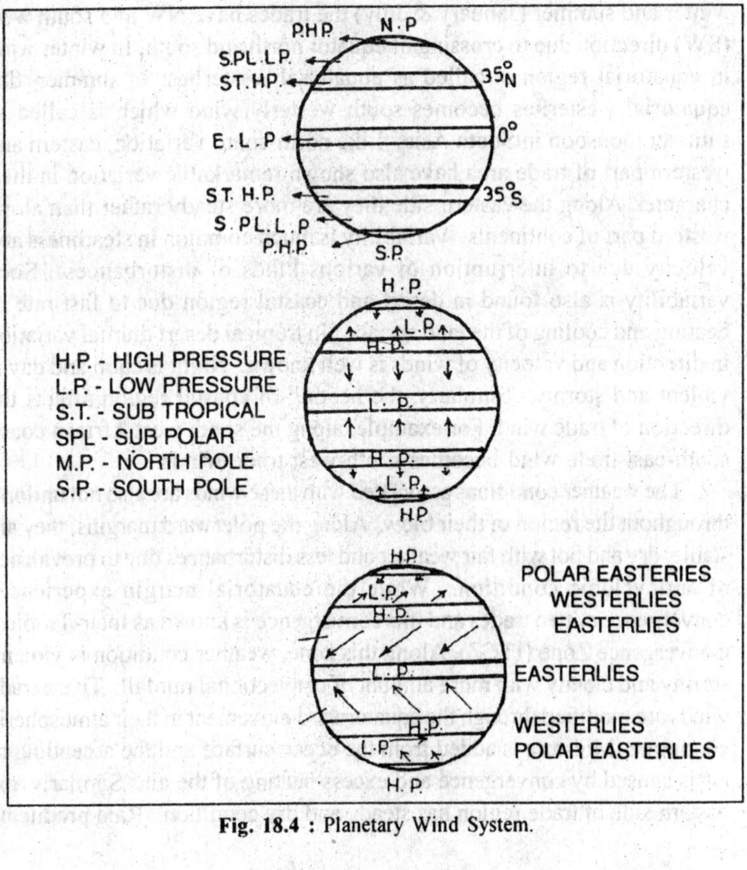

Fig. 18.4 : Planetary Wind System.

of these winds are fixed with little modification due to shifting of thermal equator during winter and summer seasons. Such winds are given as below:

Trade Winds

Trade winds blow from sub-tropical high pressure to equatorial low pressure in both hemispheres. Their direction of movement is from North-East to South-West in northern hemisphere and from South-East to North-West in Southern hemisphere. They are known as north and south trade winds respectively. In days of sailing ship, the trade wind latitudes with their fine weather, steady winds and severe storms were deservedly famed.

Moreover, the trades are the steadiest in various parts of the trade region. The variability in constancy of their direction and speed is well marked along poleward and equatorward margins. Near the equator in winter and summer (January & July) the trades have NW and south west (SW) direction due to crossing of equator north and south. In winter wind in equatorial region is called as equatorial westerlies. In summer, this equatorial westerlies becomes south westerly wind which is called as summer monsoon in South Asia. Like north-south variation, eastern and western part of trade area have also shown remarkable variation in their character. Along the eastern side they are more steady rather than along western part of continents. Variability is much common in steadiness and velocity due to interruption of various kinds of disturbances. Such variability is also found in desert and coastal region due to fast rate of heating and cooling of the land surface. In tropical desert diurnal variation in direction and velocity of winds is well known. Night is calm and day is violent and stormy. Similarly, the heating of coastal region affects the direction of trade wind. For example, along the south west African coast, south-east trade wind becomes south-west trade winds.

The weather conditions associated with these winds are also not uniform throughout the region of their blow. Along the polerward margins, they are stable, dry and hot with fair weather and less disturbances due to prevalence of anticyclonic condition. While the equatorial margin experiences convergence of two trades and this convergence is known as Inter-Tropical Convergence Zone (ITCZ). Along this zone, weather condition is violent, stormy and cloudy with more amount of convectional rainfall. These trade winds are modified through the equatorward movement in their atmospheric condition. Moisture is added from the ocean surface and the ascending of air is caused by convergence and excess heating of the air. Similarly, the eastern side of trade region has steady and dry condition. Rain producing storms are fewer due to strong subsidence condition in this region and the prevalence of land while the western part especially western side of Atlantic and Pacific oceans have variable, violent and rainy winds due to less developed subsidence of air and existence of oceanic condition.

Westerlies : The blowing of winds from sub-tropical high pressure to sub-polar low pressure is known as westerlies in middle latitudes. They are having westerly direction in both hemispheres. Their direction is from south-west to north-east in northern hemisphere and north-west to south-east in southern hemisphere. These winds are more violent and variable and they show great variation in velocity and constancy of direction. Poleward boundary is not definite but swing north and south in summer and winter respectively due to shifting of thermal equator. The polar dry winds surge into warm moist sub-tropical winds and subsequently a continuity — polar front is formed. Thus the northern region of westerlies is much disturbed with cyclone and anticyclone. Weather conditions are stormy, voilent and cloudy.

The westerlies are also showing dissimiliarity in their physical characteristics, in northern and southern hemisphere. In the former region seasonal reversal of pressure causes the wind system to be very complex due to prevalence of landmasses. These winds are very much affected by Monsoon on the eastern side of Asia and North America. Whereas the southern hemisphere exhibits stormy, extra violent and least interrupted westerlies. They are very much parellel to latitudes and move from west to east direction. They are called as Roaring forties along 40° latitude.

Polar Winds

A wind blows from polar highs to sub-polar low pressures in both hemispheres. It is known as Polar winds. Its direction is from north-east to south-west in northern hemisphere and south-east to north-west in southern hemisphere. Due to which it is also called as Polar easterlies. They are very weak and shallow. Sometimes, the westerlies cover the area of polar winds too. They are well marked in isolated trough of Aleution island in Pacific and Iceland in Atlantic ocean.

Seasonal Winds

Seasonal winds are such one which change their direction seasonally. For six months it blows from land to sea and for six months from sea to land. Monsoon is the best example of such kind of winds.

MONSOON

Monsoon is derived from one (Mausim) arabic word which means the reversal of direction of winds. Thus those winds, which are changing their direction seasonally from land to sea and *vice versa* are called as Monsoon. Monsoon is usually common in Eastern, South-Eastern and South Asian region is tropic and temperate regions. Such tendency is also prevailing along South-Eastern coast of U.S.A. and to some extent also in southern continents. But they are rather incomplete in their seasonal reversal. Hence these regions are called Pseudo Monsoon regions.

Monsoon in East Asia

The important monsoon areas in East Asia are Japan, Korea, North and South China. Winter monsoon is very strong and offshore air flow takes place from intense thermal anticyclonic high pressure in western China. It is usually dry and cold but interrupted with numerous cyclones and anticyclones of mid-Latitudes. The western coast of Japan recieves rainfall because the cold and dry continental wind gets moisture from the ocean. In summer season on shore maritime airmass strike along the coast of East Asia. But these winds are rather weaker. They have south easterly and southernly direction. Maximum rainfall takes place in East Asian countries by this monsoon.

Monsoon in South Asia

Monsoon characteristics in South Asia is due to seasonal migration of pressure belts. In summer season, South Asian countries like India, Pakistan, Bangladesh and so on, receive rainfall from SW maritime airmass, which is the extension of SE Trade wind across the equator. Summer monsoon is very strong due to high intensity of low pressure lies on the landmass. Moreover, the winter monsoon shows NE direction which is the replacement of NE trade winds. This wind has continental origin and rather weak and dry except when it passes over the water surface. Maximum rainfall occurs during summer monsoon.

Monsoon in Australia

Northern portion of Australia continent experiences monsoonal circulation. Moist northernly onshore winds, blow in Australian summer and cold and dry offshore wind blow in winter season.

Monsoon in West African Countries

The south facing coast of west Africa is dominated by monsoonal circulation within about 65 kms of coast line. Moist oceanic air flow is from south and south-west direction during summer and dry continental offshore air flow prevails in winter season. This monsoonal circulation differs from Asian monsoon system in several ways. It is considerably smaller in magnitude, shallow in depth and no planetary air-masses are involved.

Genesis of Monsoon

Question of genesis of monsoon has not been explained well till now. It is a matter of great dispute among the scientists. Many explanations have been given by meteorogists on different occasions to explain the origin and development of monsoon. A classical explanation thought that the monsoon is not more than seasonal land and see breeze. It viewed that during the summer season in northern hemisphere, the land surface is more heated. Consequently, the low pressure develops over the continental areas contrary to this, the water surface (Ocean and Sea) is rather less heated due to unequal heating and cooling of land and water. Subsequently, higher pressure persists over the oceans. Thus on shore winds from sea moves toward the land, which is called as summer monsoon. Similarly, in the winter season, due to shifting of sun towards southern hemisphere, land surface in northern hemisphere is cooled down and high pressure develops in continental regions, whereas the ocean surface records rather low pressure due to relatively higher temperature here than the land surface in the same latitude. Consequently, reverse wind circulation occurs. Winds begin to blow from land to sea, called as winter monsoon.

Moreover, the classical thermal explanations is not sufficient alone to describe such extensive and deep monsoon circulation is world especially in Asia. It is supplemented with a new explanation which is called as dynamic concept. This concept, is given by Flohn, in which he thought that the origin of monsoon is not only due to simple thermal heating and cooling of land and water surface but owing to latitudinal shifting of planetary wind belt north and south according to migration of sun seasonally. During the summer season the sun shines vertically in northern hemisphere especially on Tropic of Cancer (on 21st June). Due to intense heating of northern hemisphere, the pressure belts shifts northward and consequently winds migrate too. Inter Tropical Convergence (Equatorial low) lies beyond tropic of cancer in May and June. Hence, the south trade wind crosses the equator and changes its direction from south-east to south-west due to coriolis effect in northern hemisphere. The low pressure which is already established over continental area in south Asia. Draws the maritime winds to land. This south-west maritime airmass is described as summer monsoon.

In winter season, reverse conditions prevail. Sun shines vertically over the southern hemisphere and both pressure and wind belts migrate southwards. South trade winds regain their original poisition south of equator and the North Trade winds occupy the their original place near equator. At the same time, southern hemisphere shows low pressure, hence the winds blow from land to sea as winter monsoon. The winter monsoon is North Trade wind in south and south-east Asia (Fig. 18.5).

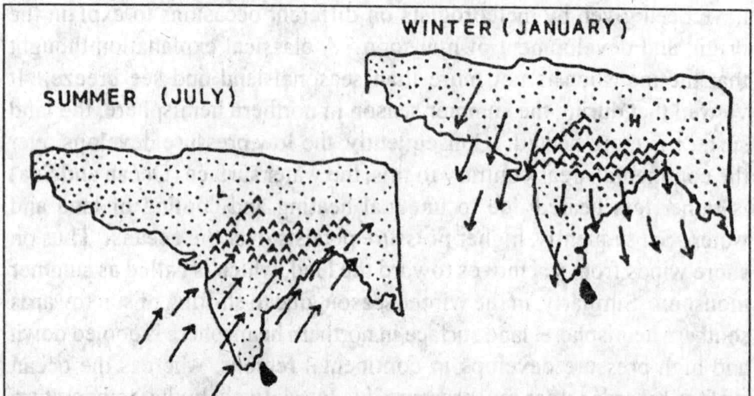

Fig. 18.5 : Summer and Winter Monsoon in India.

However, the dynamic theory is well explanatory but still needs some supplementary opinions to be more effective. Study of onset of monsoon, in south Asia, of several years indicates that the monsoon has not followed any definite date of its arrival and retreats. But some year it is delayed and some year advanced. Drought conditions are followed by flood occasionally. In this regard following views have been given time to time.

(i) Upper air circulation has been very much responsible for the origin and development of monsoon. The Jet stream, a fast blowing westerlies near tropopause, causes the delay or advance of onset of monsoon in India. The sub-tropical jet streams is bifurcated into branches in winter season, one branch lies north of Himalayan ranges and second lies south of Himalaya. The latter jet stream begins to shift polewards as the sun migrate north of equator. In the month of June due to development of ITCZ low pressure near 30°N latitude, the southern sub-tropical jet stream crosses over across the Himalaya and the south trade winds are drawn rapidly and onset of summer monsoon occurs. If, due to any reason, the jet stream remains south of Himalaya the onset of monsoon is delayed;

(ii) Elnino is another new concept to illustrate the success or failure of monsoon in different parts of the world. It is an event when abnormal pressure gradient occurs between eastern equatorial low pressure (ITC) and south Pacific sub-tropical high pressure (STH). This abnormal atmospheric condition causes weakening or reversal of such south trade winds, a flow of warm equatorial water in a easterly direction across the Pacific and other numerous varieties of repercussions in other aspects of atmospheric circulation. During 1983 the idea of Elnino was suddenly brought to world's attention by series of weather events. During a period of several months, there were crippling of droughts in Austalia, India, Indonesia, Maxico, Central America and Southern Africa. Devastating floods occurred in Western and South-Eastern U.S.A. and the occurrence of typhoon in the part of pacific at Tahiti and Hawai islands where they are normally extremely rare.

(iii) Koteshwaram supplemented his idea of thermal heating of Tibetan Plateau to the development of monsoon in summer seasons in Indian ocean. In summer season, the Tibetan highland acts as a heating engine. It is heated intensively being near the sun. The intense heating set conventional (Ascending) movement of air. This air is deflected to north and south aloft and moves as northerly and southerly upper air circulation. The southerly directed Tibetan upper air descends down in Indian ocean and supplements the air to the S. Trade wind which is known as SW Monsoon in South Asia.

Local Winds

Besides, large scale globally circulated winds, there are some other winds of local importance. They are originated due to locally atmospheric conditions and showing localized effects on character of any given area. Such winds of local importance are called as local winds.

Sea and Land Breezes

Such winds are localized in the coastal regions. During the day when sunshines, the land is heated and low pressure develops. The sea surface is relatively rather cooler than the land and high pressure persists in. Thus a landward circulation of wind is established. The movement of wind from sea to land in the day is called Sea Breeze (Fig. 18.6 (a)). Contrary to this, in the night land becomes cooler due to nucturnal radiation and the sea surface is warmer and low pressure develops. Consequently, in the night winds blows from land to sea is called as Land Breeze (Fig. 18.6 (b)).

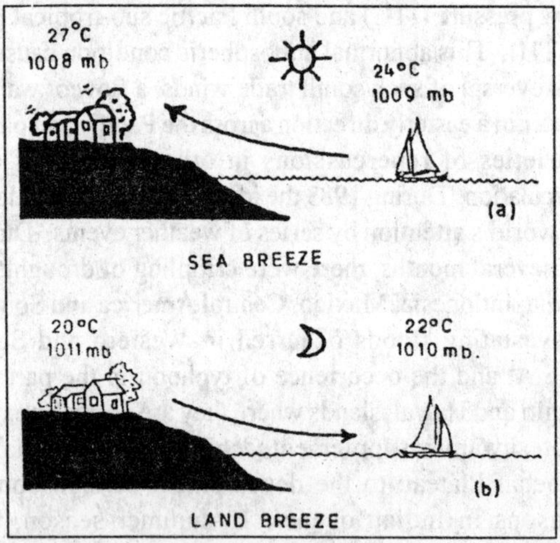

Fig. 18.6 : Sea and Land Breezes. Arrow Indicates Direction of Winds.

These winds provide relief to the coastal people from the tropical heating and they are very useful for the fisherman who go in the sea in the night and return back in the morning with sea breeze.

Valley and Mountain Breezes

Such winds are common in hilly area. It is diurnal cycle of air flow. During the day, upper slope of the valley is heated because the sun's rays first strike on the upslope. Low pressure occurs at the top of the valleys and

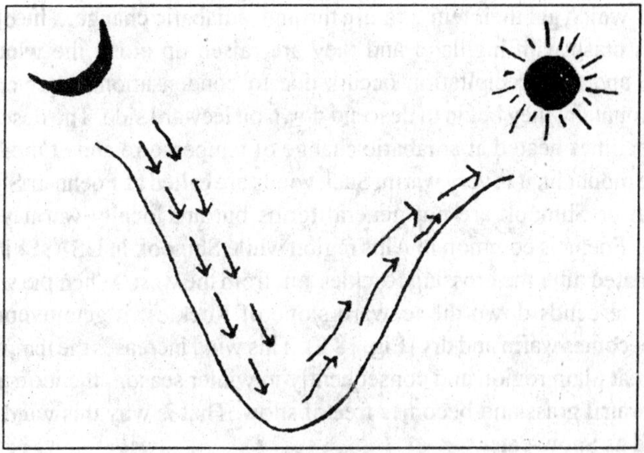

Fig. 18.7 : Valley and Mountain Breezes.
1. Valley Breeze; 2. Mountain Breeze

high pressure in its bottom. Consequently, winds blows from bottom to top of the valley called as Valley Breeze (Fig. 18.7). In the night reverse pressure condition persists. The upper slope of the valley shows high pressure due to nucturnal rediation. But the bottom remains warmer and low pressure develops here. Hence the wind blows from upper slope to bottom called as Mountain Breeze.

Foehn or Shinook

Such winds are also found in mountainous region. They are warm and dry

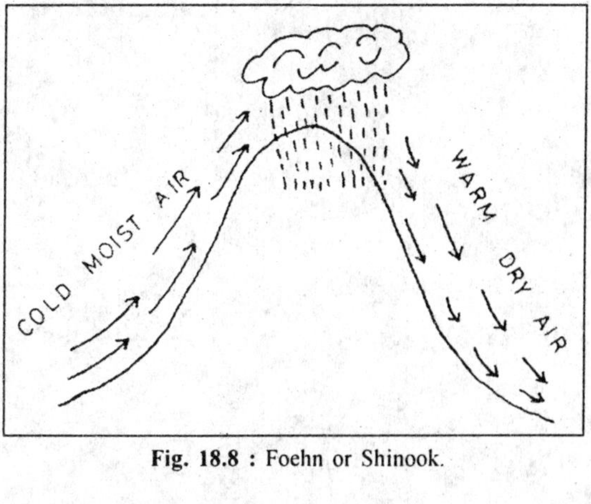

Fig. 18.8 : Foehn or Shinook.

winds which get their temperature through adiabatic change. The blowing winds clash with highland and they are raised up along the wind-ward slope and the precipitation occurs due to condensation. After crossing the mountain, they begin to descend down on leeward side. The descending air becomes heated at adiabatic change of temperature and at the bottom of the mountain it is very warm. Such winds are called as Foehn or Shinook. Foehn or Shinook are not general terms but are locally warm and dry wind. Foehn is common in Alps region while Shinook in USA. Shinook is originated after the crossing Rockies mt. from the west. When the westerly wind descends down the leeward slope of Rockies, it gets temperature and becomes warm and dry (Fig. 18.8). This wind increases the temperature in Great plain region and consequently in winter season, the ice is metled and prairil grassland becomes free of snow. That is why this wind is also called as Snow-eater.

19

AIR MASS AND FRONTS

The winds movement of air horizontally, are of various characteristics on the earth surface. They may vary in their thickness, temperature, humidity, instability or stability and the length (Distance) from time to time and from place to place. But they are distinguishable on the basis of their homogeneity and hetrogeneity in physical characters. Thus a huge body of air with homogenous physical properties like temperature and humidity (Horizontally) is described as air mass. It may vary in thickness and various strata of different temperatures and humidity may exist in any given air mass because temperature usually declines with increasing height.

Characteristics of Air Masses

The air mass as discussed above, must have the following conditions to be distinguished from the surrounding air in any region.

(i) It must have an immense body with great thickness and length;
(ii) It must have uniform physical characteristics, i.e. temperature humidity and stability horizontally; and
(iii) It must maintain its identity when travels in any contrasting region. Basic original characteristics, i.e. temperature and humidity should be retained by it for its recognition

Significance of Air Masses

From the point of view of climatology, the study of air masses is very significant. They modify the weather conditions in their invading areas. When cold air mass moves into warmer region, temperature will fall and stability of air will take place in the underlying surface. Fair and cloudiness sky may appear. Contrary to this when warm air mass enters in cold region, temperature will increase and instability will be caused in underlying air. Consequently, air will be ascending up and cloud formation and rainfall will be end result. Various kinds of fronts and atmospheric disturbances are also associated with battles of contrasting air masses in a given region. Water vapours are brought from ocean surface by maritime air masses and it is supplied to the various parts of continents by them in the form of either snowfall, rainfall, hail or sleet. Upper atmospheric air circulation is also associated with the air masses on the earth surface.

Factors which Determine the Characteristics of Air Masses

The air mass gets its characteristics and modification right from its source region to destination. Whole journey of the air mass may be affected by following factors.

Source Regions

Heavy air mass gets its origin from a suitable geographical region, which is called as source region. The source region adds the characteristics of homogenous atmospheric condition to the air mass. A suitable source region should fulfil the following conditions.

(i) It may be homogenous in physical characteristics. A large land mass or water body with uniform temperature and humidity is suitable source region;

(ii) A suitable source region should be of anti-cyclonic condition. Subsidence and divergence of air will retain the physical characteristics for longer time in wider regions. Subtropical high pressure belts are suitable source region in warm climate; and

(iii) Calm condition of air is also essential for an ideal source region. Slow and stagnant winds will remain for longer time in source region. It will absorb characteristics of the source regions upto maximum level.

Path : The air mass moves from source region to the destination through a track. This track of journey is called as the Path. It is modified by the temperature and moisture of underlying air attached with the path and *vice versa.* Thus mutual modification take place between the physical characters of air mass and the path.

Time : Every air mass takes some time to cover its journey. Duration of time is inversely related with the velocity of air (wind), modification in the air mass is also inversely proportional to the velocity.

Destination : Air mass is also modified by characteristics of the air of destination region. Warm air is modified by underlying cold surface when the former invades in latter. Temperature of lower layer of warm airmass is lowered and stability and inversion of temperature occur. Contrary to this when cold air mass enters into warm region, temperature of invader increases and consequently instability and precipitation may take place.

Modification of Air Masses

An air mass undergoes modification in its physical characteristics as moves from the source region to destination. The modification in temperature and moisture takes place in following ways.

Thermal Modification

When the air mass moves in the contrasting region, its temperature undergoes changes in the form of increase or decrease of temperature depending upon the underlying surface. For example, the warm air mass, in cold region, cools down from below and *vice versa*. Such kinds of temperature modification (Temperature addition or subtraction) is described as thermal modification. Humidity and stability of concerned air mass (invading) are also changed due to influence of underlying surface.

Dynamic Modification

This process includes the vertical movement (ascending or descending) and mixing of air by turbulence. Through vertical ascent of the air, air mass is vertically modified to great height. Temperature undergoes change at dry and wet adiabatic rates. Warm and cold airs are mixed up by voilent turbulents like cyclone, thunderstorms and torandoes. Addition and subtraction of moisture take place. Condensation of water vapours adds temperature due to release of latent heat while the evaporation subtract the temperature because of consumption of extra heat to convert liquid water into vapours. Thus vertical modification of air mass due to various kinds of vertical motion is known as dynamic modification of air mass.

Classification of Air Masses

Air masses have been classified into two following forms:

(i) Classification on the basis of source regions;
(ii) Classification on the basis of modification or nature of underlying surface.

Classification on the Basis of Source Regions

The source regions of air mass may be either land surface or water surface. Both of them may lie in either cold or warm regions. Thus the air mass is divided mainly into two forms like (i) Tropical air mass, and (ii) Polar air mass. They are represented by English capital letters T and P. Again these two have been grouped into sub-divisions separately keeping into account the nature of surface of source regions. They described as Tropical Maritime (mT), Polar Maritime (mP), Tropical Continental (cT) and Polar Continental (cP) air mass respectively.

Classification on the Basis of Modification

On the basis of thermodynamic modification of air masses under the influence of underlying surface, they can be classified in two categories, i.e. (i) Cold air mass, and (ii) Warm air mass. Cold air mass is defined as one which has lower temperature than the underlying surface. It is indicated by small "k" letter. While the warm air mass is described as one which has relatively higher temperature than the temperature of underlying surface. It is denoted by small "w" letter. These air mass may be stable and non-stable (unstable) due to different types of modification through exchange of temperature between overlying and underlying air. Thus cold and warm airmasses may be classified also as stable (s) and unstable (u) airmass.

General Classification of Air masses

After the combination both criteria discussed above, the airmass can be grouped into following types :

(1) Tropical continental warm stable Air mass (cTws)
(2) Tropical continental cold stable Air mass (cTKs)
(3) Tropical continental warm unstable Air mass (cTWu)

(4) Tropical Continental Cold Unstable Air Mass (cTku)
(5) Polar Continental Warm Stable Air Mass (cPws)
(6) Polar Continental Warm Unstable Air Mass (cPwu)
(7) Polar Continental Cold Stable Air Mass (cPks)
(8) Polar Continental Cold Unstable Air Mass (mTws)
(9) Tropical Maritime Warm Stable Air Mass (mTws)
(10) Tropical Maritime Warm Unstable Air Mass (mTwu)
(11) Tropical Maritime Cold Stable Air Mass (mTks)
(12) Tropical Maritime Cold Unstable Air Mass (mTku)
(13) Polar Maritime Warm Unstable Air Mass (mPwu)
(14) Polar Maritime Warm Stable Air Mass (mPws)
(15) Polar Maritime Cold Unstable Air Mass (mPku)
(16) Polar Maritime Cold stable Air Mass (mPks)

FRONTS

When the air masses of contrasting temperature and humidity meet together, a separating boundary is formed between them. They do not mix together and maintain their individual identity along a line with some angle subtended to the ground. Such line or plane of discontinuity is called as Front. The process of the formation of front is known as Frontogenesis. Front is a transitional zone in which mixing of air masses take place and physical characteristics change rapidly. But on the either side of the front extreme condition of temperature exists.

The fronts may move with various velocities. They may remain in stationary position for few hours. They lean to the earth making more horizontal orientation. Their gradient may vary between 1:50 to 1:500 but average is 1:150. Various kinds of weather changes are associated with arrival and departure of the fronts. However, the concept of "Front" was given by Norwegian meteorologists at the time of Second World War. They thought the air masses as armies and they fought facing from opposite direction. This line of war in military is also known as Front. A front is having following characteristics:

(i) Fairly abrupt horizontally temperature change.
(ii) Comparatively low pressure.
(iii) Inwardly blowing surface winds.
(iv) Upper surface air is updrawn by and into the upper air circulation.
(v) Increasing likelihood of condensation and precipitation from rising air.

Classification of Fronts

On the basis of aggressor (invader) air mass and the dynamic characteristics, the air mass can be grouped into four main types.

(i) Warm Front.
(ii) Cold Front.
(iii) Occluded Front.
(iv) Stationary Front.

Warm Front

When warm air mass invades into cold air mass region, former overrides the latter along a sloping line of separation. This line of separation is known as Warm Front. Warm air mass overlies the cold air and moved upward along the front due to its lightness. This front is characterised by gentle slope forming with the ground. It's average gradient is 1:200. It covers larger area and affect it by different weather condition associated with it. It is represented on map with a dark semicircles (Fig. 19.1).

Weather Condition : The arrival and retreat of the Warm Fronts cause following kinds of weather changes (Fig. 19.2).

(i) The appearance of high cloud with cirrus type forming "Holo" around sun and the moon, ahead of the warm from at a greater distance from ground point of front. This condition heralds the arrival of warm fronts at a given point;
(ii) Gradually cirrus clouds transform into stratus, strato cumulus and cumlonimbus at lower level as the front comes closer to the point of observer. Temperature increase and low pressure develops;
(iii) When the front approaches the observer, steady and continuous drizzle and rainfall occurs for longer time over larger area. The duration of shower depends upon the velocity of front; and
(iv) When a warm front passes an observer on the ground, many of the symptoms of atmospheric conflict cease out. Pressure begins to increase, sky becomes clear and wind shifts from easterly to southerly direction.

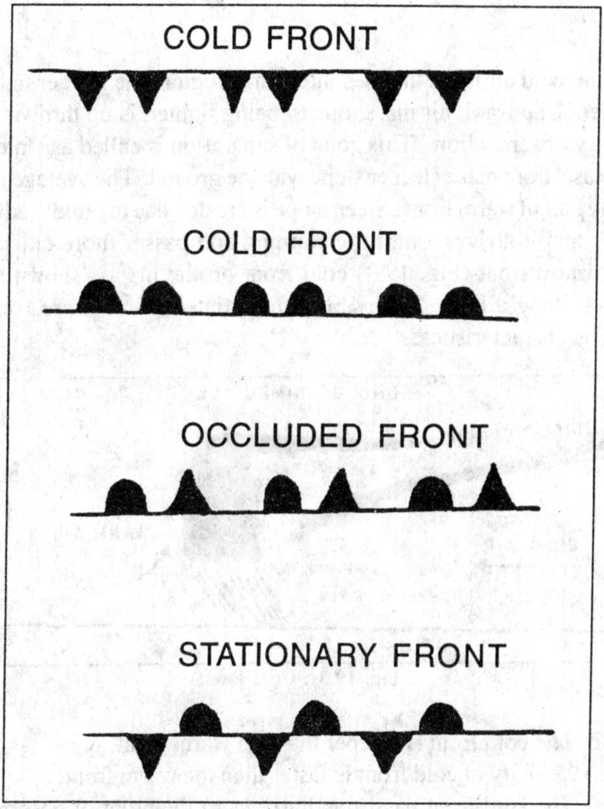

Fig. 19.1 : Representation of Fronts by Symbols.

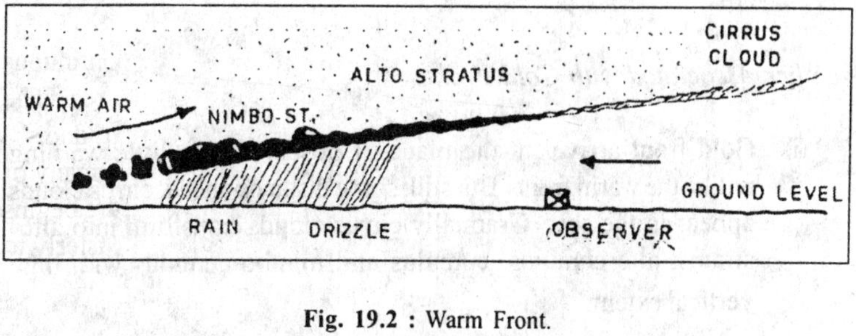

Fig. 19.2 : Warm Front.

Cold Front

When the cold air mass invades into warm region, the former undergoes the latter. The warm air mass, due to being lighter, is up thrown along a boundary of separation. This zone of separation is called as Cold Front. In this case, front make steeper slope with the ground. The average gradient is nearly half of warm front. Steep slope is created due to rapidly advancing cold air and it arrives with little warning and passes more quickly than does a warm front (Fig. 19.3) cold front on the map is shown by dark triangles along a line. It may be differentiated from warm fronts with following characteristics.

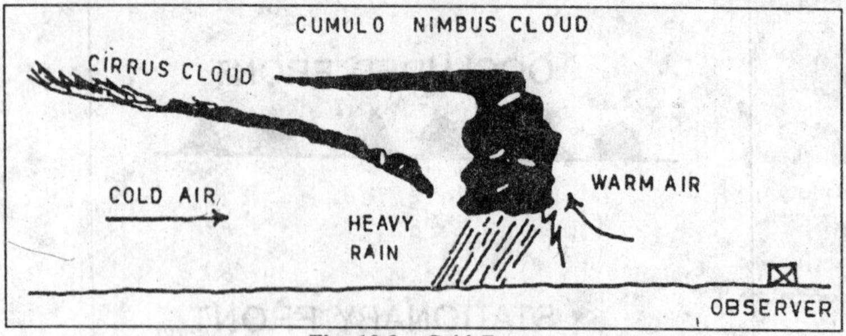

Fig. 19.3 : Cold Front.

(i) The cold front is steeper than the warm front.
(ii) Velocity of cold front is faster than the warm front.
(iii) The cold front passes and arrives to the point of observer with little warning.
(iv) Vertically high extensive cloud and turbulent and voilent weather with heavy shower of short duration are common in cold front. While the warm front exhibits extensive horizontal dense cloud with light drizzle and shower of long duration over extensive area.

Weather Associated with Cold Fronts

(i) Cold front arrives at the place of observer with little warning unlike the warm front. But still, some distance ahead, cirrus clouds appear in the sky. Gradually cirrus clouds transform into alto-stratus, alto-cumulus, cumulus and mimbo-cumulus with high vertical extent.

(ii) The cold front produce heavy precipitation, thunderstorms, hails, torandoes if warm air is unstable enough. This storm activity may extend ahead of cold front in the form of squall line of thunderstorms.

(iii) Heavy rainfall with great violent occurs over shorter area in short duration. It appears suddenly at observer creating the problem for the people.

(iv) As the cold front passes away, the temperature begins to fall and pressure increases due to presence of cold air on the ground. Consequently, there occurs improvement in visibility and fair weather. Detached cumulus and cumulonimbus clouds appear, if cold air is unstable, which produce heavy shower of rains.

Occluded Fronts

When cold front overtakes the warm front as the former moves faster than the latter, occluded front is formed. In this front three air masses are involved. Warm air is fully uprooted from the ground and upthrusted. It overlies the cool and colder air masses (Fig. 19.4). The last stage of extra

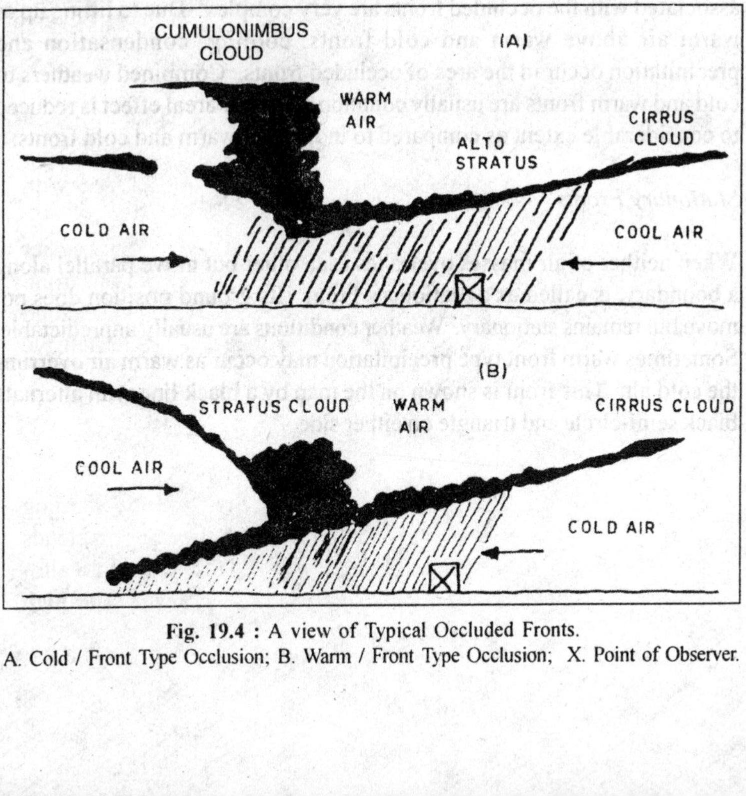

Fig. 19.4 : A view of Typical Occluded Fronts.
A. Cold / Front Type Occlusion; B. Warm / Front Type Occlusion; X. Point of Observer.

tropical cyclone presents a typical view of occlusion of cold and warm fronts. An occluded front is portrayed on the map either by an alternating red and blue line or black line with alternating semi-circles and triangles on the same side of the line.

The occluded fronts may be classified as —

(i) Warm Front occlusion, and (ii) Cold Front occlusion.

In the cold front type occlusion (Fig. 19.4a) warm air is trapped between cold air from NW and Cool air from the east and the warm air is lifted up above these two cold and cool air. Air behind cold front is colder than air underlying warm front. It is very commonly found to the east of Rockies. Contrary to this, in warm front type occlusion (Fig. 19.4b). Air behind the cold front is warmer than the air underlying the warm front. Warm air is also lifted up by cold and cool air in this type of occlusion like the cold front type. It is very common along the Pacific coast.

Weather Associated with Occluded Front : The weather conditions associated with the occluded fronts are very complex. Due to lifting up of warm air above warm and cold fronts, cooling, condensation and precipitation occur in the area of occluded fronts. Combined weathers of cold and warm fronts are usually common, but their areal effect is reduced to considerable extent as compared to individual warm and cold fronts.

Stationary Fronts

When neither of air masses overtakes each other but move parallel along a boundary, is called as a Stationary Front. Its ground position does not move but remains stationary. Weather conditions are usually unpredictable. Sometimes warm front type precipitation may occur as warm air overruns the cold air. This front is shown on the map by a black line with alternate black semi-circle and triangle on either side.

20

ATMOSPHERIC MOISTURE

Our atmosphere consists of water vapours along with other constituents like gases and dust particles. The presence of water vapours in the air is also called as moisture or humidity. The amount of moisture in the atmosphere is variable with time and space. It varies from 0.04 per cent to 4 per cent of the volume of the atmosphere. It is rather high in wet season and above the water bodies as compared to dry season and above continents or land surface. The presence of water vapours or humidity is very useful to man on the earth surface. The significance of the moisture are as given below.

(i) It control the flow of energy from the sun to the earth and from earth to space. Absorption of solar short wave radiation in the atmosphere is due to moisture in the atmosphere. Greenhouse effect is also caused by its presence;

(ii) It determines the potential of formation of various kinds of condensation like dew, fog and clouds and the amount of precipitation on the earth surface; and

(iii) Moisture is the storehouse of latent energy which derives the storms or cyclone in the atmosphere.

Source of Water Vapours in the Atmosphere

Water vapour in the atmosphere, initially, is added by cooling of hot gaseous earth. After the condensation, heavy rainfall occurred continuously for many years and the low or depressed parts of the earth were filled up with water. Again the collected water in waterbodies began to evaporate due to heating by the sun and water vapours were added to the atmosphere. Thus the evaporation of water vapour from water bodies is the foremost important source of moisture in the atmosphere. Moisture is also supplied to the atmosphere by precipitation. Uncountable drops of water are suspended in the air after precipitation. Besides, plant kingdom is not lagging behind in order to supply moisture to the atmosphere. The plant transpirates the water through their leaves by photosynthesis process. Animal including man also contribute, but insignificantly, moisture to the air through respiration. Thus Evaporespiro-transpiration is the main source of moisture in the atmosphere.

Measurement of Humidity

The humidity in the air is measured into various ways and expressed in several units. The important instruments like wet and dry thermometer and the Hair Hygrometer are used normally to determine the moisture of the air. The method and expression of humidity are discussed as follows :

Absolute Humidity

Absolute humidity refers the amount of moisture present in a given volume of the air at a given temperature. It is expressed as grams per cubic volume of air. This kind of measurement of humidity may not remain unchanged with any kind of change in volume of the air with variable temperature irrespective of any addition or subtraction of the moisture in the air. It changes due to change in volume through the expansion and contraction of the volume of air through heating and cooling respectively. For example, a given air with 2 cubic metres volume has 100 gram moisture at given temperature. Due to decrease in temperature, volume reduces to ½ size, i.e. 1 cubic metre and the absolute humidity may be expressed as 100 grams per cubic metre now against 50 grams per cubic metre earlier, though no addition of moisture is made. Hence it is not well suitable expression for meteorological study.

Specific Humidity

The second method of measuring humidity is called specific humidity. It is direct measurement of water vapour content in a given air. The weight of moisture is expressed in a given weight of the air instead of volume. In this method humidity is described as grams per kilogram. It remains unaltered with change in temperature and volume of the air. It is the best method of expression of humidity for the meteorological studies.

Relative Humidity

The most important and used method of humidity measurement is relative humidity. It is not direct measurement of the amount of moisture in a given air but it is the ratio between the present amount of moisture (AH) and the capacity of air to hold the moisture at a given temperature. It is expressed in percentage (%). When the relative humidity is 100 per cent, the air is called Saturated Air. It means the said air is not having capacity to hold further moisture. Relative humidity is variable with the change in either temperature or humidity. Change in Relative humidity takes place into following ways.

(i) When the moisture is added to the air by evapotranspiration and precipitation, the relative humidity increases up and *vice versa.*

(ii) When the capacity of air to hold moisture is changed through cooling of the air; the relative humidity increases up and when the capacity is increased through heating (increase in temperature) of the air, the relative humidity declines.

Moreover, the relative humidity practically varies in time and space. Diurnally it is higher in the morning, evening and night rather than day. Because in former three periods temperature is lower, consequently, relative humidity rises up. As the sun rises up, the temperature shoots up and relative humidity begins to decline due to increasing capacity of air to hold moisture. It is also rather high in humid tropical region than dry tropical one. For example, equatorial regions show more than 90 per cent relative humidity throughout the day rather than tropical desert, where it occurs randomely.

Some Important Features Associated with Atmospheric Humidity

Evaporation

The liquid water converts into gaseous state due to excess heating and the molecules of water escapes from water surface to the surrounding air. This kind of process of conversion of liquid water into gaseous form is called as Evaporation. The extra heat, beyond the boiling temperature which is added for vaporisation, is also moved with vapours and stored in them. This kind of heat is called as latent heat of vaporisation. That is why the vapours or steams have rather more burning efficiency than the boiled water. The latent heat is released to the atmosphere when the water vapours are condensed into liquid or solid water. The evaporation process depends on the following conditions.

(i) The rate of evaporation from a water surface is determined by temperature of surrounding air. Warmer the overlying air, higher the evaporation and *vice versa*. Thus the rate of evaporation is greater in tropical ocean than polar one;
(ii) Dryness or wetness of overlying air is next important criteria of the evaporation. If the overlying air is moist or saturated, no evaporation will take place. If the air is dry and has more capacity to hold moisture, more molecules of water scape from water surface and evaporation rate will be rather higher;
(iii) The velocity of wind or air is also determinant of the evaporation. The calm wind causes less evaporation because it becomes saturated after sometimes. Contrary to this the mobile air encourages the evaporation as the moist air moves ahead and it is replaced by dry air which has more capacity to hold moisture. Thus the rate of evaporation is directly proportional to the velocity of the wind if other factors remain constant; and
(iv) Temperature of the water surface is the main condition for the evaporation. The beginning of the process is caused due to temperature of water. Variation in temperature of the oceans has led to the variation in the amount and rate of vaporisation in different latitudinal zones.

Saturation of the Air and Dew Points

Saturation of the air refers a state of the moist air when the relative humidity reaches upto 100 per cent. Thus the saturation of the air may occur into following conditions.

(i) Through addition of more moisture in a given volume of air;
(ii) Through cooling of the air, so that the capacity of air to hold moisture is reduced; and
(iii) Thus the air becomes saturated at the different temperature depending upon the presence of actual amount of moisture in the air. The temperature, at which air is saturated and starts to condense, is called as Dew point. The dew points may be either above or below the freezing point.

Adiabatic Change of Temperature

When the parcel of air moves upward or downward, its temperature changes due to change in the volume of the air. The ascending air is expanded in volume and its temperature declines, whereas the descending air contracts in volume and the temperature rises up. This increase or decrease of ascending or descending air due to expansion or contraction in volume is described as Adiabatic Change of Temperature. This kind of change of temperature is not affected by surrounding air's temperature. Addition or subtraction of external temperature is not taking place. It is very much useful in condensation and precipitation process.

Condensation

The moistures in saturated air at dew point, begins to convert either in solid or liquid state around some suspended matters in the air. This process of conversion of water vapours into solid or liquid is called as Condensation. While the suspended matters like salts, dust particles, pollen etc., around which condensation occurs, are called as Hygroscopic nuclei.

Kinds of Condensation

Condensation may be of various forms depending upon the temperature of dew point as well as the location of its occurrence. On the basis of altitude or height above earth's surface, the condensation may be divided into following categories.

Dew

The condensation of water vapours, on the objects attached with the earth surface, is called as Dew. The drops of water may be observed on the blades of grass and stics leaves of trees etc. Formation of dew takes place into four following conditions.

(i) Long winter night in which nucturnal radiation is greater and the temperature declines to considerable extent to reach at dew point;
(ii) Calm wind causes the easy saturation of the air. Dry air does not replace the moist air due to stationary position of it;
(iii) Clear sky helps in more radiation to reduce the temperature of air attached to the earth surface; and
(iv) The temperature of dew point should be above freezing point. In the case of dew point temperature below freezing point, frost will be formed.

Fogs

Fogs are such type of condensation which takes place in lower part of the atmosphere in suspended form. They are different from the dew in such way that latter is found on the solid objects of the earth where as the former is formed in hanging form in lower part of atmosphere. Fogs are formed through the numerous process. They show a great variation in visibility depending upon the density of fogs. They are similar to cloud in their physical forms but differ in the process of formation and location. The most clouds develop due to adiabatic cooling in rising air, but rarely, uplift is involved in fog formation. Fogs have their base very near to ground. On the basis of mode of formation the fogs may be classified as follows :

Radiational Fogs

Such fogs are formed due to cooling of air closed to the earth surface by nucturnal radiation. During a long winter night, under clear sky, the terrestrial radiation is faster and outgoing energy exceeds the incoming solar radiation. Consequently, the ground becomes chilled up. Through the conduction process the overlying air becomes colder and colder and attains dew point which results in formation of fogs. They are also known as ground fogs because they occur due to chilling of the ground through terrestrial radiation. They are most common in middle and high latitudes where winter nights are very long. They are also common in valley region where as a result of air drainage, colder and havier air collects into the bottom of valley and inversion of temperature occurs. They are very short lived and tend to dissipate with sun heating during the day.

Advection Fogs

Advection fogs are such type of condensation which takes place due to horizontal movement of warm and moist air into colder region. The temperature of overlying warm air drops down due to underlying cold surface and the former gets dew point to form fogs. Such fogs are very common in coastal areas. More than 2/3 of such fogs are concentrated to the coastal area. They are relatively more long lasted than the other fogs, advection fogs are also common in interior continental location which is associated with poleward flow of cool moist humid air from lower latitudes over cold and snow capped surface.

Orographic or Upslope Fogs

Sometimes fog is created due to ascent of air along the mountain slope. The adiabatic cooling causes the formation of clouds at higher level but near the ground on gentle slope stable air may cause extensive fogs.

Precipitation or Frontal Fogs

When the front is formed, the warm air is pushed up overlies the cold air mass. Moist warm air attains dew point after ascending aloft and precipitation takes place after the condensation of water vapours. After the precipitation the underlying cold becomes saturated and condense into form of fogs. This kinds of fog is called a precipitation or Frontal Fog.

Clouds

When the moist warm air ascends up by any reason, its temperature declines at adiabatic rate. The ascending air cools down and gets saturated. Thus the air reaches at dew point. At the level of dew point condensation of water vapour takes place around the hygroscopic nuclei either in the form of water droplets or ice crystals. The water droplets and ice crystals hang together in the air. They accumulate in high altitude of troposphere. Such mass of droplets of water and ice crystals hanging in the atmosphere at higher level is described as cloud. Clouds are having uniform level at the base. Base is the level of dew point and the clouds extends up from this base, clouds are very much significant to determine the heat balance on the earth surface. Reflection, absorption, scattering and diffusion of light in the atmosphere are caused by clouds. Terrestrial radiation is also controlled by clouds. Moreover, the clouds are having various shapes and sizes and they are located at different elevations. On the basis of shapes of the clouds, they may be grouped into three forms : (i) Circus , (ii) Cumulus, and (iii) Stratus. Circus word is derived from Latin word which means a lock of hair. Such clouds are thin, whispy, composed of ice crystals rather than water particles. The stratus means spread out. Stratus cloud appears as greyish sheet or layer that covers most or all of the sky. The cumulus indicates the mass or pile of the clouds. Such clouds are massive and rounded with small flat base like cauliflower extending upward to great height.

On the basis of location of height the clouds have been grouped into 4 families.

Family A

High Clouds : Such clouds are found above 6 km from the earth surface. These clouds are thin, white and composed of tiny ice crystals due to the fact that there is low amount of moisture and low temperature at such height.

Family B

Middle Clouds : Under this family all the clouds lying between 2 to 6 km height from the earth surface. They are layered or puffy in form and composed of water unlike ice crystals in high clouds. 'Alto' word is used for this category of clouds.

Family C

Low Clouds : The clouds which develop below 2 km height are called as low clouds. These clouds occur in either patches individually or in continuous thick overcast.

Family D

Clouds of Vertical Development : When the clouds develop upward to considerable height upto 15 kms, it is categorised as clouds of vertical development. Their bases are very restricted and narrow they may lie across all three families (A to C) vertically.

However, taking both classifications in combination together, clouds have been classified into 10 categories which are internationally accepted. They are discussed briefly in Table 20.1.

Table 20.1 : Chart of Various Kinds of Clouds

Family	*Types*	*Characteristics*	*Height*
A. High Clouds	1. Cirrus	1. Composed of ice crystals instead of water droplets.	Above 6 km
		2. Thin feather like clouds of delicate and fibrous appearance.	

Contd. ...

Table 20.1 *Contd.* ...

Family	*Types*	*Characteristics*	*Height*
		3. White colour without shading.	
	2. Cirro comulus	1. They are in the form of patches of white flakes or globular masses usually very small and without shadow. 2. They are arranged in groups or lines or ripples.	
	3. Cirrostratus	1. They are thin whitist veil of clouds which causes milky sky appearance. 2. Their thickness never cover the sun and holo is formed by these clouds. 3. They are harbingers of storms.	
B. Medium Clouds	4. Alto cumulus	1. They are flattened mass of clouds arranged in layer. 2. They have wavy or linear pattern in their distribution. 3. They cause underneath shadow unlike cirro-cumulus.	2-6 kms
	5. Alto stratus	1. They are sheet clouds of grey or blue colour. 2. They have stratified or fibrous appearance and structure. 3. They are commonly followed by widespread continuous precipitation.	
C. Low Clouds	6. Strato cumulus	1. They are composed of global masses of clouds with brighter space between.	Below 2 kms

Contd. ...

Table 20.1 *Contd.* ...

Family	*Types*	*Characteristics*	*Height*
		2. The cloud masses are regularly arranged like alto-cumulus but they are much larger.	
	7. Stratus	1. It is low cloud and spreads out as fog but not rests on the ground. 2. Light drizzle is only possible of this cloud.	
	8. Nimbo-stratus	1. It is low cloud with any definite shape. 2. It has the capability to cause precipitation but not necessary precipitation occurs. 3. It is more denser than stratus without any definite lower level.	
D. Clouds with vertical Development.	9. Cumulus	1. They are vertically thick dome shaped cauliflower like clouds with well even horizontal base. 2. They are evidence of convectional current of the air. 3. They are commonly associated with fair weather.	Up to 6 kms
	10. Cumulonimbus	1. They are over-developed from of Cumulus clouds. 2. The cauliflower shape is disturbed. They spread out at top and become anvil shaped. 3. They are associated with heavy rainfall, lighting and thundering etc.	

Precipitation

The condensation process forms the various kinds of fogs and clouds. The clouds are consisting of millions of tiny droplets and ice crystals.

When these water droplets and ice crystals become so larger beyond the holding capacity of clouds in the atmospheres, they begin to fall down. under the gravitational pull of the earth, either in solid or liquid forms on the earth surface. This process of falling of water drops or ice crystals is described as precipitation.

Process of Precipitation

It must be kept in the mind that all clouds never cause precipitation necessarily but all precipitation is caused by clouds. As discussed earlier, the clouds are consisting of very tiny drops or crystals of water, precipitation cannot take place until the water droplets or ice crystals attain a larger size in which their buoyancies are exceeded by gravitational pull of the earth. Two processes, mainly, are involved for getting the ideal size of raindrops for causing precipitation.

Ice Crystal Formation : Due to buoyancy of air, the moist air pushed up to a height where temperature is below freezing point. A large number of tiny ice crystals are formed and around which millions of tiny droplets and ice crystals are attracted to form big rain drops for precipitation.

Collision of Collescence of Water Droplets Through the Passage of Precipitation : In tropical region condensation takes place at the temperature above freezing point. The water droplets of various sizes move down at different velocities. Larger the size, faster the velocity. Thus the larger droplets with faster velocity overtake the smaller droplets with slower velocity and the former attain bigger sizes. Thus due to merging of water droplets, precipitation occurs.

Forms of Precipitation

On the basis of composition whether water droplets, ice crystals or combination of both, precipitation occurs into four forms —

(i) Rainfall : When the water droplets fall as rain drops through precipitation is called as rainfall. It occurs mainly in tropical regions of low latitudes. In middle and higher latitudes, it is restricted to the warm season. Thus the rainfall is the main characteristics of warm regions and warm season.

(ii) Snow Fall : When the precipitation takes place in snow form, ice crystals, instead of water droplets, falls on the ground. They are caused when the condensation occurs at dew points below freezing temperature. It is restricted to winter season in middle and high latitudes and in high altitudes in low latitude region in the same season.

(iii) Sleet : Sleet is a kind of mixed precipitation. Small ice crystals are mixed with rain drops. Water droplets sometimes freeze up in the passage when passing through underlying cold air before reaching on the earth surface.

(iv) Hail : Hail is also similar to sleet, mixed raindrops and ice crystal, but the main difference between them is of process of formation and size of crystals. The former is consisting of big size ice crystals with many coating of ice. It is formed by turbulent storm in which tiny ice crystals are tossed up. Convective turbulent air moves to considerable height where the temperature is very low below the freezing point. Due to repeated process of tossing up and down, many layers of ice are coated around dropped ice crystals. While the latter is smaller crystal formed due to passing of raindrops through underlying cold air surface.

Types of Precipitation

As discussed earlier, cloud formation and occurrence of precipitation are associated with the adiabatic cooling of air mass. An air mass is raised up or uplifted by various ways, i.e. heating, highlands and fronts or convergence of contrasting air. On the basis of causative factor involved in raising the air mass, the precipitation may be classified as follows:

Convectional Precipitation

When an air mass is heated up it becomes lighter and moves upward to lower atmospheric pressure. Cold surrounding and air from aloft descend to take the place of heated and uplifted air. The uplifted air is cooled down by adiabatic process and it attains saturation and dew point after reaching some height. Consequently, condensation and precipitation take place. Such type of precipitation caused by raising of air by heating process is called as convectional precipitation. Such precipitation is restricted to warm season and warm regions. For example, equatorial region receives rainfall throughout the year in convectional form (Fig. 20.1).

Fig. 20.1 : Convectional Percipitation.

Orographic or Topographical Precipitation

When a blowing air mass especially warm and moist, comes across the topographic features like mountain plateaues, escarpment or high hill, is uplifted upslope. The rising air mass loses its temperature adiabatically and it reaches to the temperature of dew point. Cloud formation and precipitation take place on the windward slope of the highlands. After crossing the mountain the airmass descends down and warmed up at adiabatic rate and it ceases to precipitation on leeward slope. The opposite side of highland, where dry condition prevails, is called as rain shadow zone. Such type of precipitation caused by resistance of high land is described as orographic precipitation (Fig. 20.2). Such precipitation is

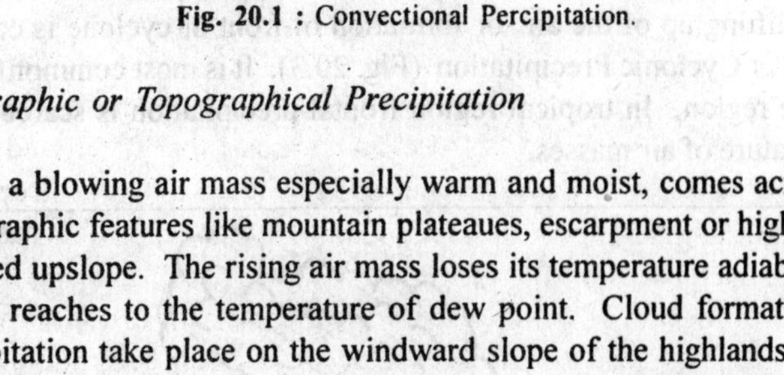

Fig. 20.2 : Orographic Percipitation.

either seasonal or periodical but it can occur in any latitudes at any time and in any season. It is more general and prolonged because the raising air mass takes place slowly along the slope. Moisture of the air and transverse location of highlead are the basic conditions of such precipitation. Along the coasts, most of precipitation is orographic type.

Frontal or Cyclonic Precipitation

When two air masses of contrasting temperature converge together, the warm and light air mass is displaced or raised up by cold and denser air mass. The overlying uplifted warm air mass cools down adiabatically and it attains the saturation condition. Thus the formation of clouds and precipitation take place. Some processes of raising of the air is also associated with the formation of cyclones. Thus precipitation which occurs due to lifting up of the air or formation of front or cyclone is called as Frontal or Cyclonic Precipitation (Fig. 20.3). It is most common in mid-latitude region. In tropical region frontal precipitation is scarce due to same nature of air masses.

Fig. 20.3 : Frontal Percipitation.

Distribution of Precipitation

The distribution of precipitation is very unevenly distributed annually as well as seasonally all over the world. The amount of precipitation is the combined effect of a number of factors. They may be included as physical characteristics, i.e. temperature and humidity of the raised air, nature of topography, distance from the oceanic or maritime influence and the latitudinal distribution of pressure. Thus the distribution of precipitation may reflect some distinguished generalized features regarding annual as well as seasonal pattern.

Annual Distribution of Precipitation

Keeping into account the amount of annual precipitation of the world, zonal character may be visualized on the basis of distribution of annual precipitation in map. The first zone which confined to tropical and equatorial regions where abundant amount of annual rainfall occurs annually. Convectional, orographic and monsoonal precipitation are common. High temperature and high moisture (due to warm oceans) are positive factors for occurrence of such region of maximum rainfall. The second region is associated with sub-tropical high pressure. It is the driest and all the tropical deserts of the world lie in this zone. Minimum precipitation in this zone is attributed to the subsidence of the air, anti-cyclonic condition and the far distance from the sea. Generally all deserts lie along the western coast of the continents where anti-cyclonic conditions are very strong and the presence of cold ocean currents led stability to moist maritime air from west direction.

The mid-latitude region is the third important zone where amount of precipitation rises up. It is the second maxima of precipitation. Interior location of continents in northern hemisphere and the eastern coast in southern hemisphere are the exception where the precipitation is scanty. The high amount of precipitation in this area (mid-latitudes) is due to high absolute humidity in the air especially in summer due to melting of ice and development of extra tropical cyclones or the polar fronts. Coastal areas also receive the precipitation due to monsoonal maritime air mass in different seasons. According to the onshore movement of the wind, the fourth zone is again considered the zone of low precipitation. This region lies polewards from 60° North and South latitude. This is the second precipitation minima of polar and sub-polar regions. Low precipitation, here is caused due to low temperature and low absolute humidity in the air. Stability of the air is very common due to heavy and dense area. Summer season is also very short not more than one month.

Seasonal Distribution of Precipitation

Seasonality of the climate is the main characteristics of almost all parts of the world except equatorial and the coastal region. Seasonals variation in the amount of precipitation is common. Summer is usually the season of wet condition due to high convectivity and moisture of air all over the world contrary to this, winter is considered dry season as it is associated with high pressure and stability of the air. Divergence of the air ceases precipitation in winter season especially in interior of the continents. Following features are the main generalization regarding the seasonal distribution of the precipitation.

(i) Shifting of thermal equator and pressure belts has lesser effect on the precipitation zones. Only mediterranean climatic region and Inter-Tropical Convergence zone are affected by this shifting. Equatorial convergence of high humidity or rainfall moves north and south in summer and winter respectively. Mediterranean region between 45° — 60° North and South latitudes along the western coasts of continent receive maximum rainfall in winter season instead of summer.

(ii) Summer season is the duration of maximum rainfall, snowfall except equatorial region, eastern coast of some islands like East Indies, Sri Lanka and eastern coast of India etc. in tropical region. The seasonal difference is well sound and coastal region show uniform seasonal precipitation.

(iii) Seasonal rhythm of the wind (monsoon) in the coastal region is not uncommon. Wettest season is determined by the direction of onshore maritime moisture laden winds.

21

ATMOSPHERIC EQUILIBRIUM

Composition of earth's atmosphere is very unique and complex. Various kinds of dry and wet gases and solid crystals are its constituents. The atmosphere is arranged into different strata with distinct physio-chemical characteristics. Air is compressible, so that atmosphere at lower level is rather more denser. It is hanged under the gravitational pull of the earth. Consequently, the atmospheric pressure decreases with altitude due to decreasing amount of gravitational pull of the earth. At the outer wedge of atmosphere the pressure is rather very less. At 5.5 km distance from grounds about half of pressure at sea level is found. Unequal heating on the earth surface by sun, the warmer and lighter air is pushed up by colder and denser surrounding air. The ascending air responds to the decreasing pressure above by expansion in its volume because air is compressible. It expands when pressure is released and contracts when pressure is forced. The expansion of rising air continues until it gets the equal pressure to the surrounding air. The outward expansion of air requires energy which is added by inherent internal energy of it. Consequently, the temperature of rising air declines and rising air cools down. Thus the decline of temperature of rising and expanding air is described as adiabatic change of temperature. In this process of change of temperature cooling of air does not take place due to subtraction of heat from it but to spreading of energy in the expanded volume of the same air. The rate of adiabatic change varies with dry or wet conditions of the parcel of convective air. Thus the vertical motion or mixing of air in upper atmosphere is associated with the relationship between laps rate, and adiabatic rate of the change of temperature. Because of variable relationships between them, ascending air is, sometimes, continuously rising up and sometimes, vertical motion is ceased and the rising air descends down to its original place. The condition of vertical mobility of the air in the atmosphere is known as Atmospheric equilibrium.

Significance of the Study of Atmospheric Equilibrium

Stability or Unstability of the Air

The study of stability or unstability of the air in metorology is very significant due to following facts :

(i) Prediction about formation of clouds, fogs and hailstorms and precipitation can be made on the basis of knowledge of the stability condition of the air in the atmosphere; and

(ii) Atmospheric pollution in upper layer is also affected by the condition of air—whether stable or instable in the atmosphere. Stable air ceases the upward movement of pollutants and they remain near the ground. While the instability causes the moving up of the pollutants in the upper atmosphere and consequently low pollution occurs near the ground.

Determinants of Atmospheric Equilibrium

There are mainly three factors which affect the condition of stability of pushed up air in the atmosphere. They are : (i) Normal laps rate, (ii) Dry adiabatic rate, and (iii) Wet abiabatic rate.

Normal Laps Rate

The temperature in lower most atmospheric layer troposphere decreases with altitudes. This rate of change of temperature in the atmosphere with increasing height is called as Normal Laps Rate. It may vary with time and space but the average rate is 6.5°C per km height from sea level.

Dry Adiabatic Rate

Warm air tends to rise above the cold air. When a parcel of the air on the earth is heated up in the day, it becomes lighter and moves upward. The surrounding colder air moves under the warm air. Due to ascending up of the air, it cools down due to expansion of its volume to adjust with the pressure of adjoining environment. The cooling of ascending air is due to spreading of its internal energy in expanded volume of it. In dry air this rate of change with height is constant, i.e. 10°C per km height. Conversely, when the rising air is descended down to the ground, the air contracts and heats up. Temperature increases at the rate of 10°C / km due to not addition of temperature from external source but because of restoration of spread energy into smaller volume from larger one. Thus the change of temperature of parcel of ascending or descending dry air, without any subtraction or addition of external temperature, due to expansion and contraction of volume respectively, is called Dry Abiabatic rate of change of temperature.

Wet Adiabatic Rate

When the moist and warm air is lifted up, it cools down initially at dry adiabatic rate, i.e. 10°C/km. After reaching at level at some height, the ascending air gets saturated and condensation takes place, consequently, latent heat of water vapours is released to air and the latter's temperature begins to reduce at relatively lower rate (dry adiabatic rate—addition of latent heat) with increasing height after the condensation level. This rate of change of temperature of an ascending air after the condensation is called as Wet Adiabatic rate. Wet adiabatic rate is not constant but may vary with the temperature at ground and moisture content of the air. Higher original ground temperature, higher amount of moisture content of ascending air exhibit very low rate of wet adiabatic rate and *vice versa*. Moreover, the air when descends down after exhausting moisture after condensation, it heats up at dry adiabatic rate and the temperature becomes rather higher on the ground against the original temperature when it was first lifted up. This increased temperature of decending air is, known as potential equivalent temperature.

Stability of Air

The heated air on the earth surface tends to uprise over the cold air as it is lighter than the latter. As the rising air is lifted up, both the environmental (atmospheric) and the ascending air's temperature decline at the normal laps and adiabatic laps rate respectively. Suppose that air on the earth surface on particular day had a temperature of 30°C, while the temperature of a parcel of uplifted air at 2 km aloft is 10°C. The comparison of the temperature derives the conclusion that the warm air (30°C) near the surface should overlie the cold air aloft (10°C), but practically it is not so, because the warm air overlies the cold air in the environment of same atmosphere pressure. Hence the vertical motion of ascending air should be observed through relationship between laps rate and adiabatic rate of change of temperature at the uniform level of elevation and pressure. When the normal laps rate is rather lower than the dry adiabatic laps rate of the ascending air, thermal resistance will be forced and the thermal buoyancy of lifting air will be stopped. Consequently, the vertical motion of air is restricted and it descends down due to being colder and heavier than the surrounding ait at a given height and pressure. For example, the temperature of air ascending and the environment on the ground are 45°C and 35°C respectively. At the elevation of 3 km, the rising air gets its temperature, equivalent to 15°C due to adiabatic cooling (10°C/km) and the environmental temperature at this height is 20°C as it reduced by normal laps rate of 5°C/km. Thus the rising air is forced to move down due to its lower temperature than the surrounding air at 3 km height. Such condition of air in the atmosphere is described as stability of the air (Fig 21.1).

Moreover, sometimes, the normal laps rate becomes very low even lesser than the wet adiabatic laps rate. In this situation the rising air is still bound to sink down even after the condensation at dew point. Such stability of air is called as Absolute stability. It means that air cannot move upward under this condition.

Effects of Stability

Meteorologically, the effects of stability of air may be reflected by following phenomena :

(i) Fogs are formed when air is ceased to move upward due to inversion of temperature. The air near the ground cools down to dew point through rapid nocturnal radiation in long winter night;
(ii) Flat cloud is a symptom of stable air. Stratus clouds may be formed and the precipitation may also occur from the clouds. But the thunder, hail and torrential rains are almost impossible in such condition;
(iii) Due to stability of the air vertical mixing or drafting of pollutants to higher level is restricted. After the sunset, the earth loses heat, the lower layer of atmosphere cools down and the temperature becomes stable. Smoke and dust remain near the ground as a strip in the evening. Pollution concentration is greater in the night than the day due to stability of air; and
(iv) The stable condition led to the subsidence and the formation of anti-cyclones on the earth surface.

Instability of Air

Instability of the air is the state of the atmosphere in which the vertical motion is continued. Air is buoyant and the surrounding air is always colder than the rising air. Instability is caused when laps rate is higher than dry and wet adiabatic rate. In Figure (21.2) line of laps rate and dry adiabatic

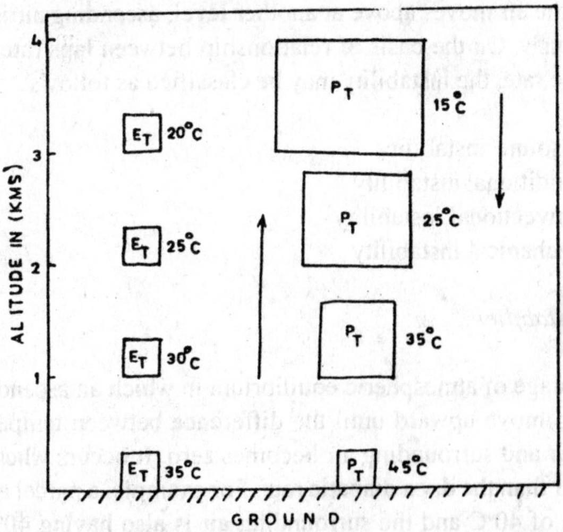

E_T = Environmental Temperature, P_T = Temperature of Parcel of air, Arrows indicate the direction of movement of parcel of air.

Fig. 21.1 : Stability of Air.

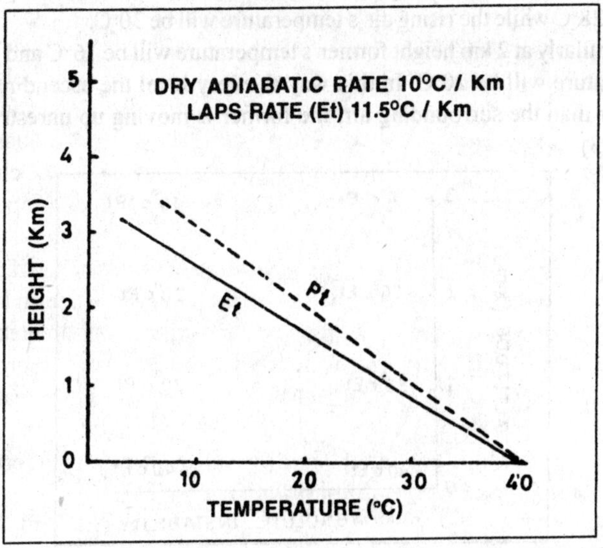

Fig. 21.2 : Instability of Air.

rate are shown. Dry adiabatic of ascending air is 10°C per km while environmental laps rate is 11.5°C/km. At every height level the temperature of surrounding environment (Et) is lesser than the temperature of ascending air (Pt). As the air moves above at another level, ascending air is moving up continuously. On the basis of relationship between laps rate, dry and wet adiabatic rate, the instability may be classified as follows:

(i) Absolute instability
(ii) Conditional instability
(iii) Convectional instability
(iv) Mechanical instability

Absolute Instability

It is a such stage of atmospheric equilibrium in which an ascending air is continued to move upward until the difference between temperature of ascending air and surrounding air becomes zero. It occurs when the laps rate is higher than the dry a diabatic rate. For example, a parcel air has the temperature of 40°C and the surrounding air is also having 40°C on the ground. Air is heated due to sunshine and becomes lighter and moves upwards. Suppose the laps rate and dry adiabatic rate are 12°C and 10°C/km respectively. At the height of one km, the environmental temperature will be 28°C while the rising air's temperature will be 30°C.

Similarly at 2 km height former's temperature will be 16°C and latter's temperature will be 20°C. In this way at every level the ascending air is warmer than the surrounding air, the former is moving up unrestrictedly (Fig21.3).

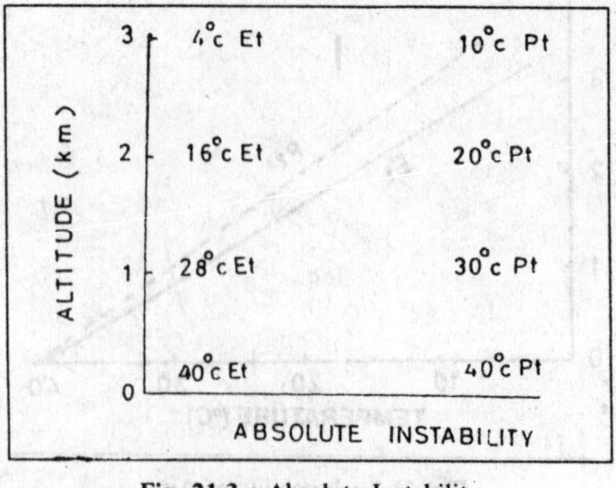

Fig. 21.3 : Absolute Instability.

Conditional Instability

In this kind of atmospheric equilibrium, an ascending moist air becomes stable at certain level of height due to higher rate of dry adiabatic rate than the laps rate. But after reaching the dew point, air is saturated and condensed. Due to condensation latent heat is released and added to stable air. When the stable air is forced up by any turbulance, the air cools up at the rate of wet adiabatic rate (which is lower than dry adiabatic rate). Thus the ascending parcel of air is moving upward continuously, if the normal laps rate becomes higher than wer adiabatic rate. For example, the moist ascending air and the surrounding air have temperature of 35°C and 30°C on the ground. Suppose normal laps and dry adiabatic rate are 7.25°C and 10°C per km. At elevation of 2 km surrounding and ascending air's temperature would be 15.25°C and 15°C respectively. In such condtion the ascending air becomes stable because it is colder and denser than surrounding air (Fig 21.4). But at this temperature, the ascending air begins to saturate and condense. Now the stable air is lifted by vertical current and starts to cool down at 5°C wet adiabatic rate. At the elevation of 3 km, the temperature of surrounding and ascending air will be 8°C and 10°C respectively and the ascending air becomes warmer and unstable. This type of instability is bound with condition of condensation of the uplifting air. Two important conditions are required to occur conditional instability.

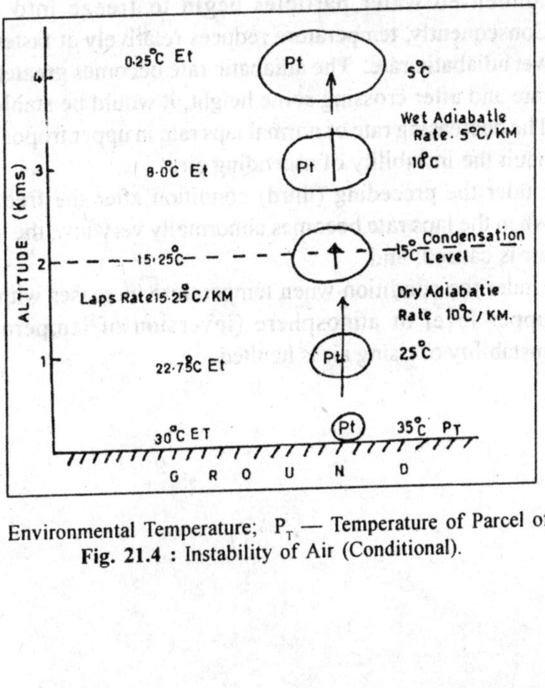

E_T — Environmental Temperature; P_T — Temperature of Parcel of Air.
Fig. 21.4 : Instability of Air (Conditional).

(i) The ascending air should be moist, (ii) Normal laps rate must lie between dry and wet adiabatic rates.

Mechanical Instability

Sometimes at higher level the normal laps rate of temperature is very high and an abnormal instable condition develops. The hot rising air is pushed up violently due to greater difference between surrounding temperature and ascending air's temperature at a given elevation. Torando, a revolving violent storm is formed under such atmospheric condition.

Conditions Under Which Instability is Haulted

The continuity of the instability is not unchecked. There are various conditions in which the unstable air is ceased and becomes stable.

(i) Absolute instability may be simply changed into stability when the ascending air is not moist or relative humidity is very low;

(ii) The moist air is ascended up to greater height through the voilent convectional currents. After the condensation, the air cools down at wet adiabatic rate and causes the conditional instability. The further tossing up, cools the air to freezing point and the condensed water particles begin to freeze into ice form. Consequently, temperature reduces relatively at faster rate than wet adiabatic rate. The adiabatic rate becomes greater than laps rate and after crossing some height, it would be stable;

(iii) The decreasing rate of normal laps rate in upper troposphere also hault the instability of ascending air;

(iv) Under the preceding (third) condition after the freezing point when the laps rate becomes abnormally very low, the stability of air is caused; and

(v) Under the condition when temperature increases with height in upper level of atmosphere (inversion of temperature), the instability of rising air is haulted.

22

ATMOSPHERIC DISTURBANCES

There are wide ranges of general circulation of the air on the surface and aloft. They are basically distributed following the permanent pressure belts in definite latitudinal extent with little seasonal modification. They are known as first order circulation and moving in same direction throughout the year. This first order circulations, in various latitude zones, are intervened and disturbed by development of various kinds of second order circulation. The second order circulation may be caused by local change in atmospheric condition as well as convergence of two contrasting first order circulation in transitional region. Upper air circulation like jet streams may also exert influence to create disturbance amidst first order circulation. Thus the development of second order circulation amidst the first order circulation (Planetary wind belts) is called as atmospheric disturbances. They are called disturbances because the normal weather conditions are very much disturbed with arrival and departure of such circulation in any kind of climatic region. These disturbances are characterised with following facts.

1. They are smaller and variable in size than the components of general circulation.
2. They are migratory and transient.
3. They have relatively short duration of persistence for only few hours, few days and few weeks.
4. They produce characteristics and predictable weathers along the track of their movements.

Classification of Atmospheric Disturbances

The origin, characteristics and associated weather conditions of atmospheric disturbances are variable with the variation in latitudinal extent. On the basis of area of their occurrences, the disturbances may be classified into two categories : (i) Extra-tropical or Mid-latitude, and (ii) Tropical Disturbances.

Extra-Tropical or Mid-Latitude Atmospheric Disturbances

The mid-latitude region is considered as highly disturbed climatic zone where weather conditions frequently change from day to day, week to week and month to month. All seasons, here, are characterised by frequent invasion of airmasses from other regions and each invading airmass brings a change in the weather. A considerable varieties of atmospheric disturbances are found in this belt but the two, i.e. (i) extra-tropical cyclone, and (ii) extra-tropical anti-cyclone of them are most important to mention.

Extra-Tropical Cyclone

The cyclone, which occurs in mid-latitude zones, between 30° to 70° latitudes in both hemispheres, is called as extra-tropical cyclone. They are very important disturbances which contribute to larger share in erratic and changeable nature of weather conditions in this region. They blow in whirling anticlockwise motion from outside to inward side following the low pressure in the centre. They usually follow the westerlies tracts and directions of movements from west to east. So the mateorologists are very much cautious about weather condition at the west from the observatory than the eastern side. These cyclones are also called as low and depressions. They cause the formation of various kinds of clouds, precipitation and squalls along the tracts which they traverse. They are very common in both summer and winter season. The extratropical cyclones are very much variable in their size and the tracks to be followed. On the basis of genesis and the location of their occurrences, they may be classified as follows :

(i) Frontal Cyclone
(ii) Tropical Depression
(iii) Thermal Lows
(iv) Lee Depression

(i) *Frontal Cyclones :* They are such cyclones which are originated due to convergence of two contrasting airmasses and the formation of fronts. They are most common in the middle latitude region.

(ii) *Tropical Depressions* : They are the penetration of tropical disturbance in mid-latitude region from the tropical region. They are weakened and ultimately, sometimes, merge with frontal cyclones exist ahead.

(iii) *Thermal Lows* : Sometimes due to intense heating of land surface, a weak low pressure develops especially in summer season. They are very weak and short-lived. Such disturbances are designated as thermal lows.

(iv) *Lee Depression* : The different frontal temperate cyclones move east and north-eastward from North Asian region in winter. They finally merge with Aleutian low and move further along the west coast of North America. Some of these cyclones cross the Rockies and redevelop leeward side of the mountain (Rockies) especially in Colorado, Texas, Alberta etc. These temperate cyclones in Canada and U.S.A. are described as Lee depression due to their genesis on leeward side.

Origin of Extra-Tropical Cyclones

The study of origin of extra-tropical cyclones is very complex and still remains unexplained. But the most common explanation of genesis of this cyclone is called as Polar Wave theory. It is believed that this type of cyclone is produced due to convergence of polar easterlies and tropical westerlies along sub-polar low pressure belts. The beginning of cyclogenesis takes place with the formation of polar wave as the consequence of contact of two contrasting air masses of heterogeneous temperature. The development of such cyclones take place through various stages. At the first stage, the wave is formed as stationary front (Polar Front) along which both air masses blow parallel in opposit direction (Fig 22.1). The polar front is intervened and disrupted due to presence of any kinds of hinderance like mountain barrier, forest and inland sea or lakes. This intervention led to the bending or meandering in polar front and consequently the warm air begins to surge into Polar region and the Cold Polar air into warmer region. Thus due to the invasion of airmasses into different contrasting temperature sectors, the bulge gradually becomes very acute and severe. Direction of winds are deflected due to coriolis force which put whirling motion of air around the low pressure at the apex of vortex. Isobar trends change from parallel to semi-circular. Two fronts-cold and warm — develop in the eastern and western edge of the cyclone.It is the third stage in which cyclone is divided in three zones like warm front, warm sector and the cold front. It is the stage when the cyclone gets its maturity.

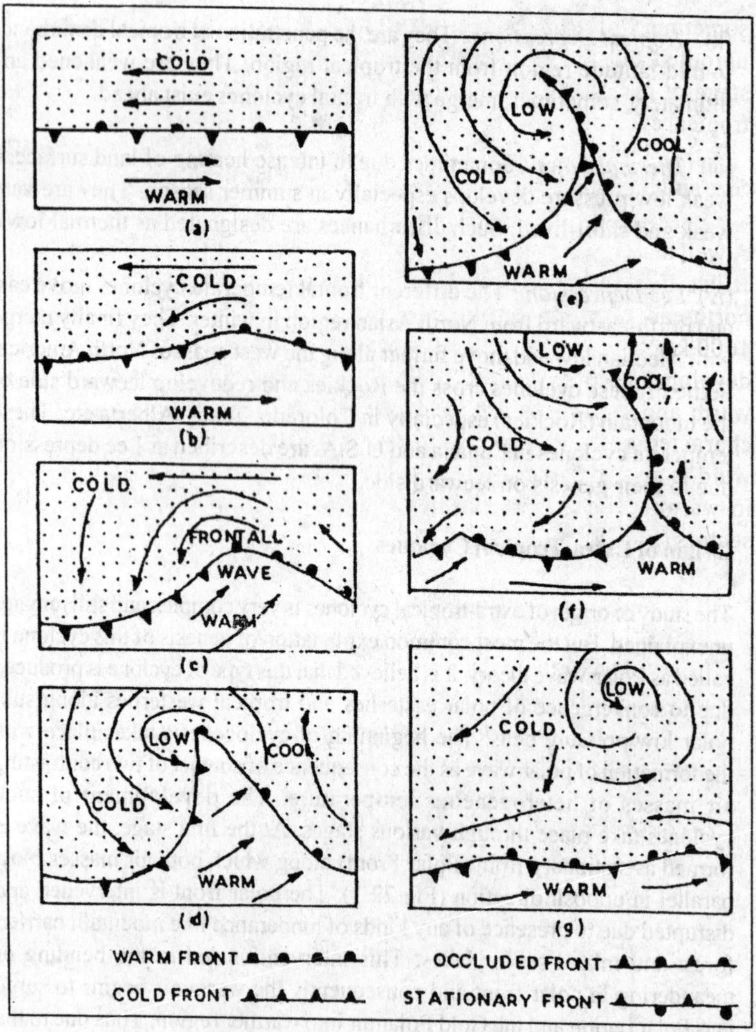

Fig. 22.1 : Stages in the Lifecycle of a Middle Latitude Cyclone as proposed by J. B. Jerknes.

Moreover, the extra-tropical cyclone moves from west to east direction with differential rate of velocity of two contrasting fronts. The velocity of cold front is rather greater than the warm front. Thus the warm sector begins to squeeze and from the top becuase at the apex the rate of movement of cold front is relatively higher. Gradually after sometime, the cold front overtakes the warm front and the latter is uprooted and thrown up aloft and occlusion of fronts takes place. This stage of cyclone is called as death or cyclolysis process of the extra-tropical cyclone.

Size and Structure of Extra-Tropical Cyclone

A well mature extra-tropical cyclone shows very extensive dimension. It has oval shape of trends of isobars with a long axis extending north-east to south-west. On average the length of diameter is about 1600 km with 990 or 1000 mb pressure in the centre. As the cyclone developes due to formation of the fronts, it is divided into two sectors warm and cold by the cold and warm fronts. These two sectors are characterized with contrasting weather conditions. The direction of wind is also different in these sectors. Southerly winds are common in warm sectors while northerly and easterly are prevalent in cold sectors (Fig 22.2).

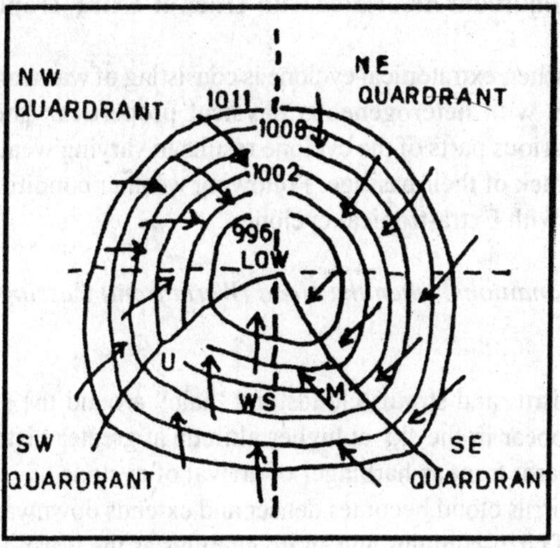

C : Cold Front; W : Warm Front; WS : Warm Sector

Fig. 22.2 : A Model of Cyclone — Temperate Cyclone (Northern Hemisphere).

Movement of Extra-Tropical Cyclone

The movement of extra-tropical cyclone is characterised with the following norms :

(i) It has counter clockwise whirling motion in northern hemisphere. Winds are blowing inward from outside;
(ii) The path of their movement is generally in the region of westerlies. They move from west to east with little modification in direction following the route of westerlies. That is why, the weather forecaster take into account the weather condition of west station rather than east in temperate region;
(iii) They move with variable rate both spatially and seasonally their velocity is lower in summer season than the winter; and
(iv) The average velocity is 35 to 40 km/hours. Extra-tropical cyclone is followed by anticyclone in mid-latitude region.

Weather Conditions Associated with Track of Extra-Tropical Cyclone

As noted earlier, extratopical cyclone is consisting of warm and cold fronts and sectors with heterogeneous physical properties. The contrasting nature of various parts of the cyclone results in varying weather condition along the track of their passage. Following weather conditions have been associated with Extra-tropical cyclone.

Weather Conditions, when the Head (Warm front) Passing Through the Observer

(a) Cirrus and stratus clouds and 'halo" around the sun and moon appear in the sky at higher altitude at greater distance from the warm front as harbinger of arrival of cyclone.
(b) Cirrus cloud becomes denser and extends downward in the form of cirro-cumulus and strato cumulus as the warm front becomes near to approach the observer. Temperature increases and low pressure develops at the observer ahead.
(c) Dense low Nimbo Cumulus and Nimbo Stratus clouds are formed when the warm front reaches to observers. Heavy and steady precipitation takes place for longer time (Fig. 22.3).

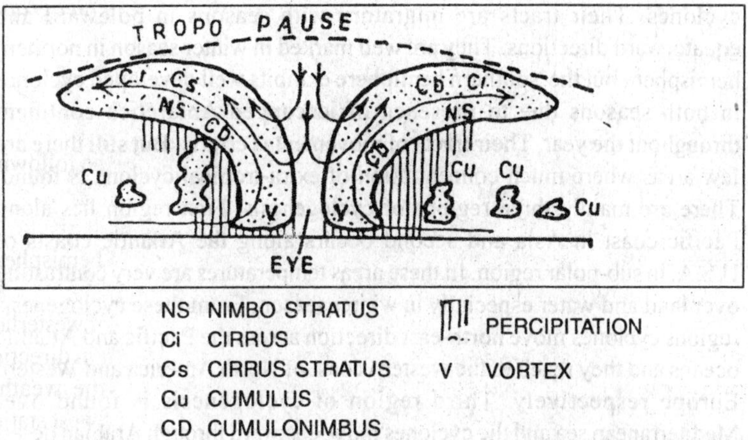

Fig. 22.3 : Structure of Tropical Cyclone.

Weather Condition in Warm Sector

When the warm sector passes through the observer, following weather condition prevails.

(a) Temperature rises abruptly, clouds are removed and relative humidity is high.
(b) Occasionally convectional precipitation occurs as downpour for short duration.

Weather Condition in Cold Front and Cold Sector

(a) As the cold front approaches the observer the cirrus clouds appear in the sky but at a shorter distance from the ground base of front.
(b) Very dense nimbo cumulus clouds in rather smaller area is formed and very heavy precipitation occurs for shorter duration due to steeper slope of the front.
(c) On the rear side of cold front as it passes the observer, temperature falls abruptly, pressure arises up and stability and inversion of temperature occur. Consequently, haze and fogs are common.

Area of Occurrence of Extra-tropical Cyclones

Mid-latitudinal zones in both hemispheres are the main area of such cyclones. Their tracts are migratory with seasons in poleward and equatorward directions. They are well marked in winter season in northern hemisphere but the southern hemiphere exhibits well developed cyclones in both seasons due to presence of icecapped Antarctica continent throughout the year. Their tracts of movement is erratic. But still there are few areas where much concentration of extra-tropical cyclone is found. There are mainly three regions of cyclogenesis. First region lies along Pacific coast in Asia and second occurs along the Atlantic coasts of U.S.A. in sub-polar region. In these areas temperatures are very contrasting over land and water especially in winter season. From these cyclogenesis regions cyclones move north-east direction across the Pacific and Atlantic oceans and they reach to the western coast of North America and Western Europe respectively. Third region of cyclogenesis is found over Mediterranean sea and the cyclones move eastward through Arabian desert to Indian sub-continent in winter season.

However, among the continents, North America shows highly concentration of extra-tropical disturbances. On the leeward side of Rockies mt. especially in Alberta, cyclones are regenerated due to frequent convergence of contrasting airmasses especially in Great Lake region. There are about seven tracts of cyclones in U.S.A. and almost all of them originate along the leeward side of Rokies mountain. They move ultimately towards North-East Coast of U.S.A. through the Great lake region.

Extra-Tropical Anticyclones

Extra-tropical anti-cyclone is a very important disturbance other than cyclone amidst general circulation of westerlies in mid-latitude regions of both hemispheres. It is a outflow of circular or whirling motion of the winds from the centre. The trends of isobars is circular or nearly circular with the highest value in the centre of the system. They blow in clockwise direction in northern hemisphere and anti-clockwise in southern hemisphere. They move and follow the tract of westerlies from west to east. Cyclones are usually preceded by anti-cyclone in such a way that the winds at the front of latter become convergent and supply wind to the former.

The genesis of anti-cyclones is generally of thermal origin. Due to radiation and conduction of heat especially in winter season the land or continental areas are chilled or frozen up and consequently high pressure develops over such region. In winter season they are well developed and extensive in northern hemisphere, Arctic high extends up to south over continental region. While in southern hemisphere, due to presence of icecapped Antarctica continent, anti-cyclones are well established throughout the year.

Moreover, the anti-cyclone is characterised with the subsidence of air aloft and divergent of air at the ground. Inversion of temperature is usually common in upper troposphere over the anti-cyclonic region. Clear sky, rainless and fair weathers are the main characteristics of such circulation. But occasionally, for sometimes, shower may occur.

TROPICAL DISTURBANCES

Tropical regions are well known with monotonous weather and climatic condition. Repetition of weather is comon phenomenon day to day, week to week and month to month unlike the mid-latitude region which constitutes the zone of dynamic changeable weather. The monotonous weather conditions, sometimes, .are disturbed due to development of some local atmospheric phenomena or disturbance. Such locally originated disturbance or perturbations are called as Tropical Disturbances. They may be grouped into two categories (i) major, and (ii) minor disturbances on the basis of their strength, effect intensity and geographical area of their location.

Major Tropical Disturbances

Tropical Cyclone or Typhoon

Tropical cyclone is intense revolving rain drenched migratory storm that occurs erratically in limited locations in low frequency. It is having inward spiralling motion with great violent velocity. The pattern of isobars is almost circular with low atmospheric pressure at the centre. It is showing counter clockwise direction in northern hemisphere and clockwise in southern hemisphere. Such cyclones are known by numerous local nomenclatures in different parts of tropical world. They are known as Hurricane in the Central and North America especially in Carribean sea and the West Indies; Typhoon in China sea and south-west Pacific ocean especially along China coasts, Japan coast, Philippines etc.; cyclone in Arabian sea and Bay of Bengal, willy willy along the the coast of Australia. These tropical cyclones are characterised with the following features.

(i) Pattern of isobars is almost circular with minimum pressure at the centre. Pressure gradient is steeper. 50-100 mb pressure difference usually occurs from central to periphery.

(ii) They are having extensive size with an average diameter between 160-1000 kms. The wind velocity is rather high not less than 119 km/hour which is minimum requirement for being hurricane. Practically, the velocity may be twice or thrice of above criteria on some occasions.

(iii) Winds from all sides converge along the edge of eye of cyclone. Eye of cyclone (central part of the cyclone) has the extension between 16 to 40 km diameter. Updraft of air occurs at the ground near the eye but divergent of air is found at the top of the cyclone (Fig. 22.4). Eye is characterised with subsidences of air.

(iv) Weather conditions, throughout the cyclonic affected area, are symmetric except in the eye. Cumulus, Nimbo Cumulus clouds with torrential rains spread out in all regions of cyclons excluding eye. Eye zone is characterized with calm and fair weather of cloudless sky.

(v) These cyclones follow the east-west direction in trade wind regions moving from ocean to continents or from low latitude to mid-latitudes. The deviation from this normal direction of movement occurs in south-west Pacific-north and north-east of New Zealand. Here the tropical cyclones move erratically from north-west to north-east as westerlies near the Fiji or Tonga island. This abnormal direction of movement is due to tropospheric westerlies (Jet stream) which extends quite equatorwards in south-west Pacific. Generally, the cyclones are steered up by upper atmospheric circulation.

(vi) They are of short lived or duration. The hurricane may be alive upto three weeks maximum. They generally occur over the tropical ocean away from the equator and begin to dissipate as move over the continents or high latitudes of low temperature due to cutting off energy supply from condensation of water vapours over the water surface.

(vii) Hurricanes are well known for the destruction of lives and properties over the affected areas. Destructions and damages are caused by high winds and torrential rains beneath the cyclones. Greater and severe destructions are experienced in sea or ocean and in coastal region through drowning of ships and boats and the surging of water over the coastal regions. Due to development of low pressure over sea surfaces a high surge of wind driven water overrides the coastal area and flood occurs.

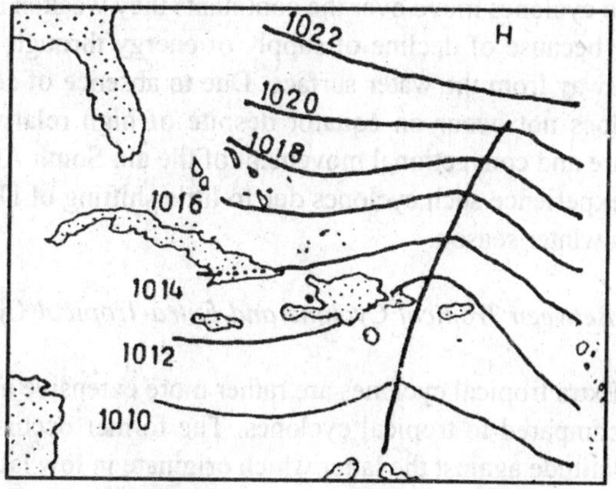

Fig. 22.4 : Tropical Disturbances in the Form of Easterly Wave.

Genesis of Tropical Cyclones

The origin of tropical cyclone is not well explained till now. It is a matter of great discussion. On the basis of their physical characteristics and the location of occurrence, some scientific arguments have been given. The tropical cyclone normally originate over the hot sea surface away from the equator. Under the following conditions such cyclones may originate and develop.

(i) Hot water surface with more than 28°C temperature.
(ii) Presence of coriolis effect to give spiral motion to the inflow winds.
(iii) Supply of latent heat of condensation as energy for development of cyclones.
(iv) Convergence of the air.

Under the above ideal conditions the western margin of oceans and seas in tropical region are the breeding place of the tropical cyclone. During the summer Inter Tropical Convergence Zone (ITCZ) lies north of equator where coriolis force is available. So over these breeding regions initially a low depression develops due to exeptionally high temperature. The low is intensified in ITCZ on account of presence of coriolis force and convergence of the air. Oceanic surface adds moisture through evaporation and the energy as latent heat of condensation.

As the cyclones move over the continents they weaken and dissipate ultimately because of decline of supply of energy through reduction of moisture away from the water surface. Due to absence of coriolis force, cyclone does not occur on equator despite of high relative humidity, temperature and convectional movement of the air. South Atlantic ocean does not experience such cyclones due to little shifting of ITCZ south of equator in winter season.

Contrast Between Tropical Cyclone and Extra-Tropical Cyclone

(i) Extra tropical cyclones are rather more extensive and violent as compared to tropical cyclones. The former occurs in the mid-latitude against the latter which originate in low latitude region;

(ii) Their direction of movement is also different. Extra Tropical cyclones follow west-east direction with the association of anticyclones in westerlies tract. While the tropical cyclones follow east-west direction without anticyclone in trade wind tract;

(iii) Extra tropical cyclones are originated due to frontal development and their lives sustain and survive due to contrast of temperature of contrasting winds. The tropical cyclones have thermal origin and the survival and sustenence of their lives depend upon the supply of latent heat of condensation;

(iv) The trends and patterns of isobars ecliptical in extratropical cyclone, while circular in tropical cyclone; and

(v) Weather conditions, are symmetric in all directions of tropical cyclone but variable in different sectors of extratropical cyclones.

Minor Tropical Disturbances

Easterly Wave

It is long weak migratory low pressure though in tropical region over the ocean or sea. The pattern of isobars is bulging in wavy form instead of circular. These waves are the initiators of the formation of well described cyclones like hurricanes. They have north south extension (Fig. 22.5), travelling from east to west in trade's path. The weather conditions differ from front to rear of the waves. At ahead of easterly waves, weather is fair and cloud less due to divergence of air while at the back or behind of the waves (crest) the weather is showery and cloudy because the air converges

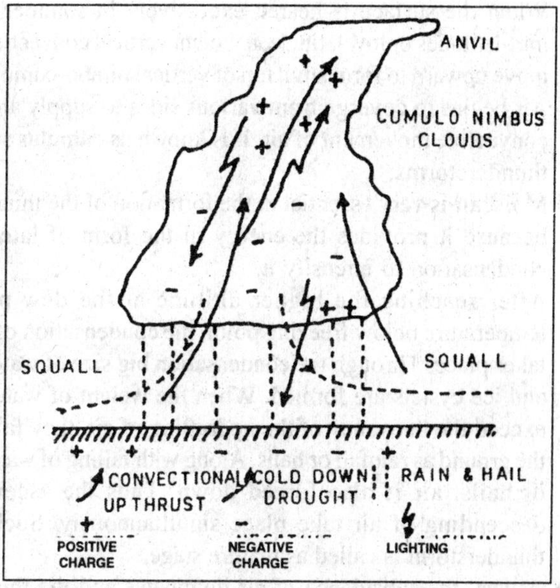

Fig. 22.5 : Thunderstorm.

here. Some of these waves develop in intensity and strength and ultimately reach to the formation of hurricanes. They may concentrate anywhere in coastal areas between 5-20° latitudes.

Weak Tropical Lows

They are very weak low pressures develop in humid tropical region usually near the zone of convergence (ITCZ). They are formed of one or two circular Isobars. The pressure gradient is very low, consequently wind velocity is very moderate. Genesis of such lows is debatable and controversial question. Main characteristic of these disturbances is related with rainfall region. They cause little showers but for larger time with uniformly overcast sky. They develop and intesify with a view to change into hurricane.

Thunderstorms

Thunderstorm is a violent convective updraft of air due to excess heating of surface of the earth. It is associated with showering, thundering, lightning and hails. Formation of thunderstorms have been characterised by following generalization.

(i) When the surface is heated excessively in summer season in mid-latitudes or low latitude, a violent vertical convective currents move upward to form anvil top of vertical nimbo-cumulus clouds. Air begins to coverge from various sides to supply and help the convective movement of air. It is known as cumulus stage of the thunderstorms;

(ii) Moist air is very essential in the formation of the thunderstorms because it provides the energy in the form of latent heat of condensation to intensify it.

(iii) After reaching the higher altitude at the dew point with temperature below freezing point, the condensation of moisture, takes place. Through the condensation big size of water droplets and ice cystals are formed. When the weight of water droplets exceed the buoyancy of the updrafting of air, they fall down on the ground as rainfall or hails. Along with falling of water droplets or hails, air is also forced down. Thus the ascending and descending of air take place simultaneously. Such stage of thunderstorm is called as mature stage.

(iv) At last, the subsidence of air dominates and the thunderstorm begins to dissipate. This stage is called dissipating stage of thunderstorms.

(v) A number of vertical cells of updraft air like chimny interspersed with cells of subsidence are common in thunderstorms.

(vi) The thunderstorms are associated with torrential local rains and hails for shorter time and in limited area where convection current move upwards. Lightning, thundering and violent squall winds are also not uncommon.

(vi) Such storms are commonly associated with high surface temperature in warmer latitudes of the earth, in warmer season in middle latitude and in warmer hours of the day.

Type of Thunderstorms

Thunderstorm may be classified into two categories on the basis of their origin. They are discussed below :

Local or Thermal Thunderstorms

These storms are originated due to excess heating of land surface with moist air. This phenomenon, usually, takes place in warm latitude, hot season and hot hours of the day. Equatorial region, equipped with high temperature and high humidity and convergent air movement, is the ideal environment for the occurrence of thunderstorms. They also occur in said latitudes but in summer season. Diurnal periodicity is main characteristics in all latitudes. Afternoon and early evening recorded maximum frequency of thunderstorms. They have local character and occur in isolation or groups with a small areal convecting.

Frontal Thunderstorms

Such storms are originated along the fronts of cyclones developed especially in mid-latitude region. It may occur at anytime irrespective of seasons, day or night. They are more severe and violent than the thermal thunderstorms. Although they occur along the warm and occluded fronts but are most frequent along the cold front of V-shaped cyclones. When the cold air meets with warm humid air in the front of cold front, the latter is upthrusted and overturned in a storm form over the edge of the front. This storm is followed by a shift of winds from south-west, south or south-east to north-west and by consequent drop in temperature. While the local or thermal storms give a little relief from the heat during the period of cloud cover and rains.

Distribution of Thunderstorms

As the thunderstorms are the function of high temperature, humidity and convergence of the air, the distribution of their occurrence is confined to few latitudinal zones. Maximum occurrences take place in equatorial doldrum region becuase of the available ideal conditions like high temperature, humidity and convergence of air along the Inter-Tropical Convergence Zone (ITCZ). Their frequency decline poleward from equator in both hemispheres due to decrease in temperature. No doubt temperate zone also exhibits occurrence of thunderstorms but only in summer season. Beyond the 60-70° latitudes poleward such weather phenomena rarely occur. The rate of decline of frequency is sharp towards sub-tropics from equator because of subsidence of the air in sub-tropics. Besides, the land areas record maximum occurrence as compared to sea surface in the same latitude due to variation in temperature. The former has rather higher temperature than the latter.

Tornado

Tornado is also a violent form of storms which occur on the land surface. It is the most destructive of all the atmospheric disturbances. Such phenomenon is characterised by following features:

(1) It is funnel shaped storm with pointed apex of very low pressure with minimum width touching the ground and wide and broader head of cumulus clouds.
(2) It is most intense vortex in nature and a deep cyclonic low pressure is surrounded by whirling cylinder of spinning with so violently that the centrifugal force creates a vacuum within the funnel.
(3) Velocity of this storm is very high due to high gradient of pressure. Within a short distance of maximum 800 metre diameter, from centre in outward, around 100 milibar difference has been recorded. The maximum wind velocity is estimated at 320 to 800 km/hour.
(4) Tornados generally are originated few metres above the ground. They become only visible when up thrown water vapours condense into dense funnel clouds moving downwards.
(5) They are short lived with maximum longevity of 8 hours. They cause devastation by uprooting off the trees and destroying the structures when the funnel sweeps along the ground.
(6) The mechanism of development of tornado is not well known. They usually occur in warm moist and instable air associated with extra-tropical cyclone. Temperature contrasts and instability of air are essential conditions. They generally occur on the land surface in the spring and early winter when temperature contrast in mid-latitude is well marked.
(7) Tornados are also associated with thunderstorm and they, sometimes, develop from the base of cumulus clouds of the thunderstorm.
(8) Tornados are commonly associated with middle latitudes and sub-tropics unlike the thunderstorm in equatorial region. More than 90 per cent tornados occur in USA because of favourable condition, viz. meeting of cT and mT air masses in flate terrain of south-east U.S.A.

23

CLASSIFICATION OF CLIMATES

The study of various elements of weather and climate in preceding chapters indicated that they are very much variable in time and space. The mean or average condition of these elements of either short or long period reflect a variation with the location of various places on the surface of the earth. It is very inconvenient and probably impossible to recognise or remember of weather or climate of every bit of earth. Thus it seems a general need to organise the climatic data with reference to time and space, in such order which may facilitate the people to understand the climatic conditions on the basis of some common characteristics. This need was felt by various climatologists, geographers and meteorologists and they made several attemps, to divide the world into various distinct climatic regions.

Study revealed that the classification of climates have been made into two ways (i) genetic, and (ii) empirical. Genetic classification indicates the methodology in which causes of climatic elements are taken into account. For example, temperature, latitude, altitude etc. which are known as determinants of the weather and climate. Greek scholars divided the world into various climatic regions on the basis of insolution distribution on different latitudes. Torrid (Tropical), temperate and Frigid (Polar) are three elementary climatic regions of the world on the basis of genetic criteria (Temperature). The area, between tropics of cancer and capricorn is described as tropical zone. The regions between Tropic of Cancer and Arctic circle in northern hemishere, and between Tropic of Capricorn and Antarctica Circle, in southern hemisphere are refered as temperate zones. While the regions between Arctic circle and North Pole, and between Antarctica Circle and South Poles are described as polar zones.

Empirical classification implies the recognition of climatic regions on the basis of observation of effect of climatic elements, on vegetation, soils, animal life and so on.

Recently many scholars made effort to classify the climates on various categories. All of them used certain criteria with a desired objectives.

Outstanding, among the modern attempts of climate classification, are those of W.P. Koppen (1900), C. Thronthwaite (1931), H. Flohn (1950), E. Kupfer (1954), B. Allissow (1956), L. Holdridge (1959) and M. Hendi (1960). They have used combination of both genetic and empirical techniques for delimitation of boundaries of main and sub-division of climatic regions.

Koppen's System of Climatic Classification

Waldimir Peter Koppen designed the system of classification of climate in 1900 A.D. and presented a modified form in 1918. His classification is the most used and widely accepted in the world among physical geographers. He adopted both genetic and empirical methodologies in his excercise. Firstly, he classified climates worldwide on the basis of vegetation types which he considered as the best expression of climate of particular region. With a view to more rational and scientific, the system devoted greater attention to the temperature precipitation, evapotranspiration and their seasonal characteristics. Thus Koppen demarcated the boundary of five major regions and some sub-regions on the basis of following criteria —

(1) Temperature
(2) Precipitation
(3) Evapotranspiration
(4) Vegetation

Major Category of Climates

Koppen divided the world into five macro or major climatic regions on the basis of temperature and evapo-transpiration and vegetation type boundary. He used English capital alphabet to designate the climate type as given below.

Climate-A (Tropical Humid Climate) : The region, where mean monthly temperature remains above 18°C throughout the year, is included in A category.

Climate-B (Dry Climate) : Such climate occurs in the regions where evapo-transpiration exceeds the precipitation. Precipitation effectiveness instead of temperature is being used as criteria. For example, 50 cm rainfall in tropical region may be considered in dry climate but same amount in sub-polar region may not be same due to low rate of evapo-transpiration in cold region.

Climate-C (Warm Humid Mid-Latitude or Temperate Climate) : Such climate is found in the region where temperature of coldest month varies between 18°C to -3°C. The mean temperature of warmest month must be above 10°C.

Climate-D (Cold Humid Mid-Latitude or Temperate Climate) : This climate covers the region where the temperature of the coldest month is below -3°C and the warmest month must have more than 10°C.

Climate-E (Polar Climate) : The average monthly temperature must be below 10°C in all the months of the year.

Sub-groups of Major Climatic Types

Koppen further recognised sub-division of climate on meso- and micro level on the basis of mainly seasonal variation of temperature and precipitation. For these subgroups small English alphabets have been used as their designation.

(A) Climate sub-groups

f, m, and *w* letters were used to denote sub-division or subgroups of A- climate. The description of letter is given below —

- *f*- implies the rainfall throughout the year above 6 cms or 2.4 inches.
- *m*- implies the driest month should have precipitation less than 6 cm but the precipitation during driest month is equal to or greater than 10 -r/25 where *r* is annual precipitation in cms.
- *w*- refers the condition in which precipitation of the driest month is less than 10 -r/25 where *r* is the annual precipitation in cm.

Hence three sub-divisions have been made in A-climate.

Af. Tropical humid or rain forest climate
Am. Monsoon climate
Aw. Savana Climate

(B) Climate Sub-group

For secondary divisions of B-climate, Koppen used capital letter W & S for precipitation and small letter *k* and *s* for temperature. The definition of these letters are given below.

S - The average annual rainfall (*R*) reported in cms exceeds *T*+14 in areas where summer is rainy season, *R* is between *T*+14 and T+7 in areas where rain is spread over the year. *R* is between *T*+14 and *T*+7 in areas where rain is spread over the year. *R* is between *T*+7 and *T* in areas with mainly winter rainfall. (*T* is the average annual temperature in °C).

W - *R* is less than *T*+14 where summer is rainy.
R is under *T*+7 where rain spreads through the year.
R is less than *T* where winter is rainy.

h - Implies average annual temperature more than 18°C

k - Implies the average annual temperature of given place is less than 18°C.

However, in combination of these alphabets four sub-groups of B or dry climate have been recognised.

1. Bwh : Tropical semi Arid climate.
2. Bwk : Mid-latitude semi-Arid climate.
3. Bsh : Tropical Desert climate.
4. Bsk : Mid-latitude Desert climate.

C - Climate Sub-dvision

C - Climate is redefined with different small letter as *s, w, f* to denote precipitation and *a, b,* and *c* for temperature. These letters are ellaborated as follows :

f - denotes moist condition throughout the year driest month precepitation is more than 3.0 cm (1.21 inches).

s - referes the summer dry. The precipitation of the wettest month of winter is three times greater than precipitation of driest month in summer when it is less than 4 cm.

w - refers winter dry season. The precipitation of wettest month of summer is 10 times greater than the precipitation of the driest month of winter reason.

a - Indicates the average temperature of warmest month is greater than 22°C, but at least 4 months over 10°C.

b - Implies that the average temperature of warmest month is less than 22°C but each of the four warmest months temperature must be above 10°C.

c - refers the average temperature of the warmest month is below 22°C but from one to three months experience 10°C or more than monthly temperature.

Hence with the combination of precipitation and temperature indicators, C- climate may be classified into following major sub-groups.

1- *Cs* (Mediterranean Climate)
2- *Cw* (China Type Climate)
3- *Cf* (West European Type Climate)

Minor sub-groups within the major sub-groups have been recognised with combination of *a, b, c*, letters at micro level (Table 23.1).

Table 23.1 : Major Climate Types

Climate Category	Symbol	Description
Tropical	Af	Tropical Rainforest, hot, rainy all seasons.
Humid	Am	Tropical Monsoon, hot seasonally excessive rainfall.
Climate	Aw	Tropical Savana, hot, seasonally dry-usually in winter.
Dry climate	Bsh	Tropical steppe; Semi-arid and hot condition
	Bsk	Mid-latitude Steppe, Semi-arid and cool.
	Bwh	Tropical Desert, Arid and hot.
	Bwk	Mid-Latitude Desert, Arid and cool.
	Cfs	Humid sub-Tropical, mild winter, moist all seasons, long hot summer.
	Cfb	Marine, mild winter, moist all seasons, short cool summer.
Warm Humid Temperature Climate	Csa	Interior Mediterranean, Mild winter short dry hot summer.

Contd.

Table 23.1 ***Contd.***

Climate Category	Symbol	Description
	Cwa	Sub-Tropical Monsoon, Mild dry winter, hot summer.
	Cwb	Tropical upland, mild dry winter short warm summer.
Cold Humid Temperate	Dfa	Humid continental; severe winter, long hot summer, moist all seasons.
Climate (snow)	Dfb	Humid continental; severe winter moist all seasons, short warm summer.
	Dfc	Subarctic severe winter, moist all seasons, short cool summer.
	Dfd	Subarctic extremely cold winter moist all seasons short summer.
	Dwa	Humid continental : severe dry winter, long hot summer.
	Dwb	Humid continental severe, dry winter, warm summer.
	Dwc	Subarctic : severe, dry winter short cool summer.
	Dwd	Subarctic : Extremely cold, dry winter, short cool summer.
Polar or Ice climate	ET	Tundra, very short summer
	EF	Permanent ice capped

Source : (1981) John Gabriel, Navarra, *Contemporary Physical Geography,* p. 381.

D- Climate sub-groups

For the major and minor sub-divisions of D-climates, *w* & *f* for precipitation boundaries and *a, b, c* for temperature boundaries have been used same as in C-climate. An addition of *d* is also used which refers to the region where the temperature of coldest month is below -38°C. Sub-groups of this climate is given below.

1. *Dfa* Humid continental (Severe winter, long hot summer).
2. *Dfb* Humid continental (Severe winter, short warm summer).
3. *Dfc* sub-arctic (Severe winter, moist all season, short cool summer).
4. *Dfd* sub-arctic (Extremely cold winter, moist all seasons, short summer).
5. *Dwa* : Humid Continental (Severe dry winter, long hot summer).
6. *Dwb* : Humid Continental (Severe dry winter, short warm summer).
7. *Dwc* : Sub-Arctic (Severe dry winter, short cool summer).
8. *Dwd* : Sub-Arctic (Extremely cold, dry winter, short cool summer).

E-Climate Sub-groups

E-climate is divided into two sub-groups on the basis of temperature variation. Capical *T* and *F* letters are used for desired objective. Explanations of the letter are given below :

T - Temperature of the warmest month is between 0-10°C.
F - The temperature of the warmest month is below 0°C.

Hence two following sub-groups have been recognised.

ET : Polar climate.
EF : Ice-capped climate.

Evaluation of Koppen's Classification of Climate

Koppen's classification is one of the most accepted worldwide scheme of climates. He has calculated the critical value of temperature and precipitation for determination of accurate boundaries among the large number of climatic regions. He followed the combined form of genetic and empirical methods to enhance utility of the scheme. Four categories of major climates excluding dry climate have been demarcated on the basis of temperature. But he is found very much sensitive and careful about B-climate. He has taken into account the precipitation effect which is the ratio between temperature, evapo-transpiration and precipitation. He has also selected the vegetation boundaries as the boudaries of climatic regions. Naturally, vegetaion is the index of climate in a combined form of temperature, precipitation, relative humidity, cloudiness, sunshine and altitudes and soils.

However, his scheme is criticized on some grounds. He did not take account all the elements of climate but only mean monthly temperature and precipitation. Different, climatic features like sunshine, cloudiness, atmospheric disturbances, air masses, atmospheric pressure etc. have not included by Koppen in the classification of climates. The use of alphabets for designation of climate is also very difficult to remember for geographers who usually designate the region with the name of regional characters.

24

CONFIGURATION OF BOTTOM OF OCEAN AND SEAS

The oceans comprise more than 75 per cent of earth's area. Like the land surface, the ocean bottoms represent the various kinds of marine features, i.e. plain, plateau, ridge, deep etc. The knowledge about the relief of ocean floors was very limited in past centuries but the development of sound recording machine enables the oceanographers to get more knowledge about the depth and topography of the ocean bottoms. By this technique sound is penetrated inside the ocean and it returns back to surface after striking the bottom. The duration of sound wave travel between surface to bottom and *vice versa* is calculated and accordingly the depths of various parts of oceans and seas are recorded. Now this eco-sounding techniques have been replaced by radar sounding and electrical echo devices to find the depth of the ocean. On the basis of collection of data, relief of land and oceans is shown by a curve line indicating the elevation and depth from the earth surface with reference to sea level. This line of curve indicating height and depth is called hypsographic curve [Fig 24.1(a) and (b)]. After the study of hysographic curve, the ocean and sea bottoms may be demarcated into various relief or depth zones as given below :

(i) Continental Shelves,
(ii) Continental Slopes,
(iii) Abyssal or Deep Sea Plains,
(iv) Ocean Deeps.

Continental Shelf

Continental shelf represents the shallow part, of the oceans and seas. It extends between low tide level and 100 fathoms depth of the sea water. It is almost uniform zone of seabed with a gentle gradient. The gradient of slope is less than 10. The width of the continental shelves vary from coast

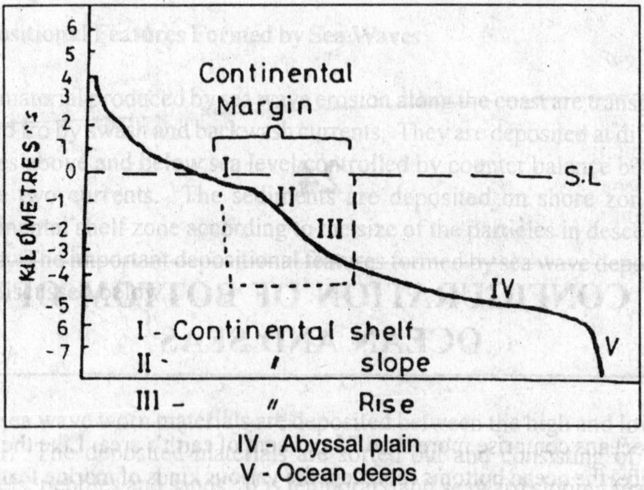

Fig. 24.1 (a) : Hypsographic Curve

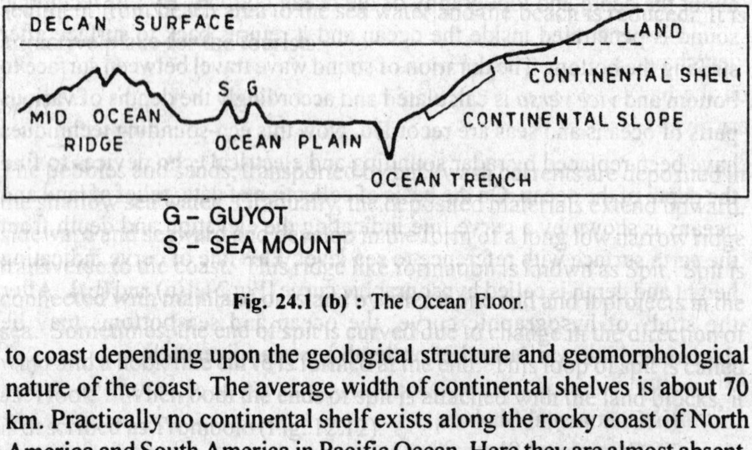

Fig. 24.1 (b) : The Ocean Floor.

to coast depending upon the geological structure and geomorphological nature of the coast. The average width of continental shelves is about 70 km. Practically no continental shelf exists along the rocky coast of North America and South America in Pacific Ocean. Here they are almost absent. Along the coast of low land like off coast of Siberia along Arctic ocean, 1500 metres wide continental shelf is recorded. Off the coast of north-west Europe, shelf is between 50 and 100 kms.

Moreover, the continental shelves are covered by terrigenous sediments mostly of sandy materials except off the glaciated coasts of higher latitudes where glacial derived gravel like debris frequently dominant. Organic deposits of marine life living on the bottom, is also found on this depth zone. Various kinds of mineral nodules are also common due to precipitation of various chemical components. The continental shelves have been grouped in following types on the basis of their formation and location.

(i) Glaciated shelf, (ii) Coral reef shelf, (iii) Riverine shelf, and
(iv) Mountainous shelf.

The area covered by this shallow depth zone varies from ocean to ocean depending upon the tectonic difference of sea coasts. The total shelf area is 27.49 million sq.km in all the oceans. The Atlantic ocean accounts for 14 million km^2, Indian ocean for 3.17 million km^2 and Pacific for 10.11 million km^2 area for shelves. Individual ocean also varies in percentage of shelf area of the total ocean area. It covers 13.3 per cent in A.O., 5.7 in P.O. and 4.2 per cent in I.O. On average it covers 7.5 per cent area of all the oceans.

The origin of continental shelves is very much disputed. Various views have been expressed regarding the possibilities of genesis of shelves on the basis of contrasting evidences persists upon them. These views are given below.

(i) The continental shelves are formed by erosion of coastal area by sea waves. Sea waves and currents erode the coast in the form of notch and cliff and abrasion plateform landward. The eroded materials are deposited in the sea and the shallow water zone is formed. Thus the landward transgression of sea through the lowering of coastal area by wave erosion, continental shelves might have been produced.

(ii) Some continental shelves have been formed by deposition of sediments, in the sea, brought by rivers. Along the mouth of rivers, the sea becomes shallower due to continuous deposition of longer periods.

(iii) According to some views, the continental shelves were formed due to change in the sea level or subsidence of costal land by tectonic movement. During the glaciation period, the sea level was rather low and coastal lands were eroded by either rivers or glaciers and a number of erosional land forms were developed during deglaciation period when ice was melted and sea level rose up to a considerable extent. Thus the coastal area came under water and the conseqently continental shelves have been formed. It is also thought that coastal area might have come under water due to subsidence of the coasts. The occurrence of canyons, valleys and depression of varying nature on the shelves are the testimony to this explanation.

(iv) The formation of continental shalves is also thought due to emergence of ocean and sea bottom. Along the coast of Norway, Labrodor and Greenland uplift has occurred. Off eastern coast of India, emergence of sea has also taken place.

Moreover, the continental shelf region is of great geographical significance. The importance of shelves are given below :

(i) Marine life is abundant due to availability of sunshine. Sun rays penetrate to the depth of the shelf and various kinds of grass, seaweeds and planktons are available. They are used as foods for marine animals. Thus zone is the richest ground of the fish due to shallow water and abundant fish foods. Newfounland, Great Britain and Japan fishing ground are confined to continental shelf.

(ii) Various kinds of minerals like petroleum, gas are also found in shelf area. More than 25 per cent of oil drilling activities are concentrated to shelf region. Most of the important ports of world are confined to this oceanic region.

Continental Slope

Continental slope is the next depth zone of ocean and sea floor to the continental shelf. It begins with outer edge of the shelf where the gradient of slope becomes steeper abruptly. The average slope of this zone is 4° but it varies from place to place. For example, off shantigo, Chile, the slope is about 45°. Along the mountainous and faulted coasts the slope is rather steeper and along the uniform and lowland coasts it is relatively gentler. The distance of slope from the coast is also variable. It depends upon the nature of coasts. In Atlantic ocean at 51°30'N continental slope starts after a distance of 75 km (50 miles) from the coast. Off the hilly coasts for example along Rokies and Andes coast, the slope begins after a short distance whereas along the coastal low land and delta region the distance between slope and coast is farther. This zone is free from the deposits due to steep slope and about 60 per cent of sediments of continental slope is mud an rest is composed of sands, gravels and organic remains. It covers smaller area of the oceans and seas. It occupies 8.5 per cent of total area of the oceans. In Atlantic ocean it occupies 7.6 per cent area, in Indian ocean 5.8 per cent area and in Pacific ocean 5.6 per cent area of ocean bottoms. The slope is intersected by a number of canyon dendritic valleys and faults. There are five types of slopes : (i) Fairly steep slope dissected by canyons, (ii) gentle slope with elongated hills and basins, (iii) faulted slopes, (iv) slopes with terraces, and (v) slopes with seamounts. The origin of continental slope is still not known, although some may be as the result of large scale earth movement called Plate tectonic. It does not always extend upto deep sea but intrrupted by wedge of deposited sediments by turbidity currents as continental rise between continental slope and deep sea plain.

Abyssal Plain or Deep Sea Plain

Beyond the continental slope and rise a broad and long featureless deeper part of the ocean is found. This deep plain is known as abyssal or deep sea plain. It is found between 2000 to 6000 metres depth. Gradient of slope is very gentle and it appears as uniform flat plain without any relief. This relief has been discorded after the deep sea sounding reports which revealed the deep sea plains possess a number of abyssal hills, Guyots, plateaus, trenches, deeps etc.

Moreover, the deep sea plain is formed by deposition of sediments of especially marine origin. Terrigenous deposits are absent due to being far away from the coasts. Ooze of Calcareous and silicious nature are common. The deposit covers the area between the hills and over the hills and consequently a featureless plain is formed in deep water. It covers the maximum area of ocean bottoms. In all the oceans the deep sea plain area covers about 82.7 per cent of total bottom area. It accounted for 85.8 per cent of total area in Atlantic ocean 90.6 per cent in Pacific ocean and 90.6 per cent in Indian ocean individually. The major part of sea plain is between 20°N to 60°S latitudes while it is almost absent between 60°-70°N latitudes.

Ocean Deeps

On the ocean bottom a large number of narrow elongated deep depressions have been recorded by the sound recorder device. Their depth is rather greater than the abyssal plain and some of them reach up to more than 9 kms below sea level. The deepest deep Mariana trench off the Guam island extends up to 11 km depth while the average depth of ocean is 4 kms. 57 deeps are known till now through the expedition in the world. 32 out of them are confined to pacific ocean. The peculiarity of the distribution of ocean deeps is that they do not occur amidst the ocean bottom but along the coast parallel to the fold mountain. Their origin is attributed to the plate tectonic movement. Along the zone of convergence of plates in Pacific ocean, the subduction of oceanic plate gave the origin of trenches and island arc.

Other Features Associated with the Bottom of Oceans

Beside above discussed major relief of ocean bottoms submarine canyons submarine ridges, Abyssal hills, seamount, guyots and atolls are also commonly found in the oceans. They are discussed below.

Submarine Canyons

A long narrow depressions are commonly found on the continental shelves and slopes. These depressions are called as Submarine canyons. They are usually perpendicular to the coast. Their profiles resemble V-shaped valley with concave rock wall. There are three groups of canyons on the basis of their location and appearance : (i) Submarine canyons which begin from continental shelves and extend to the slopes, (ii) Canyons, which are found at the mouth of rivers and extend upto continental slopes, and (iii) Dendritic canyons which is cut down in many branches.

Moreover, there are about 102 known canyons in the world. Their names are designated either with the name of rivers at which mouth they are located or by the name of the expidition which explored them. Largest canyon is found in Berring sea off Alaska. The important Canyons are the Berring, Jhemchung, Hudson, Oceanographer, Zaire, Mississippi, Indus and so on.

The origin of submarine canyons is still disputed. They were formed by various forces. Some canyons, which are located near the mouth of rivers, are viewed as the part of river valley like gorge or canyon which were cut during the glaciation period. When the sea level rose up and coastal valley submerged and changed into form of submarine canyons. Sinking of river mouths by tectonic forces is also considered as the cause of formation of canyons.

However, turbidity current, a fast flow of water consists of mud and sand in suspension, is also an important force to cut down the canyons on the bottom of oceans. This current erodes the bottom through abrasion action and also deepen the pre-existing river valleys to form the canyons. Some canyons were also produced due to diastrophism in which downfault in graben form came into being in zone of shelves and şlopes.

Abyssal Hills, Seamount and Guyot

Some isolated hills of various origin like volcanic cone, coral reef etc. are also scattered on the bottom of the oceans and seas. These hills are known as Abyssal hills. They are different from the submarine ridges in such a way that the latter are found in a chain, like mountains on the land, continously amidst of ocean bottoms. The abyssal hills which attained the height more than 1000 mt (1 km) from the floor of sea or ocean is known as seamounts. Broad and flat topped seamounts is called as Guyot. These features are exclusively concentrated to the Pacific ocean. There are about 1000 hills, seamount and Guyots in Pacific ocean.

Beside a number of atols, ring or concentratic coral reef with deep and wide lagoon, are also common. They are formed by deposition of remains of coral ployps on the marina plateform. Plate tectonic movement is the major force attributed to the origin of above discussed marine features. Along the mid-oceanic ridge, due to sea floor spreading, the volcanoes move away from the ridge and become inactive and new active volcano comes into being near the ridge. Thus the depth of ocean increases away from the ridge due to cooling and subsidence of lithosphere as the sea floor spreading moves away and spread over the ocean bottoms. Occurrence of seamounts and guyots at various depths in oceans is explained in the above discussed way.

Submarine Ridges

Like mountain chains on the land, a long continuous chain of mountains also spread out in the mid of oceans. They form the longest series of mountains, about 75000 km long, on the earth. Their width varies from few metres to hundreds and thousands of kms. Different ranges are converging and forming plateaus and some are bifurcating to form different individual ridge buttressing basins of oceans. The submarine ridges are stretching in all oceans except Pacific. Atlantic ocean is specially known for mid oceanic ridge along which sea floor spreading is very common due to diversion of the plates.

Relief-Features on Floor of Individual Oceans

In general all oceans show the identical relief features, with some exeptions.

The nature, extent and magnitude of various relief features are variables in diffferent oceans due to differences in their size and shapes, location of plate boundaries and tectonic movement.

Pacific Ocean

Pacific ocean is the largest and the deepest ocean of the world. It occupies more than 1/3rd of earth's area. It is bounded by broken coast of Asia, and Australia from the west and by coast of mountains from the east. Continental shelves cover rather less area while the deep sea plain occupies largest area of ocean bottom. The depth of sea plain is also greater than that of sea plain in other oceans. This ocean is associated with a large number of abyssal hills, seamounts, guyots, submarine rise and plateus like Alabatross near western coast of south America. Pacific ocean is basically known for the island arcs and ocean deeps or trenches. Out of 52 known oceans deeps, 32 are found in this ocean. Mariana Trench, off the island of Guam, is the deepest one. It is about 11 km deep. Kurile trench, Tonga trench Aleutian trench, Japan trench, Tucarora deep etc. are important other ocean deeps in Pacific ocean. This ocean exceptionally free from the submarine ridges. The concentration of large number of deeps and island arc from Aleutian island to New Zealand through Japan is attributed to the subduction of oceanic plate beneath the Asiatic continental plate along the margin of convergence in Pacific ocean. Volcanic eruption and earthquakes are very common phenomena in the ocean. The characteristic features of Pacific ocean are the feeble development of continental shelves, absence of any continuous middle dividing rise and abundance of deeps and trenches (Fig 24.2).

Atlantic Ocean

Atlantic is the next important ocean of the world. It occupies less than 1/6th of total area of the earth. Its shape is like "S", letter bounded by continuous landmass from the cast and west side. The northern edge is demarcated by a number of Islands, and through various straits and channels it merges into Arctic ocean. On the south it is bounded by Antarctica continent.

The study of configuration of ocean bottom of Atlantic ocean is rather more complicated. Continental shelves are broader especially in North Atlantic along the eastern coast off North Amercia and off the coast of Western Europe. In south Atlantic the continental shelves are not extensive except near the Flakland. Off the coast of Africa and Brazil it is narrower

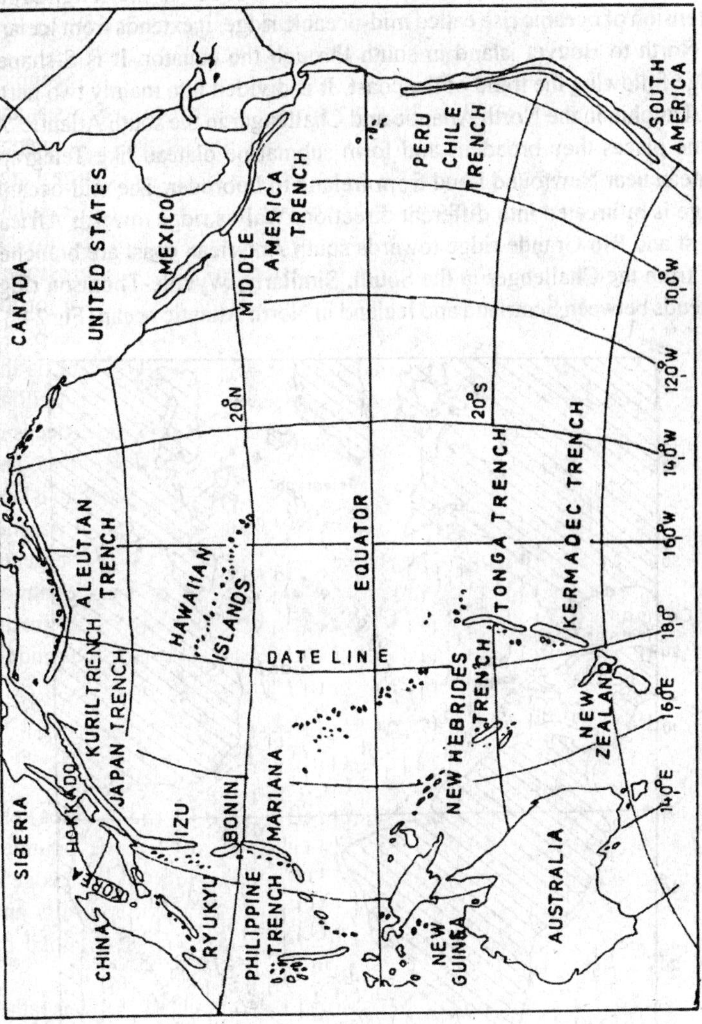

Fig. 24.2 : Distribution of Sea Trenches in Pacific Basin.

due to steep edge of plateaus.

The most important feature of Atlantic ocean is the longitudinal extension of oceanic rise called mid-oceanic ridge. It extends from Iceland in North to Bouvet island in south through the equator. It is S-shaped ridge following the trend of the coast. It is divided into mainly two parts, i.e. Dolphin in the North Atlantic and Challenger in the south Atlantic. At some places they broadens and form submarine plateau like Telegraph plateau near Newfound Land from Ireland to Labroder. The mid-oceanic ridge is bifurcated into different direction. Walvis ridge towards African coast and Rio Grande ridge towards south American coast are branched off from the Challenger in the South. Similarly, Wyville-Thomson ridge extends between Scottland and Iceland in North Atlantic ocean (Fig 24.3).

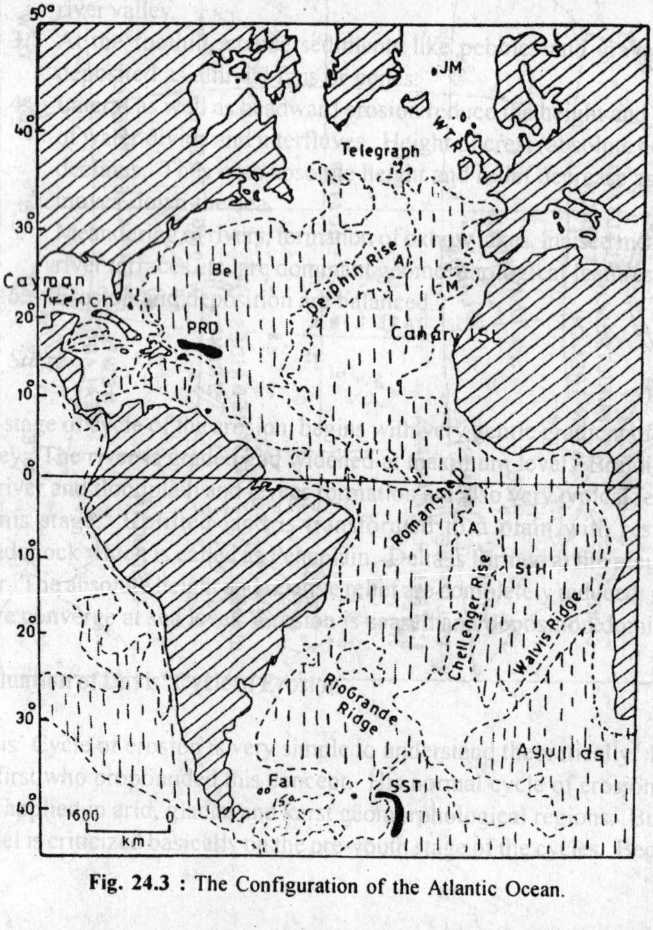

Fig. 24.3 : The Configuration of the Atlantic Ocean.

However, the Atlantic ocean is devoid of linear deeps of trenches. Only South Sandwitch, Puertorico, and Romenche deeps are found in this ocean. Besides, ocean floor spreading, along the mid-oceanic ridges, is also an exceptionally unique feature of Atlantic ocean. The Atlantic ocean is characterised by well marked continental shelves, mid-oceanic ridge and absence of deeps and trenches.

Indian Ocean

Indian Ocean is different from the Atlantic and Pacific ocean in its shape and size. It is bounded by continents from the north by Asia, from west by Africa, from east by Australia and from South by Antarctica. It does not

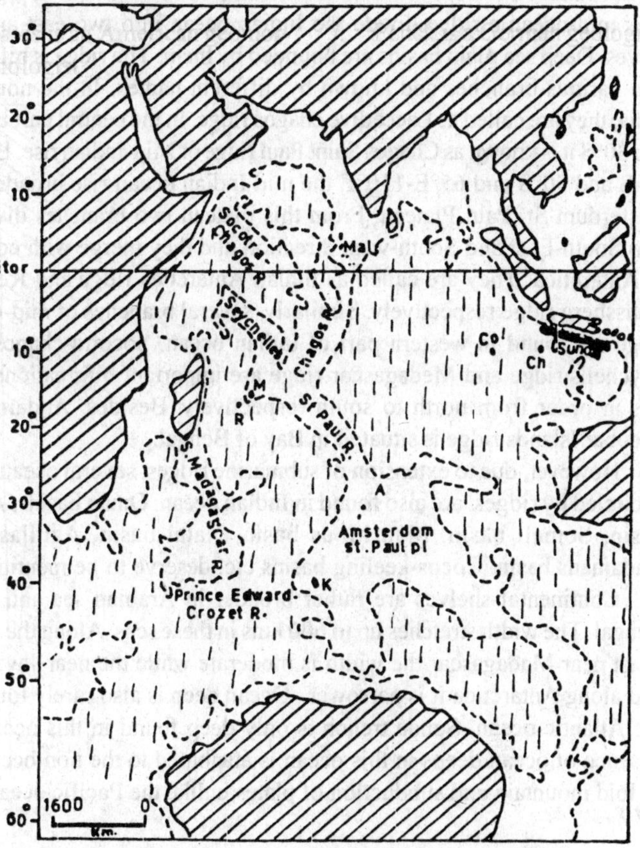

Fig. 24.4 : The Configuration of the Indian Ocean.

extend beyond tropic of cancer (Fig 24.4).

Bottom relief of Indian ocean is similar to two oceans discussed earlier. 58.8 per cent of total area is occupied by deep sea plain with depth between 4000 to 6000 mts. The average depth of the ocean is 4000 mt and this ocean may be divided into three regions : (i) Near African coast, depth is less than 6000 mts with small number of islands, (ii) The eastern part near the East Indies like Java and Sumatra, the depth is more than 6000 mt., (iii) The central part exhibits elevated bottom with less than 4000 mt. depth.

Like the Atlantic ocean, mid-oceanic ridge is also found in Indian ocean. It is continuous broad range of ridges extending from the southern tip of Indian Peninsula in North to the Antarctica continent in the south. This mid-oceanic ridge divide the Indian ocean into two east and west halves. Deep sea plain basins are bounded by them. The ridge is bifurcated into various branches and known by different names. In the north near India, they are called as Laccadive chagos ridge. In the central part between 0 to 30°S it is known as Chagos Saint Paul ridge or Mid-Indian rise. Between 30°S and 50°S and 65°E-110°E the mid Indian ocean rise broadens as a Austerdum St. Paul Plateau. From this plateau two branches divide off into South-East and South-west direction and they merge with continent of Antarctica. They are called as Indian Antarctica ridge and Kergelen-Gaussberg ridge respectively. Similarly, several branches of mid-oceanic ridge are found in western part of Indian ocean. Socotra-chagos ridge, Seychells ridge and Medagascar ridge are important bifurcations which lies in order from north to south respectively. Besides, Andaman and Nicobar Islands ridge is situated in Bay of Bengal.

However, due to extension of submarine ridges, several ocean basins, encircled by ridges, are also found in Indian ocean. Oman basins, Arabian basin, Somali basin, Mauritious basin, Natal basin, Agullas basin, Andamans basin, Cocos-keeling basins etc. deserve to be mentioned.

Continental shelves are rather broader in Arabian sea and Bay of Bengal. The width stretches up to 600 kms in these seas. Along the African coast near Madagascar the width is moderate while the near Java Island and along Antarctica it is narrower. Ocean deep is also rarely found like the Atlantic ocean. Sunda trench is only deep found in this ocean. The absence of ocean deeps in this ocean is attributed to the non-occurrence of fold mountain and subduction of plates unlike the Pacific ocean.

25

CHARACTERISTICS OF OCEAN WATER

Sea or ocean water is consisted of inumerable ingradients and show distinct physical and chemical characteristics. It is the habitat of various kinds of aquatic life, i.e. both animals and plants. The study of various aspects of ocean water is very useful for oceanographers, navigators, economists, industrialists, meteorologists, politicians and so on. In modern era ocean is seen as the storehouse of various kinds of resources which may face and solve the multifacet crisis of human being. The sea water is studied mainly in two ways : (i) Physical characteristics, and (ii) Chemical characteristics.

PHYSICAL CHARACTERISTICS

Temperature of Ocean Water

Temperature is the very important physical characterisitc of ocean water. It determines the salinity, density, plant and animal life in the ocean, circulation of water like current and waves etc. are the result of temperature in the sea. Various aspects of temperature of ocean water are discussed below.

Process of Heating and Cooling

The temperature of ocean water is the outcome of heating and cooling effect of water. The sources, which add the temperature to the sea water, are of three types, (i) heat from interior of earth, (ii) absorption of solar radiation, and (iii) kinetic energy produced by dynamic water. Interior of earth supply heat, in the form of convectional current produced due to disintegration of radioactive elements in the crust, to lower layer of ocean water. But its share in budget of heat of ocean water is insignificant. Similarly, through circulation of water-tide, current, waves, etc., some heat is produced as kinetic energy and water is consequently heated up. It is also very insignificant amount of the energy to heat the ocean water.

Insolation or solar heat is the main source of energy or heat of ocean water like the land. The absorption capacity of water is very high when sea is calm and the sun is directly overhead, about 98 per cent light penetrates the sea and only 2 per cent is reflected. Reflection increases as the angle of sunshine decreases. The absorbed heat is converted into energy and temperature of water determined. The depth and rate of penetration varies with the composition of water. Light moves at faster ate in clean water than muddy or turbidity water. Due to variation in temperature and density of water at different depth level, stratification of water persists. Lighter water overlies the cold dense water. This boundary of demarcation is known as Thermocline.

The colour of sea is also variable due to penetration and reflection of different spectrum of light at different part of the ocean. The ocean light at various depths is measured with an instrument called Photometer.

However, the absorbed heat of the ocean water by various sources is also lost to the atmosphere. Evaporation, back radiation, conduction are the important processes of cooling of ocean water. In evaporation process the heat from the water surface is used and trapped into vapour as latent heat. This latent heat releases to atmosphere after condensation of water vapours in upper part of atmosphere. The rate of evaporation and lost of energy of ocean water depend upon the (i) temperature of overlying air, (ii) velocity of air, and (iii) saturation of overlying layer of air. Calm, dry and warm overlying air cause rather higher rate of evaporation. It varies from latitude to latitude too. The maximum evaporation occurs between 20-40° N&S latitudes. In conduction heat transfers to overlying atmosphere through direct contact. Temperature is lost from warmer water to overlying air. Much of amount of heat of sea water transferred to atmosphere through the conduction process. Besides, back-radiation is also very importnat player in cooling process of sea water. Everybody radiates the energy. The amount of radiation and wavelength depend upon the temperature of the body. High temperature high amount of radiation with shorter wave length and *vice versa*. The following equation shows the heat balance of the water.

$$SR = SE + SC + SB$$

SR = Total heat received by Sea (Incoming energy from all sources)
SE = Loss of heat by evaporation (55 per cent of outgoing heat)
SC = Loss of heat by conduction (5 per cent of outgoing heat)
SB = Loss of heat by back radiation (40 per cent of outgoing radiation)

Distribution of Temperature in the Ocean

Temperature of sea water at surface as well as at depth varies from place to place and season. The variation of temperature on the water surface lies between —2°C and 30°C. The average temperature is 16°C. The differences in temperature horizontally and vertically, depends upon latitudinal distribution, location of ocean currents, prevailing winds, upwelling and sinking of water and mixing of water between open sea and marginal seas and so on. The distribution of temperature is shown by isotherm lines. The trends of isotherms of months February and August show irregularities in temperature in different places lying on the same latitude. Maximum and minimum temperature seasonally are shown in the months of August and February on the ocean surface instead of July and January as on land surface due to lag of time in process of heating and cooling of water surface. Water neither cools rapidly nor warms up fastly unlike the land surface. The following are general characteristics of temperature distribution on the water surface.

(i) Temperature decreases with increasing latitudes in both hemispheres due to latitudinal differences in the insolation/ distribution. The average annual temperature at equator is 26°C and at the latitudes of 20°, 40° and 60° it is 23°, 14° and 1°C respectively. This variation of temperature in individual ocean is markedly different as given in following table :

Table 25.1 : Latitudinal Distribution of Temperature (°C)

Latitudes	Atlantic Ocean	Pacific Ocean	Indian Ocean
70-60 N	5.64	—	—
60-50 N	8.66	5.74	—
50-40 N	13.16	9.99	—
40-30 N	20.40	18.62	—
30-20 N	24.16	23.38	26.14
20-10 N	25.81	26.42	27.23
10-0 N	26.66	27.20	27.88
70-60 S	-1.30	-1.30	-1.50
60-50 S	1.76	5.00	1.76
50-40 S	8.68	11.16	8.68
40-30 S	16.90	19.98	17.00
30-20 S	21.20	21.53	22.53
20-10 S	23.16	25.11	25.85
10-0 S	25.18	26.01	27.41

Source : Sverdrup, H.U., *Oceanography for Meteorologist,* p. 7.

(i) Table 25.1 incidcates that at every latitude the temperature is rather high in Atlantic and Indian ocean in comparison to the Pacific ocean due to the fact that former were locked by land and the latter has wide expanse and high rate of mixing of water.

(ii) Annual range of temperature in ocean is relatively low due to mixing of warm water into cold and *vice versa* through the circulation of ocean currents. The monthly maximum and minimum temperature is recorded in August and February rather than July and January due to the slow heating and cooling of water. It is affected by size and shape of water bodies. Coastal areas (marginal and enclosed seas) in tropical region show relatively higher annual range of temperature due to low mixing and land effect. North Atlantic and Pacific oceans have greater range as compared to southern hemisphere due to predomination of land in northern hemisphere.

(iii) Daily range of temperature is also very low at every latitude due to slow cooling and heating of water. It is less than 10°C. Daily maximum and minimum temperature are found at 2 p.m. and 5 a.m. respectively on the sea water. In clear sky and clam weather condition daily range of temperature is relatively higher.

(iv) In tropical region the western margins are warmer than the eastern margins due to the NE and SE trade winds and upwelling of cold water along the eastern margins. The trade winds displace the warm water from the eastern side and accumulate it along the western side. Consequently, the eastern side is cooled off and western is warmed up. Influx of cold currents as well as cold sub-surface water also reduce the temperature on the eastern margins.

(v) In higher latitude along the western side temperature is rather low as compared to the eastern side due to the effect of ocean currents and westerlies. The cold current like Labrodores Oyshio currents etc. bring the cold water and reduce the remperature along the western margin. While the warm ocean currents like Gulf Stream, Kuroshio etc. move along the eastern margins, under the influence of westerlies, consequently, temperature rises up. Between 50-60°N south of Newfoundland 0°C temperature is found while in the same latitudes along the west European coast 9-13°C temperature is recorded in February.

(vi) The isotherm lines in southern hemisphere are almost parallel to the latitudes and variation in surface temperature is inceptible. Whereas in northern hemisphere trend of isotherm lines is very irregular.

(vii) In Indian ocean highest temperature (28°C) is in the Arabian sea and Bay of Bengal. The isotherm lines in western part of ocean is parallel to the longitude instead of latitude due to effect of coastal area. The enclosed seas — Red (30°C) and Persian Gulf (33°C) record the maximum temperature due to the effect of hot desert land surrounding from three sides.

(viii) The marginal and enclosed sea in lower latitudes have higher temperature than the open sea and higher latitude. They have relatively higher temperature than the open sea due to effect of surrounding lands.

Vertical Distribution of Temperature

Unlike the land, the sun rays penetrate in water to rather more depth and the temperature upto few metres remain same as that of the surface. Beyond this narrow depth zone, the temperature decreases with the depth. But the rate of change in temperature upto few hundred metres depth is very fast in all oceans. After this, temperature decreases very slowly. Beyond 2000 mts. There is scarcely any change and the whole of the water below this depth has 35°C or 36°F temperature.

However, the rate of change of temperature with depth is not uniform. It varies from region to region. Spatial variation of temperature at different depth level is far less than variation in surface temperature. The temperature declines with depth and it occurs in the following sequence.

(i) In equatorial region initially temperature increases with depth for few metres. Due to addition of fresh water by precipitation, the surface water has low temperature and low salinity. Beneath the surface, water of high temperature and salinity occur. Again after crossing the depth of high temperature and salinity, temperature begins to fall down with the depth.

(ii) In the area of upwelling of cold water, where warm surface water is displaced the fall of temperature with depth is very low. For example off the coast of Africa and California show the same pattern of change of temperature with depth.

(iii) Sinking of cold dense water and mixing of warm surface current in high and middle latitudes affect the rate of change of temperature with depth. It becomes fast as the temperature of sub-surface water becomes very low due to convergence of cold dense water.

Similarly, due to the mixing of warm surface current, surface temperature becomes high and decrease would be fast with the depth.

(iv) In some exceptional sea like Sorgasso sea, Mediterranean sea and Red sea, temperature is higher even at greater depth as compared to the temperature of open sea at the same depth level. High amount of insolation, anti-clockwise water circulation and less mixing of water are the causes of high temperature at greater depth in Sargasso sea, while the latter two seas show the high temperature at deeper parts due to their basin like bottom, shallow water at the entrance points from the open sea (Strait of Gibralter and Strait of Bab-el Mandeb) mixing of water is checked. Hence temperature in open sea is rather less than temperature in Mediterrarean and Red sea at same depth below the bottom of straits.

(v) The enclosed or marginal seas in higher latitudes have relatively low temperature and low salinity at surface level due to addition of fresh water by melting of ice. At lower level sub-surface warm saline current moves from outside to these seas, that reversal condition (increase of temperature with depth) of temperature change with depth takes place.

SALINITY

Water is very strong solvent. It is consisting of various kinds of gases like oxygen, CO_2, Nitrogen, Hydrogen etc., chemical constituents and minerals. Sea water is only 96.5 per cent water and the rest (3.5 per cent) consists of dissolved solids as sodium chloride, magnesium chloride, potassium chloride and so on. The total mount of dissolved solids in water is referred as salinity. It is measured in gm per kg (0/00) of sea water. The amount of salinity is variable between 30-37 per cent generally in the world. The proportion of major constituents of salinity remains constant at every place. The study of salinity is very beneficial in modern time. It gives the knowledge about the chemistry of sea water which is very helpful to understand distribution of concentration of various marine life in the sea. The circulation of water like current, sinking and upwelling of water etc. are determined by variation in the salinity. Many minerals, edible items like salts, chemicals etc. may be derived by study of chemistry of sea water.

Composition of Salinity

Salinity refers the total amount of dissoved solid particles in the water. These particles are consisting of various kinds of elements. Sodium chloride is the most important salt in the sea. Besides, other small constituents are also found but in insignificant proportion. Some nutrient salts phosphate, nitrate and nitrite are also available in the sea. Nutrient salts are used by marine life in various ways. The proportion and amount of different salts are given in Table 25.2

Table 25.2 : Composition of Salinity in Sea Water

Salts	Amount (0/00)	Percentage
Sodium chloride	27.213	77.8
Magnesium chloride	3.807	10.9
Magnesium sulphate	1.658	4.7
Potassium sulphate	1.260	3.6
Calcium sulphate	0.863	2.5
Calcium Carbonate	0.123	0.3
Magnesium Bromide	0.076	0.2
	35.000	100.0

Sources of Salinity

The question of source from where salt is derived in the sea water is debatable. River is assumed as the source of salt in the sea but the analysis of river water and sea water give contrasting chemical composition. River water contains calcium in abundance while the sea water is rich in chloride. This variation in chemical composition puzzled the oceanographers. Where has calcium carbonate gone if river continuously added the salt to the sea? However, it is opined that many marine animals like corals derive calcium carbonate for building their skeleton and this chemical component is consumed in larger proportion. Sodium chloride is left unused and consequently in abundance. Moreover, the salts, initially, were added to the sea water through solution action of sea water along the sea coasts. Weathering of different rocks in coastal area also contributed to the sea water by adding salts. Thus with the continued action of river transportation (in solution) and solution action of sea water itself, salts in sea water are added throughout geological past.

Determinants of Salinity

The amount of salinity in sea water, is the combined effect of a large number of determinants. Temperature evaporation, precipitation, influx of fresh water and mixing of water etc. are the important factors of salinity. These factors are briefly discussed below —

Temperature and evaporation are directly related with the concentration of salinity. Higher temperature and higher evaporation imply higher rate of salinity, provided other factors remain constant. Sub-Tropical high pressure regions record maximum salinity if we take latitudinal distribution of salinity. Many inland seas like Dead sea, lake Van and marginal seas like Mediterranean and Red sea show relatively rather higher salinity than the open sea in the same latitudinal zone because the rate of evaporation is higher than addition of fresh water. Contrary of this high latitudinal zones have low concentration of salinity due to low temperature and evaporation.

Salinity is, however, counterbalanced with the addition of fresh water either by melting of ice, precipitation or by rivers. Addition of fresh water dilutes the salinity and concentration of salt is reduced to the given weight of sea water. For example, equatorial region shows relatively low salinity as compared to the sub-tropical high pressure, though the temperature is higher. Heavy amount of precipitation lowers the salinity here. Similarly, low salinity is also found in temperate and subpolar areas due to influx of fresh water by melting of snow. Likewise the amount of salinity is also low at the mouth of rivers as huge amount of water is added to the ocean or sea. Black sea records rather low salinity due to discharge of water by many rivers from Europe.

Above all circulation of water or mixing of water also affects the distribution of salinity. Open oceans in tropical region and cold region have relative low and high salinity respectively as compared to the marginal or enclosed seas due to less mixing of water of these bodies together. Mediterranean sea, Red sea, Persian Gulf etc. show higher rate of salinity as compared to Atlantic ocean and Arabian sea respectively. Similarly, in temperate region, the inland sea like Baltic sea exhibits lower salinity than the North sea due to less mixing of its water with that of Atlantic ocean and also due to the melting of ice accumulated from snowfalls. Gulf of Bothania records as low as 3°/00 salinity. Ocean currents have also neutralized the distribution of salinity. Warm and more saline water moves from tropical region to low saline cold water in temperate and polar region and *vice versa*. Submarine ridges also affect the salinity. On either side of ridge, different amount of salinity is reported because of its hindrance in mixing of water of both sides.

Distribution of Salinity

Average salinity of oceans and seas is 35°/00. Owing to the variation in different aforesaid determinants, the salinity varies from place to place horizontally as well as vertically, like temperature, salinity also shows latitudinal zonal distribution. It is shown on map by an imaginary line connecting the places of equal salinity. This line is known as isohaline. The isohalines and data of salinity of different latitudes indicate that the maximum salinity (36-37°/00) is found in sub-tropical high pressure regions along tropic of cancer and capricorn. Highest concentration of salt is attributed to low precipitation and high evaportaion. It decreases from this region towards both north and south. Among the equator it lies between 35-36°/00 despite of high temperature. Heavy amount of rainfall dilutes the concentration of salinity, consequently, the salinity is rather less than the sub-tropical high pressure region. Similarly, the high latitudes region also show low concentration of salinity below 35°/00 due to low temperature, low evaporation and addition of fresh water by melting of ice.

Latitudinal Distribution of Salinity

Latitudes	Salinity (°/00)
70° - 50° N	30-31
50° - 40° N	33-34
40° - 15° N	35-36
15° N - 10° S	34-35
10° N - 30° S	35-36
30° S - 50° S	34-35
50° S - 70° S	33-34

Salinity in Individual Ocean

The study of salinity in individual ocean reveals the regional differences. Atlantic ocean records highest amount of salinity (37°/00) out of all open oceans. In Atlantic ocean the following are the common features regarding to salinity distribution (Fig 25.1).

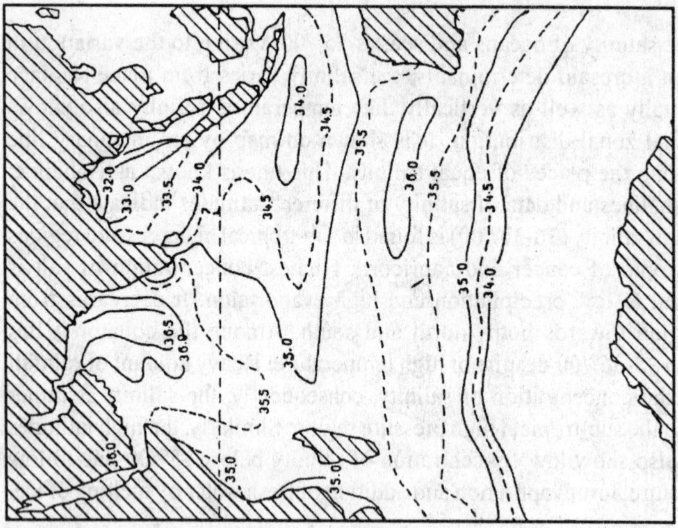

Fig. 25.2 : Surface Salinity of the Indian Ocean (in Northern Summer).

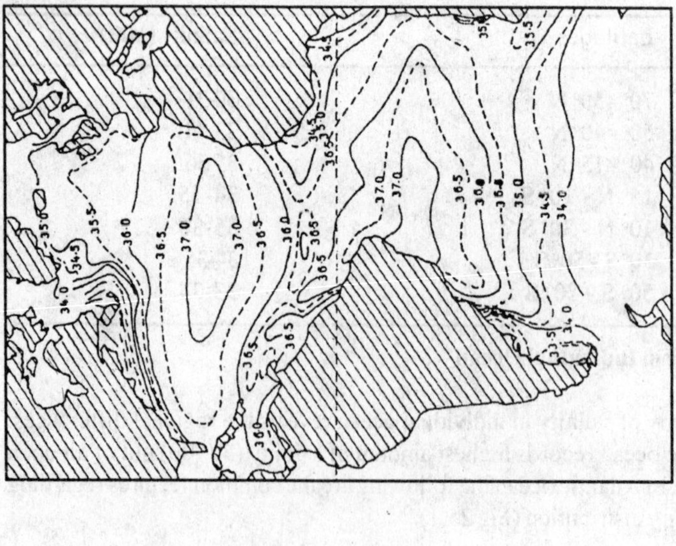

Fig. 25.1 : Surface Salinity of the Atlantic Ocean (in Northern Summer).

(i) Highest salinity (36-37°/00) is found along the western side of sub-tropical ocean in both hemispheres, instead of eastern side, due to transportation of hot and saline water from eastern coast by trade winds. Upwelling of cold water also reduces the salinity along the eastern side of ocean 25°N 40°W 15°S 25°W record Maximum around 37 per cent salinity;

(ii) In high latitude region, especially in North Atlantic ocean, the salinity is higher along the eastern coast of the ocean due to transportation of hot and more saline water from tropical region by Gulf stream current. Contrary to this, along the western side along the same latitude, the salinity is low due to influx of cold and less saline water by Labrodore cold current;

(iii) The marginal seas in tropical region record rather higher salinity and in temperate region rather low against the open ocean due to less mixing of water; and

(iv) In South Atlantic isohalines are parallel to latitudes because of pre-domination of water which shows more mixing.

Indian ocean, due to its size and land-locked by three sides, shows distinguished characteristics in terms of salinity distribution. The main features of salinity in Indian ocean are discussed below (Fig. 25.2).

(i) Isohalines are found parallel to the longitudes in North Indian ocean. 35°/00 isohalines divide the ocean into East and West halves. Western portion has more salinity than the east.

(ii) The isohalines are generally parallel to the latitudes in southern ocean due to lesser effect of Monsoon Drift.

(iii) Maximum salinity is found in the west of Australia in South Indian ocean due to dry condition.

(iv) Amount of salinity decreases from equator towards north and south in the ocean. In Bay of Bengal there is declining order towards coast due to discharge of water by rivers.

(v) In Arabian sea towards the coast salinity is rather high due to surrounding desert condition. Persian Gulf and sea have higher salinity than the open sea.

Like the other oceans, Pacific ocean also shows inequalities in distribution of salinity throughout the ocean (Fig. 25.3). It decreases from sub-tropical high pressure both north and southwards. Like Atlantic ocean, NW portion along Berring straint and Aleutian island salinity is rather

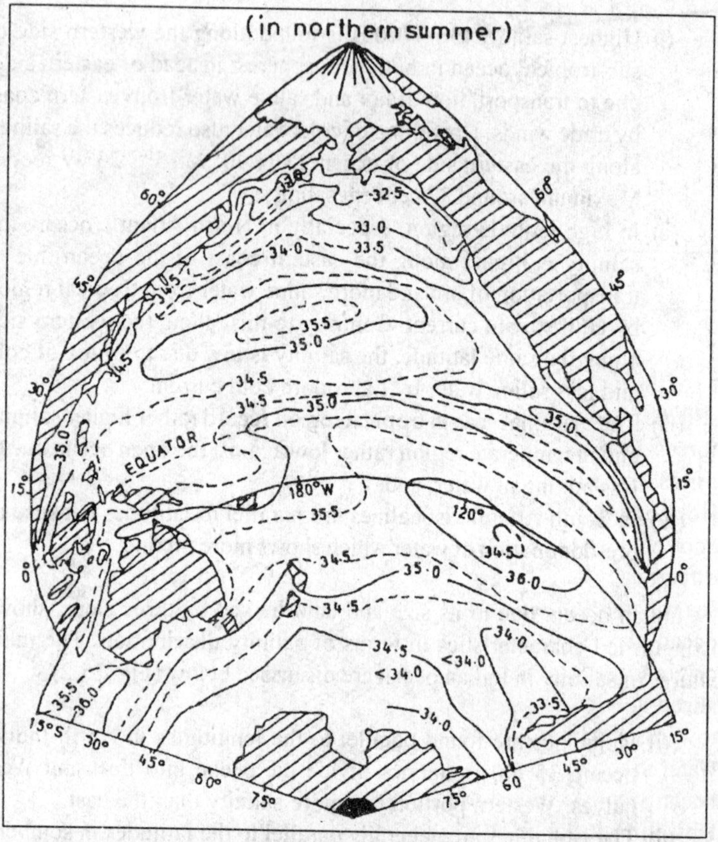

Fig. 25.3 : Surface Salinity of the Pacific Ocean (In Northern Summer).

lower than north-eastern coast of North America continent due to transport of cold water by Wyshio current in former case and warm water by Kuroshio current in latter case respectively. Upwelling of cold water has reduced the salinity along the coast of California and Peru in area of high salinity latitudinal zone. While the salinity along the eastern coast of Australia is relatively higher due to east Australian warm current.

26

CIRCULATION OF OCEAN WATER

Like the air on the earth surface, the ocean water seldom remains stand still. Its physical characters like temperature, salinity, density and external forces like sun, moon and winds have set the ocean water in various forms of motions. Mainly horizontal and vertical motions are common. Horizontal motion refers to the ocean currents and sea waves. Ocean currents are the continuous flow of huge amount of water in a definite direction either from high latitudes to low latitudes and *vice versa.* The sea waves are also a form of horizontal motion of water but it differs from ocean current in such a way that in former water moves ahead from origin to end and in the latter water does not move, but the wave moves ahead and after breaking of it in shallow depth zone, the water rushes towards the coast. While the vertical motion refers to rise and fall of water in oceans and seas. Due to attraction of the sun and the moon, the ocean water is raised up and fallen down twice a day at every point. Upwelling of cold water from sub-surface and sinking of surface water are also the form of vertical motion of ocean water.

OCEAN CURRENTS

Like the rivers on the land surface, the huge mass of ocean surface water circulate in a defined direction from equatorial region to polar regions and from polar to equatorial region. This well recognised circulation of water is called as Ocean currents. Ocean currents are of two kinds based on the place of origin. First, warm current which flows from equatorial to polar regions and the second cold current which moves from polar to equatorial region. Ocean current is also known as Drift and Stream. The ocean current which is originated by friction of winds over surface water and blowing in the direction of wind called drift. North Atlantic drift is the typical example. Sometimes direction and velocity of ocean current is determined by shape and size of coast lines along with it flows. When it flows through the opening (Strait) between islands, its velocity increases up. Such current is known as Stream current. Gulf stream warm current along the eastern coast of U.S.A. is an excellent example of stream current.

Determinants of Ocean Currents

The genesis, development and distribution of ocean currents in various oceans and seas have been the outcome of combined effects of several groups of factors lying inside and outside of oceans. These determinants are discussed below.

A. Oceanic Determinants

This group of determinants includes salinity, temperature and density of sea waters.

Temperature : The level of water surface is directly proportional to its temperature provided other factors remain constant. High temperature in equatorial region raises the level of water while the low temperature in sub- polar and polar areas lowers that water level. Consequently, warm ocean currents in all oceans move from hot equatorial to cold Polar Regions. Contrary to this sub-surface currents moves from Polar Region to equatorward as cold water sinks down in latter region and upwells in former region.

Salinity : Salinity is also an important determinant of ocean currents. High saline water has rather low level of surface water and low saline water shows high level of water. Consequently, the current moves from low salinity water to high salinity water. For example, ocean current from Atlantic ocean to Mediterranean sea flows due to variation in salinity of these water bodies. In turn, a sub surface current flows from Mediterranean to Atlantic ocean. Many cold currents such as Canary, Venezuella, West Australian current, Humbolt current etc. are flowing from low saline cold water to high saline subtropical or tropical water.

Density : Like the high saline water, high density water has low level of water surface. Salinity and density are positively interrelated. Consequently, ocean current moves from low dense water to high dense water following the slope of water surface.

B. Atmospheric Determinants

Evaporation, insolation, pressure and prevailing winds are ex-oceanic factors, which determine the trend and direction of ocean currents.

Evaporation and insolation are associated factors. Higher the insolation, higher the evaporation if the other factors remain constant. Transparency of atmosphere, cloudiness, temperature and humidity of overlying air and the velocity of winds are various elements which determine the rate of evaporation in the sea water. Clear sky, high insolation, calm winds and dry overlying air favour high evaporation. Rate of evaporation and level of water surface are inversely related. Thus the ocean current blows from the water of low evaporation to the water of high evaporation.

Prevailing winds play very important role in determination of direction of ocean currents. The direction of ocean currents in equatorial region from east to west and in temperate region from west to east is due to the effect of trade and westerlies planetary winds respectively. North Indian ocean shows great effect of Monsoon wind on the reversal of direction of ocean currents seasonally. Counter equatorial current vanishes from Indian ocean during south-west monsoon due to merging of former with the latter (Fig 26.3a). Similarly, the ocean current also follow the anti-pressure gradient in determining its direction. It flows from low pressure area to high-pressure area as the level of surface is higher in low pressure region and *vice versa*.

Insolation rate and pressure are inversely related. High amount of insolation leads the high temperature and level of ocean water. Similarly, the low pressure also results in high level of water. In this way, the ocean current will move from high temperature and low pressure to low temperature and high pressure if the other factors remain constant. For example, ocean current moves from Baltic Sea (Low pressure) to North sea (High Pressure). Equatorial currents move from equator of high insolation to polar region of low insolation.

Shape and size of the Coast

The direction of ocean currents in every ocean is modified by the nature and form of the coast. For example, westward south equatorial current is obstructed by presence of projection of land (Cape San Roque) of South America and it is divided into two branches—one goes northwards and mix with north equatorial current; and second moves southwards along the eastern coast of South America in the name of Brazil Current (Fig.26.1). Similar conditions are found in Pacific as well as in Indian ocean. East Australian Current in Pacific ocean and the seasonal ocean current in Indian ocean are the example of currents which follow the coastline.

Factors Associated with the Earth

Gravitational Force and rotation of the earth are two important determinants of ocean currents. The amount of gravitation force as well as Coriolis force (One to rotation of the earth) play key roles in determination of the direction of the current. The general trend from south to north in northern hemisphere and from north to south in southern hemisphere is due to high gravitational force at poles and low at equator. The whirling motion of the currents in both hemispheres is due to rotation of the earth. The currents move clockwise in northern hemisphere and anticlockwise in southern hemisphere.

General Pattern and Characteristics of Ocean Currents

The study and analysis of pattern of ocean currents in all the oceans indicate the following characteristics.

(i) The ocean currents move in circular motion forming a cell in all oceans. For example, North Equatorial Current, Gulf Stream Current, North Atlantic Drift and Canary Cold Current encircle the water in Atlantic ocean. The enclosed water is the zone of calm water which is known as 'Sargasso Sea'. Similarly, North Equatorial current, Kyu-Roshio Current, and California Current form a circle in Pacific ocean (Fig.26.2).

(ii) A counter equatorial current is the common phenomena, in Indian, Pacific and Atlantic oceans, between N and S Equatorial current. In Indian ocean summer season does not experience counter equatorial current exceptionally.

(iii) Warm currents move towards higher latitudes and cold current towards the low latitudes.

(iv) Cold water of lesser density moves into warmer region along the eastern coast of continents in higher latitudes.

(v) Warm currents and cold currents are found in lower latitudes along the eastern and western coast of continents respectively.

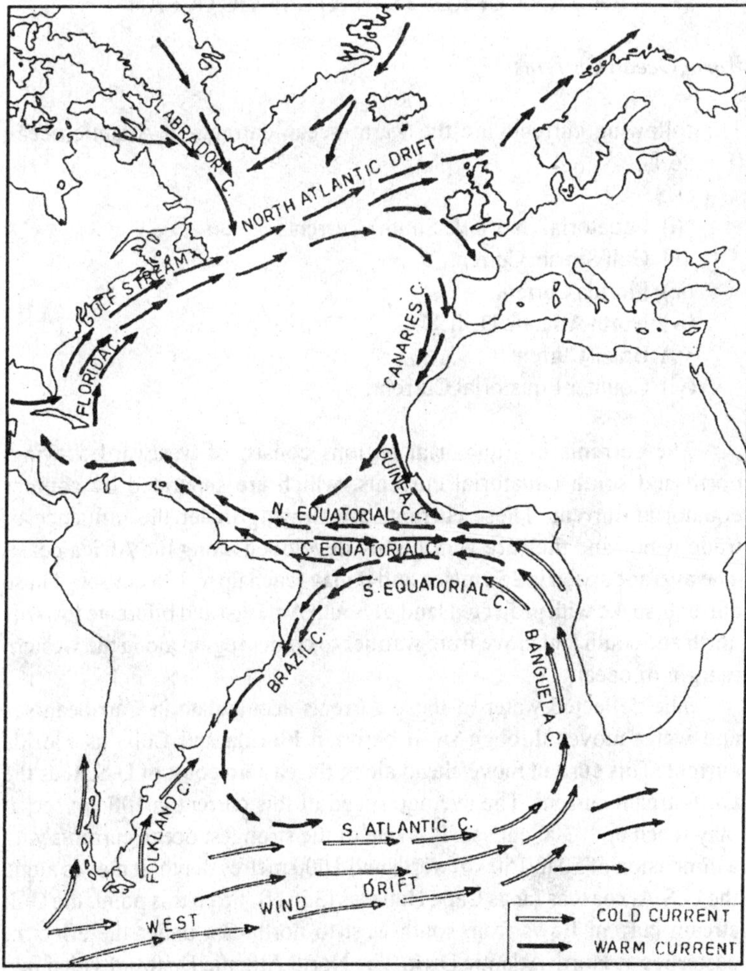

Fig. 26.1 : Currents of the Atlantic Ocean.

(vi) The high latitudes have warm currents along the eastern coast and cold currents along the western coast of continent.

Distribution of Ocean Currents

A large number of ocean currents of various categories are found in different oceans. A brief discussion of ocean currents in individual ocean is discussed below —

OCEAN CURRENTS IN ATLANTIC OCEAN

Warm Ocean Currents

The following currents are the warm ocean currents in Atlantic ocean (Fig. 26.1).

(i) Equatorial (North & South) Currents.
(ii) Gulf Stream Current.
(iii) Florida Current.
(iv) North Atlantic Drift.
(v) Brazil Current.
(vi) Counter Equatorial Current.

The currents in Equatorial regions consist of westward flowing, north and south equatorial currents, which are separated by counter equatorial current. These currents are flowing under the influence of trade winds and they are very highly developed along the Africa coast. The average speed is 25 cm/sec and it may reach up to 125 cm/sec. These currents strike with projected land of South America and bifurcate towards north and south and move from warmer to colder region along the western margin of ocean.

The deflected water of these currents accumulate in Carribean sea and water moves through strait between Florida and Cuba as Florida current. This current move ahead along the eastern coast of U.S.A. as the Gulf-stream current. The average speed of this current is 100 cm/sec. It may reach upto 300 cm/sec. It is one of the strongest ocean currents with a dimension of 50 to 150 km width and 1000 metres depth. It moves along the U.S.A. coast as far as Cape Hatteras (35° N). From this point, the Gulf stream current flows from south-west to north-east under the effect of easterlies as North Atlantic Drift. The North Atlantic Drift is divided into three branches-one moves eastwards to Great Britain, northwards to Arctic sea and southwards as Canary current. Thus these currents form a ring within which no perceptible current is found. It is area of seaweeds and called as "Sargasso Sea".

Moreover, the south equatorial current strikes with Cape Sao Rogue of South America and is divided into two branches — one of which moves southwards along the coast of Brazil as Brazilian current. Near 30°S it comes under the influence of Westerlies and moves from south-west to north-east to merge with cold wind drift, called as South Atlantic Drift or current. Along the western coast of Africa it moves northwards as Venzuela Cold current.

Due to flow of equatorial currents from east to west, the water of eastern side is displaced and accumulate along the western side. Thus a current from west to east flows between north and south equatorial current, as a supplementary to compensate the displaced water along the eastern margin of the ocean, is known as Counter Equatorial current (Fig.26.2).

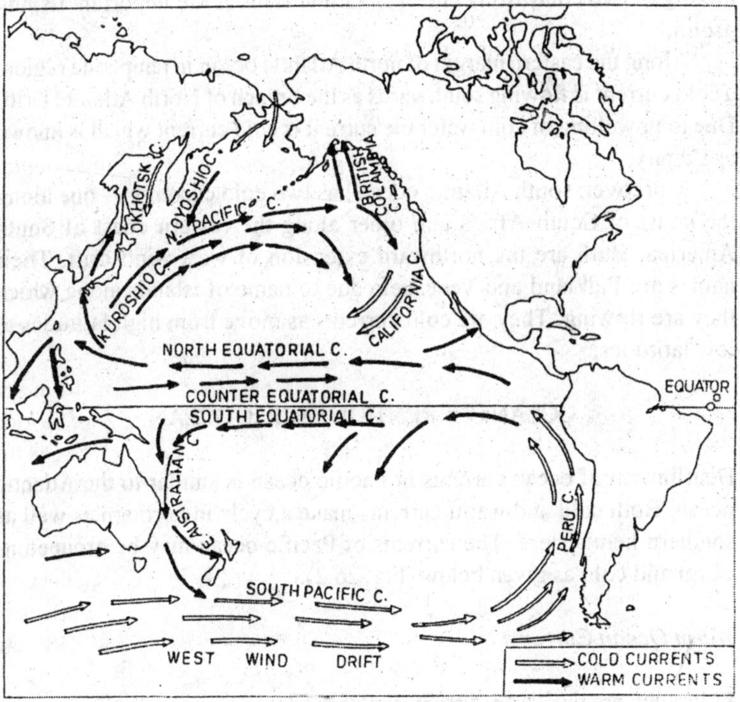

Fig. 26.2 : Currents of the Pacific Ocean.

Cold Ocean Currents

The following are the cold currents in Atlantic ocean.

(i) Labrodor current
(ii) Canary current

(iii) Venezuela current
(iv) Falkland current.

In the north Atlantic ocean, a ocean current which flows from Bay of Baffin through Davis strait southwards is known as Labrodor current. This current is originated due to prevailing north-east polar easterlies in sub-polar area. It moves along the Labrodor and Canadian coast and merges with Gulf Stream-east of 50° W near Newfoundland. The water of warm Gulf stream and cold Labrodor is differentiated by cold wall. Due to mixing of two contrasting currents, Grand Bank is the important fishing ground.

Along the eastern margin of north Atlantic ocean in temperate region, a cold current is flowing southwards as the branch of North Atlantic Drift. Due to upwelling of cold water the current is cold current which is known as Canary.

Moreover, south Atlantic ocean has two cold currents — one along the coast of South Africa and other along the eastern coast of South America. Both are the northward extension of west wind drift. Their names are Falkland and Venezuela due to name of islands along which they are flowing. They are cold currents as move from high latitudes to low latitudes.

OCEAN CURRENTS IN PACIFIC OCEAN

Distribution of ocean currents in Pacific ocean is similar to the Atlantic ocean. Both cold and warm currents make a cycle in northern as well as southern hemisphere. The currents of Pacific ocean may be grouped as warm and cold as given below (Fig. 26.2).

Warm Ocean Currents

Following are the warm currents in Pacific ocean.

(i) North Equatorial Current,
(ii) South Equatorial Current,
(iii) Counter Equatorial Current,
(iv) East Australian Current,
(v) Kuroshio Current, and
(vi) North Pacific Drift.

Cold Ocean Currents

Following are the important Cold currents in Pacific ocean

(i) Okhotosk current,
(ii) Oyashio current,
(iii) California current, and
(iv) Peru or Humbolt current.

Like the Atlantic ocean, in the north and south of equator westwards flowing ocean currents, under the influence of trade winds, are also common in Pacific ocean. Their volume is rather large as compared to these currents in Atlantic due to the absence of land obstruct. The north equatorial current strikes with Philippines and Taiwan island and move along the China coast from south to north. Under the influence of westerlies, it follow the direction from south-west to north-east. It is known as Kuroshio current near Japan and north Pacific drift off north-west coast of N. America continent. From the Berring strait cold water of lesser density flows from the polar area and mix with Okhotosk cold current and they form together Oyashio cold current off coast of Hokkaido. Along the north-east coast of Japan due to converging Oyashio cold current and Kuroshio warm current, world's biggest fishing ground has developed. The water of cold current sinks down below the warm water of North Pacific Drift and turn southwards along the California coast as California Cold current. Thus all these currents form a circular motion (cycle) in north Pacific.

South Equatorial current moves from east to west in South Trade wind area and it strikes with eastern coast of Australia and a southernly branch moves along the Queensland as East Australian warm ocean current. This current again turns towards New Zealand from Tasmania as South Pacific Drift. It merges with west wind drift and moves northwards along the Peru coast. Hence this current is known as Peru or Humbolt current. This area is rich in fishfood as Guano a valuable source of fertilizer. This cold current ultimately mix with south Equatorial current and consequently, a circle is formed in South Pacific. Moreover, a counter equatorial current from west to east is also found between north and south equatorial currents.

OCEAN CURRENTS IN INDIAN OCEAN

The ocean currents in Indian ocean especially northern hemisphere are

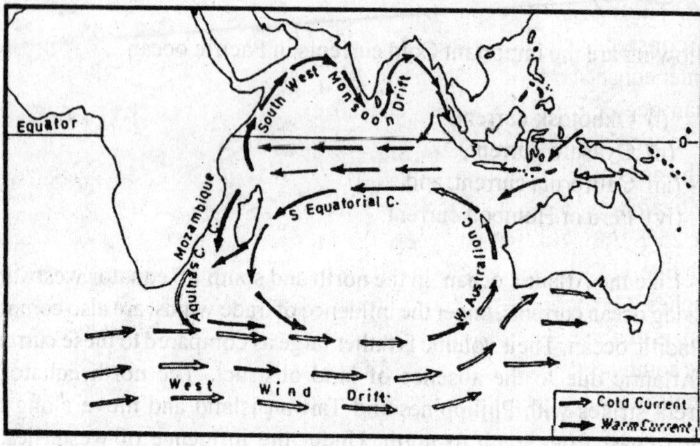

Fig. 26.3 (a) : Currents of the Indian Ocean (Summer).

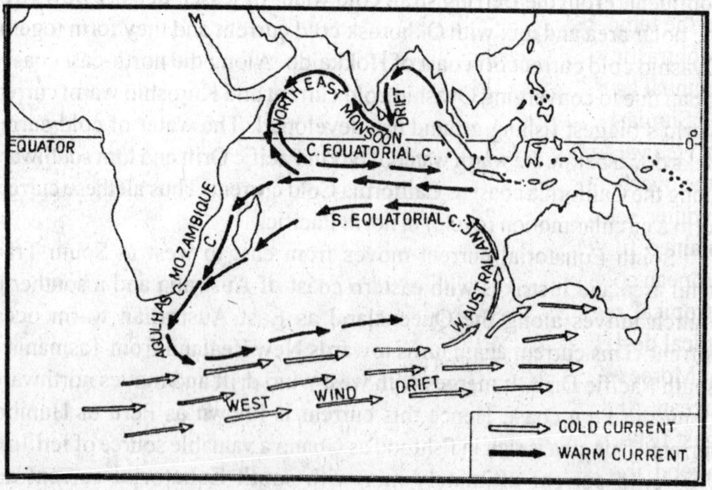

Fig. 26.3 (b) : Indian Ocean : Distribution of Ocean Currents (Winter).

well controlled by seasonal change of monsoon winds. North equational current flow, from east to west or from north-east to south-west along the coast under the influence of North-East Monsoon. It is known as north-east Monsoon Drift. Contrary to this, in the summer season it comes the effect of S.W. monsoon and direction becomes from south-west to north-east in Arabian sea and Bay of Bengal. Here, it is known as S.W. Monsoon Drift. The south-west wind drift merges with counter equatorial current and consequently in summer season, Indian ocean remains without counter equatorial current exceptionally (Fig.26.3a). Moreover, in southern Indian ocean, south equatorial current exists, and well developed from east to west direction. It strikes with eastern coast of Africa and moves southwards passing through Mozambique and Aguilhas which are called as Mozambique and Agulhas warm currents respectively. These currents, after 25°S, turn towards, east under westerlies and known as West Wind Drift like in P.O. and A.O. At 110°E, this drift is bifurcated into two branches. One branch goes northwards along the west coast of Australia which is described as West Australian cold current. Second branch moves at the south coast of Australia and mix in P.O. Thus in the Indian ocean all currents are warm except the west Australian cold current.

Effects of Ocean Currents

Ocean currents play very important role in determining the coastal climate and the temperature of ocean and sea surface water. The mixing of water of contrasting temperature, moderate the sea surface temperature. Range of annual as well as daily temperature is very low.

Similarly, the warm currents increase the temperature of cold region and *vice versa.* The cold currents in warm region retard the precipitation as the overlying air become stable. The North Atlantic ocean is open even in winter season for the Navigation than other region in same latitude remained frozen, due to presence of North Atlantic Drift. Hence West European countries show cool temperate climate. While the cold currents in tropical region have caused the desertification in coastal lands. All tropical deserts are located near cold currents.

Moreover, ocean currents have also contributed to economic values of ocean. At the place of convergence of warm and cold currents, world's largest fisheries have been developed. Cold currents, bring Plankton, the fish food, from the cold region in abundance. For example, the ocean near Newfoundland and Japan are having much fish due to convergence of two warm and cold currents.

TIDES

Beside the ocean currents, there is also vertical rise and fall of water after a fixed interval everyday under the attraction of the sun and the moon.

This periodical rise and fall of ocean water is called as Tide. It is very useful not only in navigation but also provides an enormous potential of energy in the coastal regions. Coastal geomorphology and the economic life are well affected by tidal currents. The study of tide is very complex as it has great variation in frequency, magnitudes and the height due to variation in depth of water, coastal geomorphology, nature of bottom topography and the variation in relative position of the sun and the moon with reference to earth.

Genesis of Ocean Tides

The tides are originated due to attraction of the sun and the moon exerted on the earth surface. The attraction between two bodies is inversely proportional to the square of distance between them. Thus the attraction of the sun to the earth is lesser (46% of the moon) than that of the moon. Though the sun is nearly 30 million times more massive than the moon. The former is 490 times farther away from the earth than the moon. Thus the gravitational pull of the sun towards the particles of water on the earth appears negligible as compared to the attraction of the moon.

Moreover, the tides twice occur at every point of water surface of ocean and sea, one of the place beneath the moon and other at opposite side of the earth. The tide on the former place is due to attraction of the moon and at latter place is attributed to the centrifugal force of the earth. Suppose (Fig.26.4) the earth is surrounded by water from all sides with uniform depth. Two forces, gravitational force of the moon and the centrifugal force on the earth, are working together and exert their effect in varying magnitude. The gravitational force decreases as the distance increases from the moon. Thus at the A point maximum, at c medium and b minimum attraction is experienced by the moon. The centrifugal force is the tendency of the earth due to

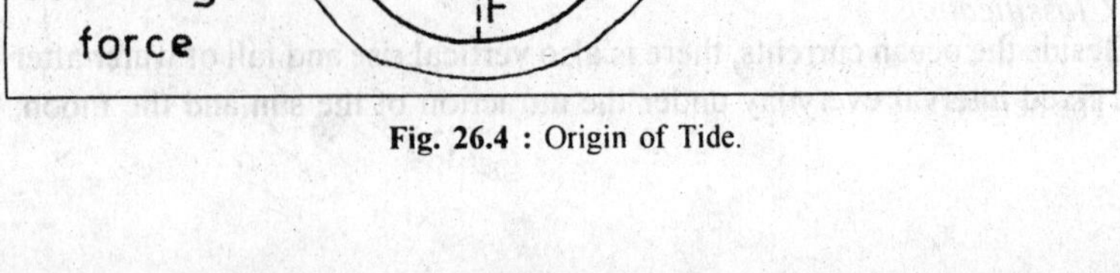

Fig. 26.4 : Origin of Tide.

rotation to throw objects outward from the centre and it is uniform at every point of the earth. Thus at A water is bulged out due to excess attraction over the centrifugal force. At the centre both forces are balanced and no effect is felt. At the b point the centrifugal force exceeds the gravitational force and the water is bulged. Thus two high tides occur at a time. Similarly, at the time of bulging out of water, the level of water falls down on the other side at E and F due to dragging of water towards the rise of water. Thus two low tides also occur at a time.

Type of Tides

Tides vary in their frequency, direction of movement and height from place to place and time to time. Hence they may grouped into various types on the various criteria.

On the Basis of Frequency of Occurrence in a Day

The tides may be classified into following categories on the basis of their frequency during 24 hours at same place.

(i) *Semi-Diurnal Tides :* They occur twice a day with uniform height. Such are common in Atlantic ocean along the USA coast, northern coast of Venezuela, Brazil, south eastern coast of Argentina, whole coast of Africa, Europe and British Isles.

(ii) *Mixed Tides :* Such tides occur two times in a lunar day with unequal magnitudes. They have the interval of 12 hrs 25 m. They are common in Pacific ocean along the coast of USA, along the Asian coast of Arabian sea of Indian oceans and along the eastern coast of S. America in Atlantic ocean.

(iii) *Diurnal Tides :* In this category only one high and one low tide occurs in a day. They are predominant in Gulf of Tonkin and Gulf of Mexico, south-west of Australia in Indian ocean, Gulf of Thailand, along northern coast of Java, Sumatra, southern coast of Borneo, the northern coast of New Guinea.

Classification of Tides on the Basis of Height of High Tides

The height of rise of water (high tide) varies with time and space. Narrowness of bays, shallowness of water and the mutual attraction of the sun and moon on the earth surface are important factors attributed to the variation in the height of tides. Out of these, the combined attraction of the sun and the moon and their locations with reference to earth are very much responsible for twice maximum rise and twice minimum rise of water in a lunar month. Thus the tides are classified into two groups as given below.

Spring Tides

These tides record the highest point of rise of water on the day when the sun, the earth and the moon lie in straight line. Thus the gravitational force of the sun and the moon act in combined form. Consequently, the sea water rises up to the maximum height. Such tides are known as Spring Tides. On the first moon and full moon such tides occur (Fig.26.5).

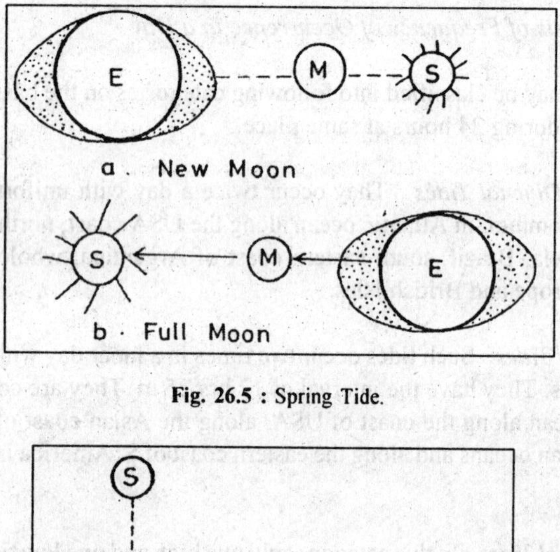

Fig. 26.5 : Spring Tide.

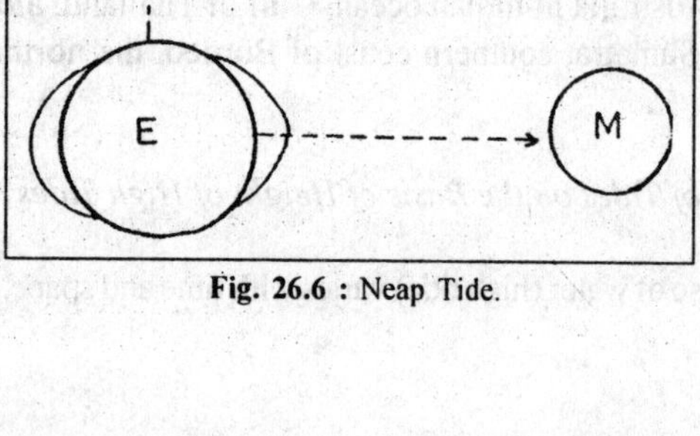

Fig. 26.6 : Neap Tide.

Neap Tides

On the days of first quarter and the third quarter of the lunar month, the sun and the moon lie at right angle with reference to earth (Fig.26.6). In this the attraction force of these bodies is divided and counter balanced and subsequently minimum pulling force is exerted on the surface (water) of the earth and consequently the water rises up to minimum height. Such tides are called as Neap Tides.

Duration of Tides

Tides, more or less, repeat themselves at each location on successive lunar days. The occurrence of tide does take at the exact time of appearing of the moon beneath on a not take particular place on the surface of the earth but it follows time lag up to few hours. This time is fixed for each location and is controlled by friction between ocean bottom and water. It is also due to blocking up of flow of water by landmass prevalent in the ocean or along the ocean.

Moreover, the duration of two successive high and low tides is 12 hours 25 minutes for each. As a rule the moon revolves the earth. Every point should experience tide after the internal of 12 hours (Tide occurs twice a day at every point). But in practice, due to variation in the velocity of the earth and the moon on their own axis, the duration of coming a meridian beneath the moon is 24 hours 50 minutes. This duration is also known as lunar day. 26.6 explains the reason why two successive after interval of 12 hr.25 m instead of 12 hours. The velocity per hour of earth is relatively slower as compared to that of moon. Suppose at X place on the earth was beneath M1 (Moon) one day before (Fig. 26.7). Next day the moon reaches at M2 and Y point on the earth comes beneath M2 after 24 hours. A point does not lie beneath M2 because Moon has moved little ahead due to rather high velocity. Thus the earth has to take few minutes (approximately 50 minutes) to bring the X point beneath the M2. Thus the same Maridian on the earth comes beneath the moon after 24 hours 50 minutes instead of 24 hours and consequently, the tides occur after every 12 hours 25 minutes.

Tidal Bore

When tidal waves rush to the coast, a large number of rivers get contact with the progressive waves. These tidal waves with the force, enter the

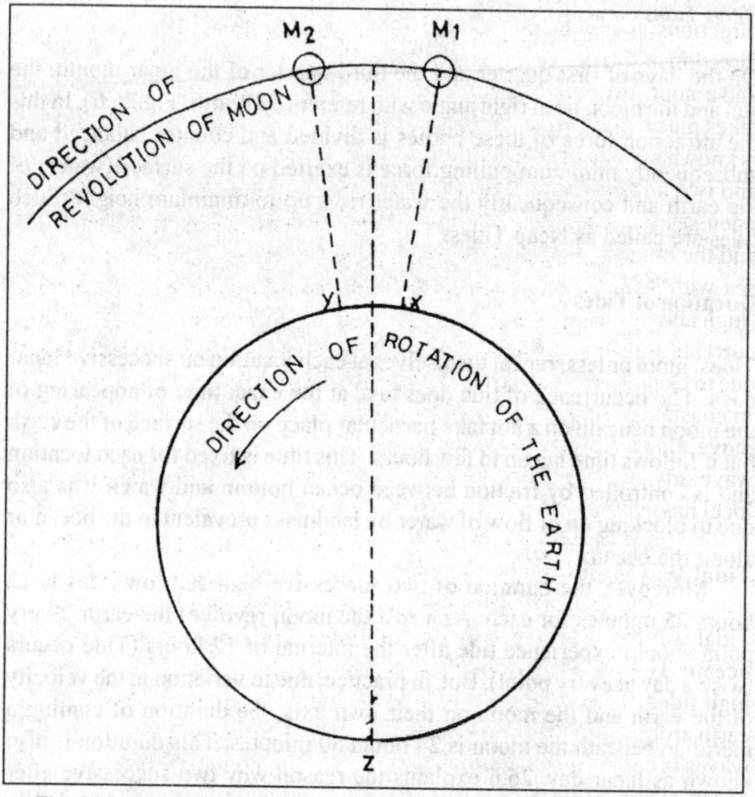

Fig. 26.7 : To show why the time intervals of consecutive Tide is 12h 26m.

estuaries or channels of rivers up slope as a wall of water. The high water of tidal waves appear as standing wall of water against the river flow the channel. This wall like wavy water is called as Tidal Bore. The speed, velocity and magnitude of the bore, depend upon slope, depth and flow of river and the amplitude, range and force of tidal waves involved in the process. Amazon, Hudson, Missori-Mississippi river etc. are examples of experiencing tidal bores. A spectacular bore occurs in Tsientang river in China, in Bay of Fundy and British and French rivers.

Tidal Currents

Occurrence of tides in sea or ocean is associated with this horizontal motion of water. As water rises and piles up, the water from different directions is drawn up and a horizontal flow of water takes place in the direction of tidal waves. Similarly, at the time of low tide water retreats back and a counter flow of water in opposite direction towards place of low tide takes place. Such type of horizontal flow of water due to tidal phenomena is known as Tidal Current. These currents occur between two hours before and two hours after of high tide and low tide. They are not perceptible in open sea because high water and low water are thousands of km. away and the range of tide is also very low in open sea. Contemporary to it the sea with shallow depth have high range of high and low tides and they (high and low tides) are found at relatively shorter distance. Consequently, currents are much perceptible. Open sea shows swirling motion of current due to Corlios effect and clockwise current is found in northern hemisphere. In coastal area they may be parallel tides occur to coast when tidal waves move along the coast, while they may be transverse, as when the tidal wave advance at right angle to the coast. These currents are helpful in local navigation.

Tidal Waves and Their Movement

Tidal wave is determined by the magnitude of high tide and low tide in ocean. The wavelength is the distance between either two successive high tides or two successive low tides. It is the half of circumference of earth around the latitudes of tides. The direction of the movement of these waves is controlled by general direction of rotation of the moon, depth of ocean, size and shape of landmass and ocean bottom relief. The occurrence of high tides in same time at various places is shown by a line on the map. This line is called as Co-tidal line. Distribution of co-tidal lines indicates that tidal waves follow the movement of the moon from east to west in open sea especially in Arctic ocean. But in Atlantic, Pacific and Indian ocean, the wave travel from south to north. In Atlantic ocean as they move northwards, convexity to north increases up due to narrowness of the ocean. Along the western coast, the tidal waves wash the coast from the west. The tidal waves move around the amphidromic points, counter clockwise in northern hemisphere and clockwise in southern hemisphere because of coriolis effect. Amphidromic point is placed where no tide occurs due to effect of moon.

27

OCEAN DEPOSITS

The floor of ocean and sea is not only beds of rocks but various kinds of loose materials cover the bottom like soils on the land surface. If we could drain the ocean, we would see vast areas covered mostly by very fine grained muddy sediments. The sediments include the unconsolidated materials of beaches, salts, marshes, bays and ocean deeps. These sediments are known as marine or ocean deposits in oceanography. They are derived from land area through the wearing , tearing and transportation by various geomorphological agents like rivers, glaciers, winds and sea waves. Marine life, both animal, and plants, also contributes to deposit by decomposition of their remains after death. Thus the marine deposits consist of both organic and inorganic materials.

Significance of Ocean Deposits

The study of ocean deposits is very useful for human being due to following facts :

(i) Ocean deposit is not only the accumulation of some fragments of rocks or shells of animals and plant but also the storehouse of minerals and energy resource like petroleum and gas. The exploitation of these resources is based on appropriate study of the deposits; and

(ii) The marine deposits tell us the geological history of origin of ocean bottom. The presence of fossils is the recorder of age of sediments and in turn the origin of oceans. Climatic and chemical history of sea water is also trapped in them.

CLASSIFICATION OF OCEAN DEPOSITS

Marine deposits have been classified in many ways on the basis of location, composition, texture and source of origin of sediments found in the ocean. On the basis of location, (i) littoral deposits between high and low tide line, (ii) shallow water deposits between low tide line and 100 fathom depth line, (iii) bytha deposits on the continental slope, and (iv) abyssal deposits over deep sea plain and deeps have been recognised. But the most important and convenient classification of marine deposits is decided on the basis of chemical composition of sediments (Fig 27.1). Both organic and inorganic deposits are common. Their concentration of occurrence vary with the depth of ocean but absolute demarcation is difficult. Thus the marine deposits may be classified as (i) Terrigenous deposits, and (ii) Pelagic deposits.

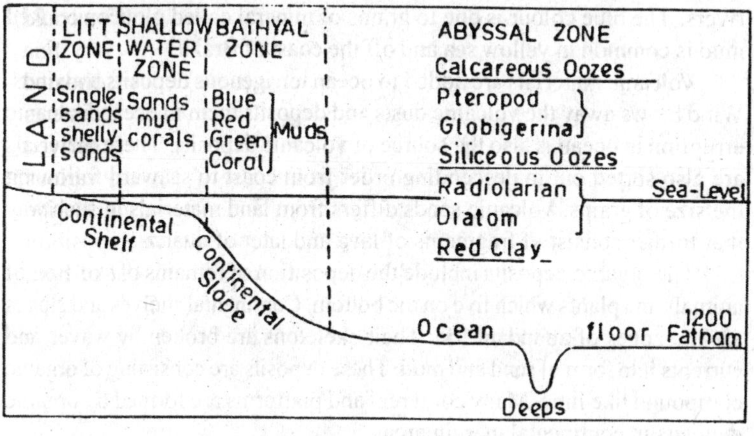

Fig. 27.1 : Generalised Profile of the Ocean Floor showing the General Distribution of Marine Deposits.

Terrigenous Deposits

The continental margin zone of ocean bottom is covered by solid unconsolidated loose particles which have been derived mainly from the land area by agents of erosion, i.e. rivers, winds, glaciers and sea waves. Remains of skeleton and shells of marine life living on the ocean bottom are also deposited on the bottom. Such deposits, which are spreading on the continental shelves, slopes, are described as terrigenous deposits. It includes following groups of sediments.

(i) Materials derived from wear and tear of land.

(ii) The remain of plants and animals which live on the bottom of the sea.

(iii) Volcanic materials.

These materials derived from land constituted the larger proportion of terrigenous deposits. The eroded materials by wear and tear of rocks by river, glacier and winds are of various sizes. They are found on the sea bed in sorted form. Bigger particles like pebbles are along the outer margin and finer like sand and mud are common inward side of edge of continental shelf or slope. Some fines may also cross the continental margin occasionally. These materials consist of mainly quartz. The mud (finest particles) is of various colours and described as red mud, green mud and blue mud. Blue mud is most common and widely spread type. Green mud is found in the area of continental slope of rocky coasts of absence of rivers. The blue colour is due to grains of mineral called glouconite. Red mud is common in yellow sea and off the coast of Brazil.

Volcanic materials are added to ocean terrigenous deposits by winds. Wind blows away the volcanic dusts and deposits them in ocean. Volcanic erruption in ocean is also the source of volcanic deposits. These materials are also sorted out in descending order from coast to seaward following the size of grains. Volcanic sands differs from land materials in such way that former consist of fragments of lava and later of quartz.

The organic deposits include the deposition of remains of skeleton of animals and plants which live on the bottom. Continental shelves and slopes are the zones of abundant life. Their skeletons are broken by waves and currents into form of sand and mud. These deposits are consisting of organic compound like lime. Many coral reef and platforms are formed by organic deposits in continental margin area.

Pelagic Deposits

Pelagic term is used to denote the deep sea water marine environment. In this zone terrigenous materials are rarely found due to being farther from the coast. Sediments deposited are of mainly marine origin. Thus the marine deposit, which have been derived from marine life floating on the surface and some finest volcanic ash derived from oceanic valcanoes are known as pelagic deposits.

The pelagic deposits include following groups of sediments based on their origin and chemical composition.

(i) *Organic Deposits*

(a) Calcareous ooze

(b) Siliceous ooze

(ii) *Inorganic Deposits*

(a) Red clay

(b) Chemical precipitates

Organic Deposits

The animals and plants float on the surface of ocean water and their shell and skeletons are deposited on the bottom of deep sea in the form of very fine and loose form called as Ooze. These oozes vary in chemical composition depending upon the biological organism whether they are having calcium or silica dominated shells or skeletons. Thus the pelagic organic deposits may be grouped as (i) Calcareous oozes, and (ii) Siliceous oozes.

Calcareous Oozes

Calcareous ooze refers the sediments formed by disintegration and decomposition of calcareous marine animal and plants. Petropods and globegerina are the important life of calcium content. This ooze does not occur below 5000 mt depth because the solubility of calcium under high pressure and low temperature is very high. Thus calcareous oozes can be put into —

(i) Petropod ooze (ii) Globegerina ooze

Petropod oozes are derived from the cells of multicellular animal known as Petropods. They are found in warm shallow water but away from the coast especially in Atlantic oceans. Ridges and plateaus rising from deep sea water are the suitable places of such ooze. Warm and minimum annual range of temperature are ideal conditions. Mid-Atlantic ridge in tropical region has the highest concentration of petropod ooze.

Globegerina ooze is derived from the single celled Foraminifera which is rich in calcium carbonate. At many places limestones have been formed by this ooze. It spreads over in Indian ocean, Atlantic ocean and South Pacific ocean especially in warm and temperate region. Exceptionally this ooze is also restricted by 5000 mt. depth. At some places it extends up to shallow water zone where terrigeneous deposit does not mask the ooze.

Siliceous Oozes

Diatom and rediolaria are two important marine organisms rich in silica in their shell and skeleton. There are two types of siliceous ooze : (i) Diatom ooze, and (ii) Radiolarian ooze. These oozes do not dissolve easily. Thus they are found at rather greater depth than the calcareous ooze.

(i) Diatom ooze: Diatom is microscopic marine plant which is widely spread in the ocean. Many herbivores depend upon them. It is also known as Sea grass. Diatoms are siliceous and best flourish in cold deep water. It is found in broader belt out of terrigenous deposits of Antarctica continent and narrower strips in northern Pacific.

Radiolarian Siliceous Ooze : Radiolarian is the minute marine animal rich in silica. They are commonly found in warm and deep water. Equatorial Pacific and Indian oceans are important area of such deposits. They are absent in Atlantic ocean. Deeper parts of ocean like deep sea plain and trenches are covered by Radiolarian ooze.

Inorganic Deposits

Besides the organic materials, inorganic materials like volcanic dust, dissolved calcareous matters, various precipitates falling from atmosphere and meteorite dust are also deposited on the ocean bottom of deep water. To maintain distinguish between organic and inorganic is very difficult. There are two groups of inorganic marine deposits (Fig.27.2).

Red Clay

Red clay is most widely spread pelagic deposit. It composed of fine decomposed volcanic materials, finest particles of terrigenous clay, meteorite dusts and inorganic precipitates. They may be also unrecognised fine biological sediments. It consists of mainly hydrate of aluminium, oxide of iron. It is thought to be product of disintegration of land minerals. Red clay contain calcium (6.7%), siliceous organism (2.39%), minerals (5.56%) and fine washing or hydrated silicate of alumina (85.35%). Radioactive elements are also found in red clay. It occurs generally in deeper part of all oceans and in areas of low biological activity. It accounts for 36.1 per cent of total marine deposits. 15.9 million square km area in Atlantic ocean, 143 million km^2 in Pacific ocean and 16 million km^2 in Indian ocean is covered

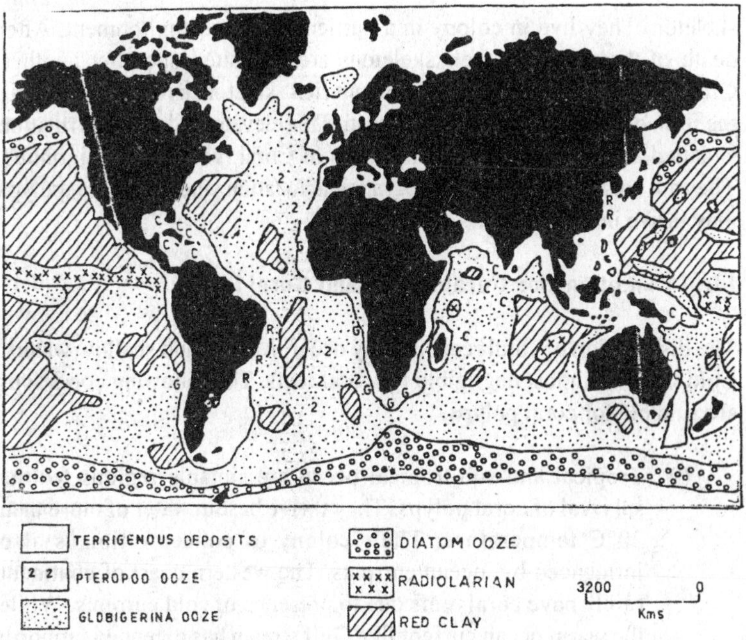

Fig. 27.2 : The World Distribution of Marine Deposits based on a map by Sir John Murray, The Ocean (London 1928).

by red clay.

Chemical Precipitate

Besides the red clay, the ocean also contains the materials that are formed from chemical solution in water and deposited in the form of chemical precipitates on the floor of the sea. Phospharite nodules, manganese nodules are important inorganic materials in the ocean. Phospharites are found on the continental shelves and slope. They are especially abundant off the coast of California, Southern U.S., South Africa and Australia. Manganese nodules are found in very small fraction of ocean deposits but they are economically very significant in future. Copper, cobalt nickels and various other valuable minerals are included.

CORAL REEF

Coral polyps is a minute marine organism, consisting of calcium in its skeleton. They live in colony in a particular marine environment. After death of these polyps, their skeletons are deposited one after another. Gradually, due to deposition of calcium rich skeleton, a rocky structure in the form of ridge is formed in the ocean. This elevated ridge like formation is called as Coral reef. The distribution of Coral reefs is not ubiquitous. They are confined in an ideal geographical location where the coral polyps may thrive in abundance.

Ideal Conditions for Coral Polyps and Coral Reefs

The distribution of coral reefs usually occur in warm and shallow waters. Cold water is almost negligible for coral reefs. The following conditions are suitable for coral polyps.

(i) Tropical and sub-tropical ocean are suitable regions for the survival of coral polyps. They thrive best in water of more than 20°C temperature. Their colony of concentration is also influenced by ocean currents. The western coast of continent rarely have coral reefs due to presence of cold currents. While the warm ocean current like Gulf stream has extended far north the concentration of coral polyps along the eastern coast of U.S.A;

(ii) Shallow water is another suitability of abode of coral polyps. 30 fathoms (60 metres) is the ideal depth above which coral polyps best survive. Thus the coral reefs are usually found on the continent shelves area. Some coral reefs are located in deep sea water about which many explanations are given; and

(iii) Muddy and sedimented water is harmful for coral polyps. Clean water with high salinity is ideal condition for growth and development of coral polyps. Along the mouth of rivers, their number is low or absent. This is the reason why the coral reef grows seaward sides instead of landward side. Waves and currents along seaward side continuously removes the mud or sediments. Fresh water is also not suitable.

Kinds of Coral Reefs

On the basis of location and structure of reefs they may be divided into three categories.

Fringing Reef

Along the coast of island or continent, initially, a coral reef is formed. It is attached to the shoreline or sometimes separated by a narrow lagoon. Thus a longitudinal ridge like coral formation along the coast of island or mainland or around the volcanic hills is known as Fringing Reef (Fig.27.3).

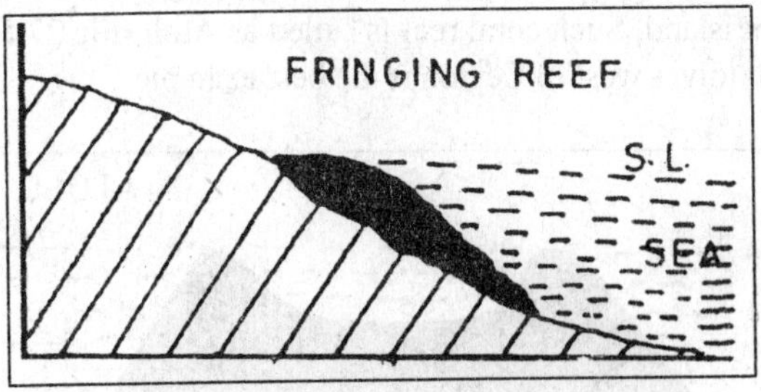

Fig. 27.3 : Fringing Reef.

Barrier Reefs

They are formed after the growth of fringing reefs. All the reefs grow outward (sea side) and upward side as they provide clean saline water and foodlike plankton which are required by Coral polyps. Thus fringing reef gradually moves away from the coast and broader legoon is formed. Such reef is known as Barrier (Fig.27.4). It is broken at some places to have the connection between open sea and legoon. Great barrier reef along the Queensland coast in Australia continent is the most excellent example of barrier reef.

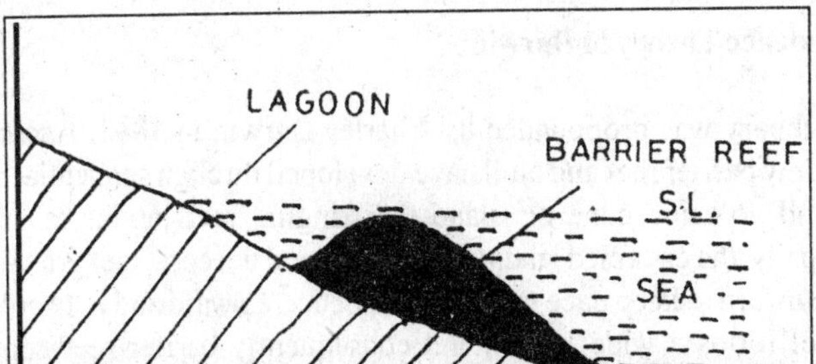

Fig. 27.4 : Barrier Reef.

Atoll

Atoll is a ring like coral reef formed around island initially as fringing reef. With the passage of the time it grows away from the coast and barrier reef is formed. At the last the encircled island disappears from the scene as it comes under the water either by subsidence and rising of sea level. Thus the barrier or fringing reefs are left as a ring like ridge enclosing a lagoon without an island. Such coral reef is called as Atoll (Fig.27.5). Suvadiva atoll in Maldives west of Ceylon is the best example.

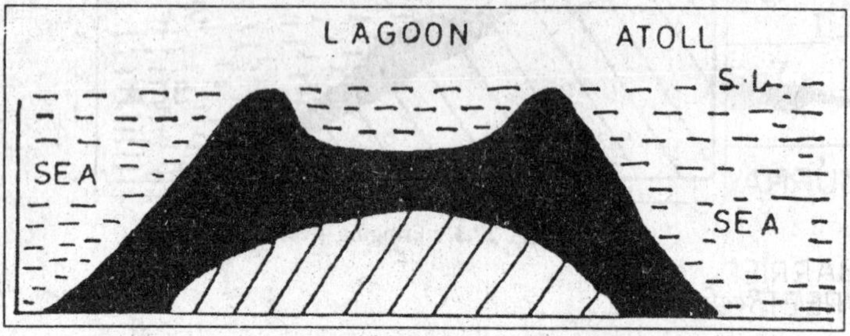

Fig. 27.5 : Atoll.

Origin of Coral Reefs

The distribution of coral reefs in shallow as well as in deep water has puzzled the oceanographers. The ideal conditions of coral polyps are warm, clean, saline and shallow water but occurrence of some barrier reefs and atolls in deeper part of ocean in unfavourable condition led the various views on the origin of coral reefs (Fig. 27.6). The various views and theories of genesis and development of reef are discussed below.

Subsidence Theory of Darwin

This theory was propounded by Charles Darwin in 1842. According to this view barrier reef and atoll have developed through sequential process. Around volcanic cone or island the fringing reef begins to originate. Gradually, the encircled landmass subsides and the coral reef grows upward and outward to keep pace with the subsidence, upward and outward growth of reef forms a wide lagoon and consequently barrier reef comes into being. At last, the island or volcanic cone subsides under water and Atoll is formed. Thus the barrier reefs and atolls have been evolved through

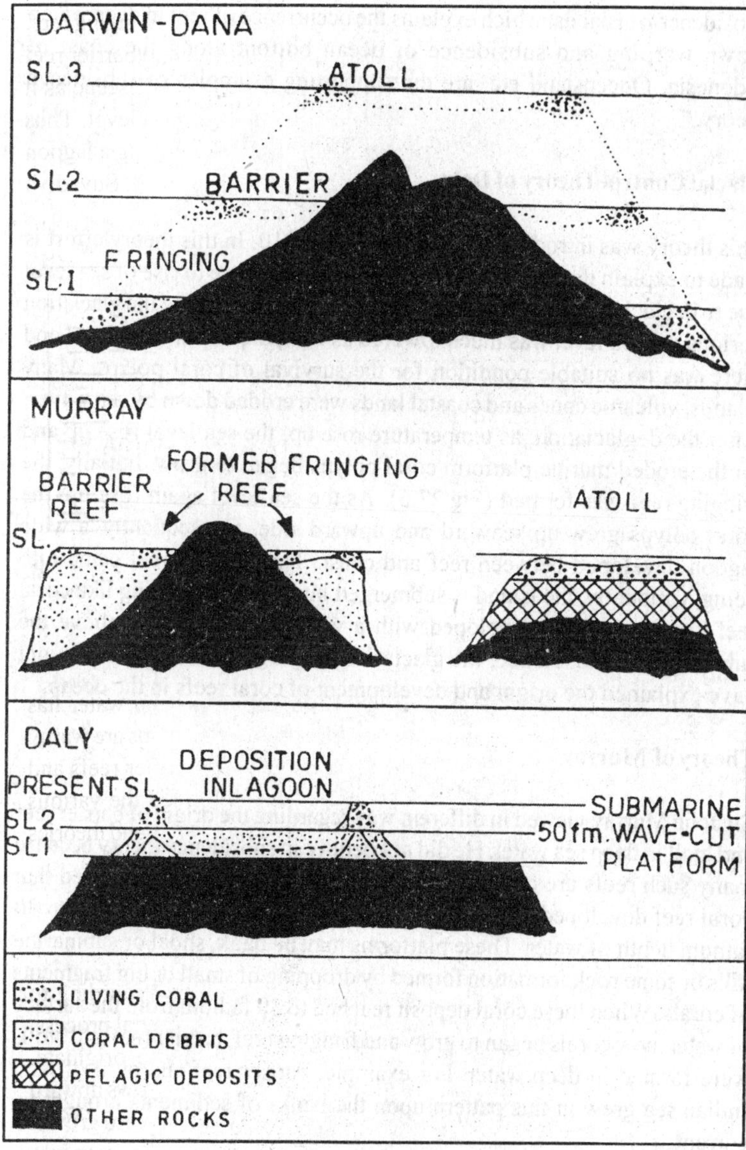

Fig. 27.6 : Theories of Atoll Formation.

subsidence of islands, which explains the occurrence of reef in deep water. Down warping and subsidence of ocean bottom along the coast of Indonesia, Queensland etc. are the supporting examples of subsidence theory.

Glacial Control Theory of Daly

This theory was introduced by R.A. Daly in 1910. In this theory effort is made to explain the origin of coral reefs with reference to rise of sea level due to deglaciation process in the past. It explains that during the glaciation periods the sea level was much lowered as compared to present level and there was no suitable condition for the survival of coral polyps. Many islands, volcanic cones and coastal lands were eroded down by sea waves. After the deglaciation, as temperature rose up, the sea level rose up and on the eroded marine platform coral polyps began to grow. Initially, the fringing reef was formed (Fig.27.6). As the sea level again rose up, the coral polyps grew up seaward and upward side. Consequently, a wide lagoon developed between reef and coast. Thus barrier reef came into being. Gradually, the island is submerged into water and a ring like coral reef known as Atoll developed with a wide circular lagoon above the submerged island. Hence, the glaciation and deglaciation process might have explained the origin and development of coral reefs in the ocean.

Theory of Murray

Sir John Murray viewed in different way regarding the origin of barrier reef and atoll in deep sea water. He did not believe in subsidence theory because many such reefs are common in non-subsidence region. He viewed that coral reef developed on the marine platforms which have elevated to 30 fathom depth of water. These platforms may be bank, shoal or submarine hills or some rock formation formed by dropping of small or big fragments of corals. When these coral deposit reached to 30 fathom from the surface of water, new corals began to grow and fringing reef, barrier reef and atolls were formed in deep water. For example, barrier reef and atoll of west Indian sea grew in this pattern upon the banks of sediments brought by current.

GLOSSARY

Ablation : The process of carrying away or removing now generally restricted to the loss of ice by glacier, mainly by surface melting and evaporation.

Abrasion : The wearing away of part of earth's surface by action of winds river, glacier etc., utilizing moving debris, i.e. sand, pebbles and gravels.

Absolute Humidity : The presence of moisture in a given volume of air is called as absolute humidity. It is usually expressed as grams per cubic metre of air.

Absolute instability : Air that has laps rate greater than dry adiabatic rate.

Absolute stability : Air that has laps rate less than wet adiabatic rate.

Abyssal Deposit : The pelagic deposits on the abyssal plains is described as abyssal deposits.

Abyssal Hill : An isolated hill on abyssal plain with more than 1000 mts. Height from the ocean bottom.

Abyssal Plain : It is the deep sea plain which extends from outer edge of continental slope of ocean bottom. It has uniform surface with greater depth covered by Pelagic deposits.

Adiabatic Temperature Change : Cooling or warming of air caused, when air is allowed to expand or compressed, not because heat is added or subtracted.

Advection Fog : A fog which is formed when warm air moves over cold surface and is thereby cooled to dew point. It is very common near Grand Bank (Newfoundland) due to moving of warm moist air of Gulf Stream over the cold air of Labrodor current.

Aeolian : Relating to or caused by winds. The erosional and depositional work done by blowing winds are known as Aeolian erosion.

Aggradation : The worm materials, produced by agents of denudation are deposited at certain suitable places. This process of deposition of sediments is called as Aggradation.

Air mass : A large body of air, usually, 1600 kms or more across, that is characterised by homogenous physical properties, i.e. temperature and humidity at any given altitude.

Air Pressure : The pressure exerted by the weight of a column of a air above a given point.

Albedo : The reflection of shortwave radiation from atmosphere and the earth to the space is described as Albedo.

Aleution Low : A large cell of low pressure centred over the Aleution islands of North Pacific during the winter.

Alluvial Fan or Cone : The deposit of sediments laid down, by swift flowing stream as it enters a plain or open valley along the foothill forming a conical shape. This pebble studded conical shaped deposit is known as Alluvial Cone or Fan.

Alluvial Plain : A level tract bordering a river on which alluvium is deposited. It includes flood plain, delta and alluvial Fan.

Alpine Glaciers : The glacier which flows in pre-existing river valley in mountainous region is known as Alpine or valley glacier. It is usually found at higher elevation of mountains in tropical region and lower elevation in cold region.

Altocumulus Cloud : A type of medium cloud which is generally consists of masses of small, relatively thin, globular patches, sometimes so close together that their edges join.

Altostratus Cloud: A type of medium cloud in the form of a continuous sheet or veil, which is sometimes thin, sometimes so thick as completely obscure the sun or moon.

Anemometer : An instrument used for measuring the velocity and direction of the winds.

Aneroid Barometer : An instrument used for measuring air pressure that, consists of evacuated metal chambers that are very sensitive to the variation in the air pressure.

Antarctic Circle : The parallel or line of latitude 66°32'S, commonly assumed to be 66° 30'S . Owing to inclination of earth's axis, the sun does not set here one day (22 December) of the year. Similarly, one day (21 June) sun does not rise too along this line.

Antecedent Drainage : A river system in which the original direction of flow is maintained across the land where upliftment has taken place.

Anticline : The crest of a fold is known as Anticline.

Anticyclone : A whirling motion of winds in which high pressure is located at centre. They have circular or oval isobars and winds blow outwards, clockwise in northern hemisphere and anticlockwise in southern hemisphere.

Antipodes : The places situated on the opposite side of the earth, so that a straightline can be drawn through the centre. Generally lands antipode to oceans or seas with few exception.

Aphelion : The position of the earth on its orbit when the maximum distance 152000000 km. is found between the earth and the sun.

Aquifer : A porous and permeable layer of rocks which accumulate underground water and allow water to percolate down.

Arctic Circle : The parallel or line of latitude at 66°32'N, commonly assumed to be 66°30'N. Owing to the inclination of earth's axis, the sun does not set on one day (21st June) of the year and the sun does not rise on one day (22nd December) of the year.

Arenaceous : A term applied to sedimentary rocks which contain sand, e.g. sand stones.

Argillaceous : A term applied to sedimentary rocks containing a substantial proportion of clay minerals, i.e. shale. They are commonly formed into water.

Artesian well : A type of well in which water is coming out continuously from underground reservoir under hydraulic pressure of water. For the construction of this well initially an outlet is dug, by man in overlying impervious rock, upto the water table. Such wells are commonly found in the area where rock beds are in synclinal or basin form and an aquifer is bounded between non-permeable rocks.

Asteroid : Small planets are, between orbit of Mars and Jupiter, revolving around the sun.

Atoll : A type of Coral reef which is in circular form enclosing a broad Lagoon without any island within it.

Annual March of Temperature : The annual cycle of air temperature changes. Usually the lowest monthly mean occurs a month or two months after the winter solstice while the highest monthly mean temperature commonly occurs a month or two months after the summer solstice. January and July generally represent the extreme months.

Annual Range of Temperature : The difference between the mean monthly temperature of the warmest and the coldest month of the year.

Bar : A ridge like formation in the sea along the shoreline formed by deposition of sediments produced by sea wave erosion and deposition.

Barchan : A crescent shaped sand dune with the horns on leeward side and convex slope windward side. It is formed when winds are steady for longer time in the same direction with low velocity.

Burrier Reef : A long narrow ridge with wider lagoon, along the coast, formed by deposition of skeletons of dead coral polyps especially in warm and shallow oceans. They develop seawards and upwards gradually. Great Barrier Reef off Australian coast in Pacific is an excellent example.

Base-Level : The lowest level upto which a river can cut its valley vertically. It is determined by the water level of lake or sea where the river discharges its water.

Barometer : An instrument to measure the atmospheric pressure.

Batholith : A dome shaped discordant mass of igneous rock formed by intrusion of magma. It is usually very massive, often extending over hundreds of km in width. It is seated at greater depth attached with, sometimes, magma chamber.

Basket of Eggs : A glaciated topography with the presence of a large number of drumlines together is described as basket of eggs.

Beach : A strip of land or terrace bordering the sea, sloping gently seaward from the land. It is area which lies between high and low tide water level. It may consist of shingle and sand or mud.

Beaufort Scale : The series of number devised by Admiral Beaufort at the beginning of 19th century to differentiate approximately between various wind strengths. The scale varies from 0 to 12 indicating calm or motionless air and hurricane respectively.

Beheading of River : When two rivers flow on either side of watershed, they make headward erosion at differential rates. The stronger river cuts rather deep valley than the weaker one. After cutting across the watershed, these rivers meet together at certain point. The stronger and deeper river draws the water of weaker river lying relatively at higher level. Thus the channel or flow of water of latter stream is captured by former one. Consequently, the weaker is delinked from its down stream valley. This process of capturing of flow of water of weaker stream by stronger one is described as beheading of the river.

Bergschrund : The gap or crevasses left round the upper rim of glacier ice within a cirque as the ice moves downwards, thus separating the latter from the rock wall beyond. It usually provides a difficult obstacle to any attempt to climb from snowfield to the mountain peak. It plays very important role in the formation of cirques.

Blind Valley : In Karst region, a valley either dry or still containing a stream which ends suddenly in caverns, cave or sink holes.

Blizzard : A storm of powdery snow, sometimes, with small ice crystals, is found in cold region of Canada and North U.S.A. It is a cold wave, often, disturbs the calm of winter anticyclone, coming with northerly wind of gale strength in rear of an eastward moving depression. It is very dangerous in Prairies (U.S.A.).

Block Mountain : It is an uplifted block of rocks between two or more fault lines. Black Forest in Germany is the example of Block mountain.

Blow Hole : A hole near sea shore which has been formed in the roof of cave and through which air and possibly water is forced by rising tide. A sound like whistle is produced from this hole.

Bore : A high tidal wave experienced in the narrow streams estuary, advancing upstream like a wall of water. It is produced by sudden retardation of normal tidal wave when it reaches shallow water and meets the river current.

Braided River : A network of small interlaced streams derived from single stream. It was formed after the original stream had deposited sediments, dividing and forming new channels and then these channels had joined up many times.

Breaker : When the sea wave reaches in shallow waterzone, its wavelength is reduced and height increased up. The crest of the wave breaks up as the trough touches with ocean bottom due to shallow depth. After breaking the crest, a current of water moves towards the coast. This landward moving current is known as *breaker.*

Calorie : The amount of heat required to raise the temperature of 1 gram of water by 1°C.

Canyon : It is deep and narrow river valley formed in the mountainous region by youth river. Such type of valley or depression is also common in the seas and oceans. They are described as submarine canyons.

Catchment Basin : The area from which a given drainage system receive the water.

Caverns : Underground caves or depression with intact roof are very common in Karst to topographic region. They are usually known as caverns. Sometimes, caverns are interlinked from each other by swallows.

Centrifugal Force : It is the force on the rotating surface due to which a particle tends to move outwards from centre to periphery.

Chinook : The name applied to a foehn wind in the Rocky Mountains.

Circle of Illumination : The line (great circle) separating day light from darkness on the earth.

Cirrus Cloud: A type of high cloud which is thin delicate ice crystal cloud often appearing as veil-like patches or thin whispy fibres.

Cirques : It is a armed chair like depression, along the slope of mountain, formed by erosion of valley glaciers.

Cliff : A high and extremely steep rock face approaching vertical either inland or along a coastline.

Climate : A description of aggregate weather conditions of longer time and larger area.

Clouds : A form of condensation, at higher altitude, consisting of suspended water droplets and ice crystals.

Cloud of Vertical Displacement : Cloud that have their bases in low height but extend upwards in middle and high cloud regions.

Cold Front : The discontinuity, at the forward edge of advancing cold air mass, along which warm air mass is uplifted.

Condensation : The process of conversion of water vapours into water droplets or ice crystals at the dew point is called as Condensation. Dew, fog, mist and clouds are different forms of condensation.

Conditional in Stability : Moist air with a Laps rate between dry and wet adiabatic rate.

Conduction : The transfer of heat from one body to another through movement of heat from one particle to another. This process continued until the both bodies get uniform temperature.

Continental Air mass : The air mass which originates over land area which is normally dry.

Continental shelf : A shallow depth zone, less than 100 fathoms deep encircles coasts in all oceans is described as continental shelf. It's width varies from coast to coast in different oceans. Gradient of slope is very gentle. Economically this zone is very significant.

Contours : The imaginary line drawn on map connecting the places of equal height from sea level.

Coriolis Force : The deflective force of the earth's rotation on all free moving objects including the atmosphere and the ocean. Deflection is to right in northern hemisphere and to the left in southern hemisphere.

Corrasion : A type of process of wearing and tearing of rocks in which transported debris erodes the rocks in the way by friction or strikes and simultaneously, they (debris), themselves, are also torn into finer sizes.

Corrosion : It is a form of erosion by river in which some soluble rock dissolves in the water and is removed from their original places in the form of solution.

Convection : A method of transmission of energy in which heat is transferred through the movement of particles. Such process is common in equatorial region where air is heated and moves upwards. Cold air moves from different sides to take the place of displaced warm air.

Cumulus Cloud: A type of cloud with cauliflower appearance and flat base.

Cumulus stage : The initial stage of thunderstorm development in which formation of cumulo-nimbus cloud dominates by strong updrift of air.

Current (Ocean) : The flow of huge amount of water in a definite directions in all oceans is described as ocean currents. They may be divided into warm and cold currents.

Cyclone : A strom with whirling motion in which low pressure exits in the centre. Winds blow from periphery to centre. It moves anti-clockwise in northern hemisphere and clockwise in southern hemisphere.

Daily Range of Temperature : The difference between maximum and minimum temperature of a day.

Daily Mean : The mean temperature for a day that is determined by averaging the 24 hourly reading or more commonly, by averaging the maximum and minimum temperature for a day.

Degasing: During the process of cooling of molten state of the earth, different kinds of gases were released and formed an atmosphere. The process of escaping the gases from hot gaseous surface of the earth is known as Degasing.

Dendritic Pattern : A drainage pattern with tree like appearance is called as dendritic pattern. Upstream is joined by a number of tributaries like roots of trees. Middle course of river look like a stem without any tributary but the lower valley again birfurcated into various distributaries. Thus whole network of river's tributaries and distributaries look like a tree with roots, stem and branches.

Dew Point : The temperature at which moist air is saturated and condensation of water vapours takes place.

Diatom Ooze : It is siliceous type ocean deposit of marine vegetable origin. It consists of siliceous parts of microscopic diatom plants or algae which usually inhabit the cold water of the world. There is broad belt of diatom ooze extending round the earth in the southern ocean.

Distributary : A branch or outlet which leaves the main stream and does not rejoin it, carrying its water to sea or lake is called distributary.

Discontinuity : When two airmass of contrasting temperature converge together, a line of separation is formed between them above which warm lighter air is displaced off by cold air. This line of separation is termed as discontinuity or front.

Doldrum : The equatorial belt of calms or light variable wind lying between two trade winds.

Doline : A closed hollow in karst region formed by solution action of the limestone near the surface. It is usually formed by collescence of a number of sink holes together in a given area.

Drumlin : An elongated hill or ridge of boulder clay, usually oval and shaped like half eggs. It is formed by glacial deposits and has gentle slope leeward side and steep slope on the front side.

Dry Adiabatic : The rate of cooling and heating of ascending or descending air due to expansion or contraction respectively. The rate of dry adiabatic rate is 10°C/km.

Dry Climate : A climate in which evapo-transpiration exceeds the precipitation is described as dry climate.

Dust Bowl : A depression like formation in desert area. It is formed by wind action in which sands are blown away by deflation process and gradually a depression is formed. Sometimes, it reaches to water table and water comes out. Consequently, an oasis came into being. This wind formed depression is known as dust bowl.

Earthquake : The shaking of earth due to any reason is described as earthquake. Earthquake usually occurrs due to tectonic movement like faulting, volcanism and plate tectonic movement. Weak zones of the earth are very much prone to the earthquakes.

Earth Rotation : The movement of the earth on its axis is described as earth's rotation. It completes one rotation in 24 hours.

Eolian : The action which takes place by wind is termed as eolian. The geomorphological features originated by wind erosion, especially in desert region, described as eolian topography. Peclastal rocks, yardang, zugen, sand dune etc. are examples of this topography.

Epi Centre : The place on the earth's surface where first shock of earthquakes is felt. It lies vertically above the focus.

Equatorial Low : A quasi continuous belt of low pressure lying near the equator and between the subtropical highs. It is also called as doldrum.

Equatorial (E) Airmass : A warm hot airmass which originates in tropics near equator.

Equinox : The point in time when the vertical rays of the sun are striking the equator. March 21 is the vernal or spring equinox in northern hemisphere and September 22 or 23 is the autumnal equinox in the northern hemisphere. Length of day and night on these days are equal all over the world.

Erosion : The wearing away of the land surface by various natural agents like water, winds, glaciers and sea waves is termed as erosion or denudation.

Escarpment : An inland cliff or steep slopes which is formed by either erosion or faulting, termed as Escarpment.

Esker : A long narrow ridge of sand and gravel, which was once the bed or stream flowing beneath or in the ice of a glacier left behind when the ice melted.

Estuary : The mouth of river where tidal effects are evident and where fresh and sea water mix up. If the land consists of soft rock, the funnel like shape of estuary may be due to scouring action of the river and the tide but in most cases it is due to subsidence of coastal low land. The river valley has been flooded by the sea and the tides have maintained rather than actually created the shape of estuary.

Evaporation : The process of change of water into gaseous forms is known as evaporation. Evaporation rate varies from region to region and from time to time. High temperature, dry winds with rather more velocity are the conducive conditions for the evaporation.

Extrusive Igneous Rock : A rock which is formed by cooling and solidification of Lava on the earth's surface.

Fault : The displacement of broken block of the rock along the fracture line (Fault Plane) is known as Fault. A fault consists of hanging wall, foot wall and fault plane. It is formed commonly, by tensional force.

Fetch : At sea, the length of open water across which the wind is blowing.

Fiord : A long narrow inlet into the sea coast with, more or less, steep sides. It is found along the edges of mountainous regions which were once heavily glaciated. Usually fiord is extremely deep but becomes shallower towards the mouth, possibly because the glacier lost some of its power of erosion as it melted or because the terminal moraines blocked the end.

Firn : The granular substance, intermediate between snow and ice which is formed as the snow accumulated at the head of glaciated valley is being transformed into glacier ice. As the glacier moves down, the firn gradually becomes welded into a completely crystalline mass of ice. Firn is also known as Neve.

Fissur Eruption : The ejection of molten materials from beneath the earth's surface through cracks or joint or fissures of the rocks. Shield type volcano or the ocean floor spreading is formed by fissure eruption.

Flood Plain : A uniform monotonous zone, either side river valley, formed by deposition of alluvium brought by river during flooding period.

Focus : It is the point of origin of seismic waves which lies beneath the earth's surface or crust at different depth levels.

Fluvial topography : The geomorphological topography formed by denudational works of water (river) formed as fluvial topography.

Focus : It is the point of origin of seismic waves which lies beneath the earth's surface or crust at different depth levels.

Foehn Wind : A hot local wind which is found along the leeward or rain shadow zone of the highland. A moist air ascends up along the mountain and cools down at wet adeabatic rate. After reaching dew point, condensation and precipitation take place on windward slopes. After crossing the highland the air descends down on leeward slope and warms up at dry adibatic rate. Such winds in U.S.A. known as shinook and in Europe as Foehn.

Fog : A kind of condensation which takes place in the lower part of troposphere. It is suspended and hanging in the atmosphere unlike the "dew" which condense on the solid objects on the earth. Calm wind, clear sky and long winter night are favourable factors for the formation of fog.

Freezing Point : The temperature at which water begins to convert into ice. 32 degree F or 0° degree C is known as freezing point.

Frigid Zone : A temperature zone lying between subtropical high and poles in both hemispheres.

Fringing Reef : A coral reef (ridge like formation), occurs along the coast, is called as Fringing reef. It is formed by deposition of skeleton of coral polyps. A very narrow lagoon is found between reef and the coast.

Front : A very important meteorological event which originated when two contrasting airmasses from opposite directions converge together. Warm air is lifted up by cold air and a separation link, substended to certain angle with ground, between cold and warm air is formed. This line of discontinuity is known as front.

Frontogenesis : The process of formation of fronts.

Frontolysis : The process of dying of the fronts is called as frontolysis.

Fumarole : A hole in earth's crust from which steam and gases, such as CO_2 are emitted under pressure. It is frequently found in Volcanic region.

Geanticline : An anticline on large scale extending over several miles.

Geosyncline : A large shallow depression in which sedimentation and subsidence have taken place over a longer periods. It is bounded by two forelands from opposite side. This is considered as place from which fold mountains have been evolved by compressional forces.

Geostrophic Winds : A wind, usually above a height of 600 metres, that blows parallel to the isobars due to absence of any kind of friction.

Geyser : It is a kind of hot spring in which water with steam is ejecting out at regular intervals. It is usually found in the volcanic regions. Yellow Stone Park of USA is an excellent example of hot spring and geyser.

Glaciated Deposits : The plucked rock debris by glacier is deposited as the ice begins to melt. It is found in very irregular and assorting form.

Such deposit is described as till or drift.

Glacier : The moving ice is known as glacier. It occurs in high latitude and high altitudes region where temperature remains below freezing point.

Globigerina Ooze : Pelagic ocean deposits consisting of calcareous content derived from remains of foraminifera (the globigerina). This ooze is the most wide spread type ocean deposits in Atlantic ocean, Indian ocean and southern Pacific ocean. It is most developed at the depth of 1500 to 2500 fathoms.

Gneiss : One of an extensive series of rock with a bonded structure of metamorphic type. It varies widely in chemical composition and usually characterized by the most conspicuous dark mineral present. The banding due to the separation of light and dark minerals in crude layers, is the most charactertic feature of a typical gneiss.

Graben : A Rift valley formed due to subsidence of block of rock between two fault lines i.e. Dead Sea.

Graded River : A stage of river when the down cutting of river valley reaches to the base level.

Green House Effect : The transmission of short wave radiation and selective absorption of long wave radiation by atmosphere usually by CO_2 and water vapours.

Green Mud : A marine deposit which contain much fine clay and also some grains of minerals, gluconite, a silicate of iron and potassium from which it derives its green colour. Green mud is found particularly on the continental slope off tropical coast where large rivers enter the sea.

Ground Water : The water beneath the earth surface is known as underground water.

Ground Water Table : Upper limit of underground water in the zone of saturation beneath the earth's surface.

Guyaot : A flat topped seamount in the ocean.

Hail : Precipitation in the form of hard, round pellets or irregular lump of ice that may have concentric shells, formed by the successive freezing of layers of water. Such form of precipitation occurs when the water droplets are crossed up and down by turbulent storms.

Halo : A ring of light which surrounds the moon or sun when latter is seen through the ice crystals of thin veil of high cloud. When the solar halo is formed, it is rarely seen with the necked eye owing to the brightness of the sun. The commonest type is one in which the angular distance between moon and the halo is 22°. It is formed by refraction of light through the ice crystals of high clouds.

Hanging Valley : A tributary valley which enters a main valley from a considerable height above the latter, so that a stream flowing down the main valley as water fall or rapids. Such valleys are formed by glaciers when a number of small glaciers join the main glacier. Cutting down of main valley is rather higher than the small tributaries flowing at some height from the main glacier. After melting of the glaciers such valley appears like hanging valley.

Heat Budget : Balance between the amount of incoming solar (short wave) radiation and outgoing earth's long wave (terrestrial) radiation.

High : The anticyclonic condition in which high pressure lies at the centre and successive decreases towards the periphery.

Hook : A type of spit which on seaward side, is curved under the influence of blowing winds or ocean or sea wave currents.

Horn : A pyramid like peak amidst a number of cirques developed from different sides of a mountain.

Hot Spring : The ejection of hot water from natural outlets.

Humidity : The presence of moisture in a given volume of air is termed as humidity.

Humus : The decomposed forms of remains of animals and plants in the top layer of the soil. It enriches the soil fertility. It gives soil black and brown colour.

Hydrology : The Science which studies about the distribution, genesis and characterstics of water in atmosphere, lithosphere and biosphere.

Hydrological Cycle : The exchange of water among atmosphere, lithosphere and biosphere in the forms of solid, liquid and gases.

Hygrometer : An instrument to measure the humidity in the air.

Hypabyssal Rock : An intrusive igneous rock which is formed by cooling and solidifying magma beneath the earth's surface at very shallow depth.

Hypsographic Curve : A diagramatic curve in which height and depth of land and ocean respectively are shown by vertical distance above and below the horizontal line. The area of different altitudes is represented along the length of the horizontal line. Depth and height are extremely exaggrated.

Ice bergs : A mass of land ice which has been broken off or calved from the end of glaciers or from an ice shelf and afloats in the sea. When a glacier enters the sea, the ice is buoyed up by the water and portion of the glacier is easily broken off and floats away.

Ice Capped Climate : A type of climate in which temperature is below freezing point throughout the year.

Ice field : Snow accumulates at the top of hills above snow line after snow fall. Gradually it becomes very extensive and compact as the source of glaciers. This place of accumulated ice is known as Ice field.

Incised Meander : A new meander cut by rejuvenated river in the old meanders.

Insolation : Incoming short wave solar radiation, which intercepts on the earth's surface, travelling at the rate of 300000 km per second. It reaches the ground in approximately 8 minutes. The amount of insolation varies with latitude angle of sun rays, duration of sun shine and transparancy of the atmosphere. The amount of insolation received by per square cm area per minute is 1.98 calories which is described as solar constant.

Inter Tropical Convergence (ITC) : North and south trade winds converage near the equator, the area of low atmospheric pressure. This zone of convergence is termed as Inter-tropical convergence zone which oscillates north and south of equator with rhythm of the seasons.

Interfluve : A ridge between two rivers.

Internal Drainage : The river which discharge its water in lakes but not in the sea or ocean.

Instability : When a parcel of air is continuously rising up as its temperature remains above the environmental temperature at every level of height when laps rate is greater than dry a diabatic rate.

Insular Climate : A type of climate in which temperature is very moderate throughout the year due to oceanic influence. Such climate is also known as island climate because it is mainly found in Islands. For example, Japan, Great Britain.

Intrusive Rock : The igneous rocks which are formed beneath the earth's surface after the cooling and solidification of intruded magma in the rocks.

Inversion of Temperature : It is a atmospheric condition when temperature increases with height instead of falling a normal laps rate.

Inversion Relief : A type of relief in which land that was once at a relatively high level, i.e. anticline, has been eroded so severely that it is now at a lower level than land previously below it, i.e. neighbouring syncline.

Ionosphere : A ionized layer of atmosphere lying above mesosphere extending from 80 km to 400 kms from the sea level. Radiowaves are reflected from this layer to the earth surface, so that electronic media of communication would have been developed in modern era.

Isobar : The line connecting the places of uniform atmospheric pressure reduced to sea level.

Isotherm : An imaginary line drawn by connecting the places of equal temperature reduced to sea level.

Isohaline : The line connecting the places of equal salinity in the sea and oceans.

Isobath : A line connecting the places of uniform depth.

Isohyte : A line connecting the places having equal rainfall.

Isoseismal line : Line joining the places of uniform seismic intensity.

Jet Stream : A fast blowing wind near tropopause that meander in relatively narrow belts.

Joint : Due to tensional force, the rocks are broken up and a gap is formed. This gap is described as fracture or joint.

Kennelly Heaviside Layer : A layer in the lower portion of ionosphere in which most of the significant reflection of long radiowaves occur.

Ketabatic Wind : The flow of cold dense air down slope under the influence of gravity. The direction of flow is controlled largely by topography.

Laccolith : A concordant intrusive rocks which occurs between two layers of rocks beneath the earth's surface. It is a dome shaped with uniform base at shallow depth. It is formed when a volcanic eruption is miscarried and arrested below the surface. Due to movement of Magma, beds of rocks are folded up and within the anticlinal zone magma is intruded and solidified in the form of Laccolith.

Lagoon : A narrow belt of sea or ocean between coast and bars or islands is termed as lagoon.

Land Breeze : The winds blowing from land to sea during the night when the land becomes cooled down by nucturnal radiation. Consequently, a high pressure develops on the land and low pressure on the sea surface. So wind flows from land to sea. It is more convenient for fishermen who saril their boats in the sea for catching the fish.

Laps Rate : The temperature falls at the rate of 6.5°C/km with the altitude. This rate of decreasing temperature is termed as laps rate.

Lapies : The surface of limestone rocks in the karst region which are grooved and fluted owing to solution by rainwater containing CO_2. Due to solution action, the area is converted into ridge and furrow topography which is described as lapies.

Lava : When the magma reaches on the surface of the earth by volcanic eruption, is termed as Lava.

Leaching : In humid region, the soluble soil minerals are washed down with water to the subsoil. This process is known as leaching process of soil formation.

Levees : It is ridge like formation which occur on the flood plain. It is formed by deposition of coarse sediments on the either side of river channel during the flood period.

Lithosphere : The realm of rock upto few kms depth in the crust is called lithosphere.

Loess Plain : A plain of loose soils, formed by deposition of sediments brought by wind, is described as loess plain.

Low Cloud : The clouds below 2 km height.

Magma : The molten state of the rocks beneath the earth's surface.

Mantle : The weathered loose materials at the top of the rock, is described as mantle. Mantle is also intermediate zone of interior of the earth.

Maritime Airmass : The airmass which originates from the ocean's surface.

Mediterranean Climates : The climate of areas surrounding the Mediterranean sea is known as Mediterranean climate The main characteristics of the climate is that the winter is rainiest season. Winter is cool and summer warm.

Mesosphere : Third layer of atmosphere lying above the stratosphere. It extends from 50 km to 80 km height. Temperature decreases with altitudes in this layer.

Metamorphic Rock : The rock which formed by alteration in the physical and chemical characteristics of any rock inside the earth's crust under the influence of heat and pressure. They are crystalline and foliated rocks.

Millibar : The standard unit of pressure measurement used by the National Weather Service. One millibar (mb) equal to 100 Newton per square metre.

Mistral : A cold north west wind that blows into the western Mediterranean basin from higher elevation to the north.

Monsoon : Seasonal reversal of wind direction associated with large continents especially Asia. In winter wind blows from land to sea, in summer it blows from sea to land.

Mountain & Valley breeze : Daily upvalley winds, and nightly down valley winds commonly blow in mountain valleys.

Nabula : Cloud of Intersteller gases and particles.

Nappe : A huge recumbent fold with both limbs practically horizontal which has been forced to a great distance, i.e. several kms over the underlying formation, thus covering like a cloth. It is named after the French word "Nappe" meaning table cloth.

Natural Bridge : It is remnant of collapsed roof of caverns in limestone regions.

Neap Tide : A tide of minimum height which occurs on ¼ and ¾ of moon when the sun and the moon make right angle to the earth. Consequently the gravitational pull of them is neutralized and the water is drawn to low height.

Normal Fault : A type of fault in which hanging wall moves down along the fault plane and heave is directly proportional to the throw. It is formed by tensional force of the earth movement.

Northeaster : The term applied to describe the weather associated with an incursion of mp air into the north-east from Atlantic ocean. It is a strong north-east winds with freezing or near freezing temperature. The possibility of precipitation makes this an unwelcomed wind.

Nunatak : The isolated hills amidst the continental glacier.

Oasis : A depression or lake in the desert which is surrounded by settlement and the vegetation.

Occluded Front : A front formed when a cold front overtakes the warn front.

Ocean Current : The flow of huge amount of water from tropical to polar and *vice versa* in a definite direction.

Orogenesis : Process of formation of mountains by folding and faulting.

Orographic Precipitation : The precipitation which takes place due to rising of air along the mountain slope.

Outwash Plain : A flat plain formed by deposition of fluvio-glacial deposits after the terminal moraines.

Overfold : A type of fold in which both limbs overlie each other with horizontal orientation.

Oxbow Lake : A crescent shaped lake formed by cutting off meander from the main channel of the river.

Pangea : Supercontinent supposedly composed of all the continental crust of the Earth and later fragmented into Laurasia and Gondwana.

Peat : A dark brown residum produced by the partial decomposition and disintegration of mosses, sedges, trees and other plants that grow in marshes.

Pedocal : The soils, rich in calcium, found in arid, semi-arid and sub-humid climatic conditions in different parts of the world. Calcification is the dominent soil forming regime.

Pelagic deposit : The ocean deposits which covers the deep sea plain area. It consists of mainly the sediments, organic and inorganic of marine origin.

Peneplain : A flat topography, with monodnock, formed after the denudation of highland by agents of erosion.

Perched Water : Groundwater that is unconfined and separated from an underlying underground water by on impermeable bed in the zone of aeration.

Permeability : A character of the rock by which the flow of water downwards is determined. The amount of permeability is direct proportional to the inter-connection of pores of the rock.

Photosynthesis : The process of forming the plant food with the help of sunlight and water is described as photosynthesis.

Piedmont Plain : A plain consists of coarse rock fragments with large porosity alsong the foothills described as piedmont plain. It is formed by combination of a number of alluvial fans.

Plate Tectonic : A theory to explain the drifting of continents and the oceans developed recently by geologists. The whole world is divided into six plates which either converge or diverge or move parallel to each other. Earthquakes, volanic eruption, formation of fold mountains and sea trenches are different geological phenomena on the earth surface caused by plate movement.

Plutonic Rocks : The intrusive igneous rock located at greater depth.

Porosity : The space between the grains of rock particle is described as prorosity of the rocks. Porosity is direct proportional to the size of the particles. Coarser the size, more porous rock and *vice versa*.

Pot Hole : A deep narrow depression on the bottom of river valley formed by abrasion action of transported pebbles which are caught in any depression on the valley floor. Through whirling motion of water and caught rock fragments, the whole becomes deeper and deeper by grinding process and it develops as pot hole.

Polar Easterlies : The easterly winds blow from poles to sub-polar low pressure in both hemisheres.

Polar Front : After convergence of polar Easterlies and Westerlies, a front is formed. This front is called as polar front.

Precipitation : The falling of water droplets, and ice crystals from the clouds under the effect of gravitational force is termed as precipitation.

Pressure Gradient : The change of atmospheric pressure per unit distance in declining order is described as pressure gradient.

Radiation : The emission of energy in defferent wavelengths is termed as radiation. Everybody radiates energy. The amount and wavelength or radiations depend upon the temperature of the body. Solar and terrestrial radiation control the heat balance and the temperature on the earth's surface.

Radiation Fogs : Fog resulting from radiation cooling of the ground and adjacent air. It is primarily morning and evening phenomena.

Rain Shadow : A dry area on the leeward side of a mountain.

Range of Temparature : Difference between maximum and minimum temperature.

Rejuvenation of the river : Restarting of cycle of erosion due to either upliftment of upstream or falling of sea level.

Relict Mountain : Hill or mountain which have been denuded and lowered by agents of erosion over a long duration of time. Aravali mountain is a good example of relict mountain.

Relative Humidity : The ratio between the absolute humidity and the capacity of air to hold the maximum humidity. It is expressed as percentage.

Relief : The difference between highland and lowland in any area.

Reverse Fault : A fault in which hanging wall move up in relation to foot wall and the heave decreases as the throw increases up. It is formed by compressional force.

Richtar Scale : An instrument to measure the intensity of earthquakes.

Rift Valley : A valley formed by subsidence of the rocks between two fractures. It is also known as Graben.

Run off : After the precipitation, the water flows down the slope, this flow of water is known as run off.

Salinity : The presence of salt per kilogram water in ocean or sea is formed as salinity. It is expressed as °/00.

Saltation : Transport of sediments in which the particles are moved forward in a series of short and intermittent bounces.

Saturation of Air : A moist air with 100% relative humidity is termed as saturated air.

Sea Breeze : A coastal local wind that blows from sea to land, caused by temperature difference when the sea surface is colder than the adjacent land.

Sea Floor Spreading : The ocean floor is spreading away from the mid-oceanic ridges. The driving force for this phenomena is the convective current originated in the mantle of the earth.

Sea Level : The level of surface of the ocean.

Sea Mount : An isolated hill in the deep ocean bottom rising more than 1000 mts. From the floor.

Seismic Waves : The vibrations produced from the earthquakes inside the earth from the Focus are called as seismic waves.

Shore line : The inter-section of a specified plane of water especially mean high water with the shore which is a line that changes with tide or water level.

Soil profile : The cross-section of soil from bed rock to top soil is termed as soil profile. It consists of three layers or horizons –A, B and C.

Soil Structure : The pattern and design of arrangement of particles of soils into various aggregates of groups.

Solar Constant : The rate at which energy from the sun is received just outside the earth's atmosphere on one square centimetre area in one minute. It is about 1.98 calories.

Solstice : Two instances during the year when the earth is so located in its orbit that the inclination (above 23) of polar axis is towards the run. The days are 21st June and 22nd December for north and south pole respectively. They are known as summer as well as winter solstice in northern hemisphere. On these days highest and lowest duration of days are experienced in the same hemisphere respectively.

Specific Humidity : The presence of moisture in given weight of the air. It is expressed as grams/kilogram.

Spit : A long narrow ridge of sands projecting towards sea from the mainland. It is formed by deposition of sediments brought by backwash current. Spit is located at transverse to the coast unlike the bar.

Spring : The ejection of water from underground through the natural outlets.

Stalactite : A finger like deposition of calcium on the roof of caverns (Karst region) is termed as Stalactite. It is hanging down and formed after evaporation of water drops attached to the roof of caverns.

Stalagmite : An upward moving finger like formation formed by deposition of calcium after evaporation of water drops found on the floor of the caverns in limestone region.

Stationary Front : A front which remains immobile. Warm and cold air masses blow parallel along the front.

Stratosphere : The second layer of atmosphere above the troposphere. Temperature increases with height. It extends from tropopause to 50 kms height from sea level.

S-Waves : An earthquake wave which travel in deeper part of the earth. It cannot travel through liquid or gaseous state of medium. It is also known as secondary or transverse waves.

Talus : Coarse and angular rock fragments derived from and accumulated at the base of a cliff or steep rock slope.

Tarn : A depression filled with water that occurs at the bottom of cirques formed by glacial erosion.

Terminal Morains : The glacial deposits at the mouth of glaciers.

Terrigenous Sediment : The ocean deposits of shallow water which is derived from terrestrial (earth) materials by agents of denudations.

Thrust Fault : A low angle fault along which the hanging wall has moved up relative to the foot wall.

Thunderstorm : A storm produced by cumulonimbus cloud and always accompanied by lightning and thunder. It is of relatively short duration but casually associated with strong wind gusts, heavy rains and sometimes hails.

Torando : A violently rotating column of the air attended by a funnel shaped or tabular cloud extending downward from a cumulonimbus cloud.

Trade Wind : The winds which blow in the tropics from NE in northern hemisphere and SE in southern hemisphere.

Troposphere : The lowest layer of the atmosphere in which different kinds of weather phenomena are confined. It is the closest to the earth. Temperature decreases with altitude at the normal laps rate of 6.5°C/km.

Trombolo : A hooked spit in the sea or ocean.

Trough : An elongated region of low pressure.

Tythes Sea : A geosyncline between Angaraland and Gondwana Land, from which the mighty Himalayas were uplifted. Mediterrerean sea is the remanant tythes sea.

Ultra Violet : Radiation with wavelength from 0.2 to 0.4 micrometres.

Underground Stream : The flow of underground water from one cavern to another is described as sub-surface stream. Sometimes, a surface stream disappear after joining the sink hole and reappear as spring at another place.

Unstable Air : Air that does not resist vertical displacement, if it is uplifted. Its temperature does not cool as fast as that of surrounding air and, hence, it will continue to rise on its own way.

U-shaped Valley : A type of valley with a broad floor and steep walls produced by valley glacier's erosion.

Vadose Water : The water in the zone of aeration.

Valley Breeze : A gentle wind blowing up valley or mountain slope due to heating of upper part of valley in the day.

Valley glacier : The glacier moves down the mountain slope or valley.

V. Shaped Valley : A typical form of valley formed by river erosion.

Volcanism : The process of eruption of lava, gases and streams from the underground through a vent.

Wave Cut Bench : A level or nearly level narrow platform that is produced by wave erosion. It extends outward from the base of wave cut cliff.

Warm Air Mass : The air mass originated from tropical source regions.

Warm Front : A gently sloping frontal surface in which there is active movement of warm air over the cold air.

Warm Type Occluded Front : A front that forms when the air behind cold front is warmer than the air underlying the warm front it is overtaking.

Water Table : The upper level of underground water.

Weather : The average conditions of weather elements of shorter period and smaller area.

Westerlies : The wind blows from sub-tropical high pressure to the sub-polar region in both hemisphere. It blows from S W to NE in northern hemisphere NW to SE in the southern hemisphere.

Wet Adiabatic Rate : The adiabatic cooling or warming of saturated air. The rate of temperature change is variable but it is always less than the dry adiabtic rate.

Yardang : Steep sided under cut, overhanging ridges of rock, separated by corridors which have been cut by wind borne sand. They are roughly parallel and tie in the direction of dominant wind.

Youth Stage of River : The stage of river erosion in which down cutting is dominant due to steep slope. Waterfall, rapids and pot holes have been formed in this stage of river's cycle of erosion.

Zone of Saturation : A sub-surface zone in which water fills the pores and it becomes saturated.

REFERENCES

Chapter-1

Gregory, D. (1978), Ideology, *Science and Human Geography,* Hutchinson, London.

Harvey, D. (1969), *Explanation in Geography,* London.

Harvey, M. E. and Holly, B. P. (1983), *Theme in Geographic Thought,* London.

Husain, M. (1988), *Evolution of Geographical Thought,* Rawat Publications, Jaipur.

Husain, M. (1994), *Human Geography,* Rawat Publication, Jaipur and New Delhi.

Jackson, P. and Smit, S. J. (1984), *Exploring Social Geography,* Allen and Unwin Co., London.

James, P. E. (1972), *All Possible World : A History of Geographical Ideas,* New York.

King, L.W. (1969), *Statistical Analysis in Geography,* Englewood Cliff.

Smith, D.M. (1977), *Human Geography, A Welfare Approach,* Edward Arnold Co., London.

Taylor, G. (1957), *Geography in the Twentieth Century,* London.

Wooldridge, S. W. and East, W. G. (1951), *The Spirit and Purpose of Geography,* London.

Chapter-2

Boris Levin, *Origin of the Earth and the Planet,* Foreign Language Publishing House, Moscow.

Harold, C. Urey (1968), *Origin of the Earth in the Study of the Earth,* ed. J. F. White, Prentice Hall of India, New Delhi.

Long Well, C. R. Knopf, A. Flint, R. F. (1946), *A Text Book of Geology,* John Wiley & Sons, London.

Mukerjee, P. K. (1976), *A Text Book of Geography,* World Press Private Ltd., Calcutta.

Singh, S. (1993), *Physical Geography,* Prayag Pustak Bhawan, Allahabad.

Smart, W. M. (1951), *The Origin of the Earth,* Cambridge University Press, London.

Wooldridge and Morgan (1959), *An Outline of Geomorphology,* Longmans, London.

Chapter-3

Enayat, A. (1982), *Physical Geography,* Kalyani Publishers, New Delhi.

Holmes, A. (1965), *Principles of Physical Geology,* Thomas Nelson & Sons, London.

Lee, L. Don and Judson, S. (1954), *Physical Geology,* Prentice Hall, London.

Lehmann, I (1968), *Interior of the Earth As Revealed by Earthquakes, in Study of the Earth, Reading in Geological Science* ed. White, Prentice Hall of India, New Delhi.

Mukerjee, P. K. (1976), *A Text Book of Geology.*

Steer, J. A. (1964), *Unstable Earth,* Kalyani Publishers, New Delhi.

Chapter-4

Ahmad, E. (1982), *Physical Geography,* Kalyani Publishers, New Delhi.

Hallom, A. (1973), *A Revolution in the Earth Science from Continental Drift to Plate Tectonies,* Larendon Press, Oxford.

Mukerjee, P. K. (1976), *A Text Book of Geology,* The World Press Pvt. Ltd., Calcutta.

Singh, S. (1993), *Physical Geography,* Prayag Pustak Bhavan, Allahabad.

Steer, J. A. (1964), *The Unstable Earth,* Kalyani Publishers, New Delhi.

Wooldridge, S. W. and Morgan, R. S. (1937), *The Physical Basis of Geography, An outline of Geomorphology,* Longman Green & Co., London.

Chapter-5

Ahmad, E (1982), *Physical Geography,* Kalyani Publishers, New Delhi.

Fletcher, G. L. and Wolfe, C. W. (1959), *Earth Science,* D. C. Heath and Co., Baston.

Mukerjee, P. K. (1976), *A Text Book of Geology.*

Petter, Harris (1973), *The composition of the Earth in Understanding the Earth ed.* Gass Smith and Wilson, Open University Press, Sussex.

Strahler, N.A. (1965), *Introduction to Physical Geography*, John Wiley & Sons, New York.

Chapter-6

Ahmad, E. (1982), *Physical Geography,* Kalyani Publishers, New Delhi.

Bradshaw, M. J., Abott, A. J. and Gelsthorpe, A.P. (1978), *The Earth's Changing Surface,* The English Language Book Society and Hodder and Stoughton Ltd., Edinburgh.

Hallen, A. (1973), *A Revolution in Earth Science : From Continental Drift to Plate Tectonies.,* Lorendon Press, Oxford.

Holmes, A.(1965), *Principles of Physical Geology,* Thomas Nelson & Sons. London.

Horrock, N. K. (1981), *Physical Geography and Climatology,* Longman Group Ltd., England.

Mukerjee, P. K. (1976), *A Text Book of Geology,* The World Press Private Ltd., Calcutta.

Singh, S.(1993), *Physical Geography,* Prayag Pustak Bhawan, Allahabad.

Stamp, L.D. (1956), *The Earth's Crust,* George G. Harrap & Co. Ltd., London.

Steer, J.A. (1964), *Unstable Earth,* Kalyani Publishers, New Delhi.

Strahler, A.N. (1963), *The Earth Science,* Harper and Row, New York.

Chapters-8, 9, 10, 11, 12, 13

Easterbrook, D. J. (1969), *Principle of Geomorphology,* McGraw Hill Book Co., London.

Guilcher, A. (1958), *Coastal and Submarine Morphology,* Muthuen and Co. Ltd., London.

Horrocks, N. K. (1981), *Physical Geography and Climatology,* Longman Group Ltd., England.

Mukerjee, P. K., *A Text Book of Geology.*

Singh, S. (1993), *Physical Geography,* Prayag Pustak Bhawan, Allahabad.

Strahler, N. A. (1965), *Introduction to Physical Geography,* John Wiley and Sons Inc., New York.

Tarr, B. S. and Engleen, O. D. (1948), *New Physical Geography,* The Macmillan Co, New York.

Thormsbury, D. W. (1969), *Principles of Geomorphology,* John Wiley and Sons Inc., New York.

Wooldridge, S. W. & Morgan, R. S. (1937), *A Physical Basis of Geography, A Outline of Geomorphology,* Longman Green & Co., London.

Chapter-14

Jenny, H. (1941), *Factor of Soil Formation,* Macgraw Hill Co., New York.

Strahler, A. N. (1965), *Introduction to Physical Geography,* John Wiley & Sons Inc., London.

Chapters-15, 16, 17, 18, 19, 20, 21, 22, 23

Barry, R. G. and Chorley, R. J. (1982), *Atmosphere Weather and Climate,* Methuen, London.

Critchfield, H. (1961), *General Climatology,* Prentice Hall, New York.

Das, P. K. (1968), *The Monsoon,* Arnold, London.

Gates, E. S. (1979), *Meteorology and Climatology,* George, H. Harrap & Co. Ltd., London.

Kellaway, G. P. (1957), *A Background of Physical Geography,* Macmillan and Co., London.

Kendrew, W. G. (1953), *The Climates of the Continents,* Oxford University Press, London.

Koeppe, C. E. and Delong, G. C. (1958), *Weather and Climate,* Macgraw Hill Book Co., New York.

Koppen, G. (1954), *Kalima der Erde* (Map) Justu Perthes, Darmstadt Germany, American Distributor, A. J. Nystrom and Co., Chicago.

Lutgens, F. K. and Tarbuk, E. J. (1979), *The Atmosphere,* Prentice Hall, New Jersey.

Reihl, H. (1954), *Tropical Meteorology,* McGraw Hill Book Co., New York.

Taylor, F. G. (1965), *Elementary Meteorology,* Prentice Hall Inc., New Jersey.

Trewartha, G. T. (1953), *An Introduction to Climate,* McGraw Hill Co., New York.

Chapters-24, 25, 26

Bhatt, J. J. (1978), *Oceanography,* D. Van Nostra and Co., New York.

Guilcher, A. (1958), *Coastal and Submarine Morphology,* Muthuen & Co. Ltd., London.

King, C.A.M. (1962), *Oceanography for Geographers,* Adwin Arnold Publishers Ltd., London.

Monkhouse, F. J. (1958), *The Principles of Physical Geography,* University of London Press, London.

Sharma, R. C. and Vatal, M. (1962), *Oceanography for Geographers,* Chaitanya Publishing House, Allahabad.

BIBLIOGRAPHY

Ahmad, E. (1982), *Physical Geography,* Kalyani Publishers, New Delhi.

Barry, R. G. and Chorley, R. J. (1982), *Atmosphere, Weather and Climate,* Methuen Press, London.

Bhatt, J. J. (1978), *Oceanography,* D. Van Nostrand Co., New York.

Boris Levin (1960), *The Origin of the Earth and Planets,* Foreign Language Publishing House, Moscow.

Bradshaw, M. J., Abott, A. J. and Gelsthrope, A. P. (1978), *The Earth's Changing Surface,* The English Language Book Society and Hodder and Stoughton Ltd., Edinburgh.

Bunnet, R. B. (1965), *Geography in Diagrams,* Longmans Green and Co., London.

Butzer, K. W. (1976), *Geomorphology From the Earth,* Harper and Row Publisher, New York and London.

Byers, H. R. (1959), *General Meteorology,* McGraw Hill Book Co., New York.

Critchfield, H. (1961), *General Clematology,* Prentice Hall, New York.

Don, W. L. (1951), *Mateorology with Marine Applications,* McGraw Hill Book Co., New York.

Easterbrook, D. J. (1969), *Principles of Geomorphology,* McGraw Hill Book Company, London.

Finch, V. C. *et al.* (1949), *Elements of Geography,* McGraw Hill, New York.

Fleteher, G. L. and Wolfe, C. W. (1959), *Earth Science,* D. C. Health & Co., Boston.

Floro, S. D. (1954), *Torandoes of United States,* University of Oklahoma Press Norman, Okla.

Frankline, W. G. (1970), *Introduction to Meteorology,* John Wiley and Sons Inc., London.

Freeman, T. W. (1961), *A Hundred years of Geography,* Duckworth.

Gates, E. S. (1979), *Meteorology and Climatology,* George G. Harrap and Co. Ltd., London.

Geiger, R. (1950), *The Climate Near the Ground,* Harvard University Press, Cambridge.

George Gamow (1962), *Biography of the Earth,* Macmillan, London.

Guilcher, A. (1958), *Coastal and Submarine Morphology*, Muthuen & Co. Ltd., London.

Hallam, A. (1973), *A Revolution in the Earth Science, From Continental Drift to Plate Tectonie,* Lorendon Press, Oxford.

Harold, C. Urey (1968), *Origin of the Earth in Study of the Earth,* Ed. J. F. White, Prentice Hall of India, New Delhi.

Holmes, A. (1965), *Principles of Physical Geology,* Thomas Nelson and Sons, London.

Horrocks, M. K. (1981), *Physical Geography & Climatology,* Longman Group Limited, England.

James, P. E. (1966), *A Geography of Man,* Blaisdall Publishing Co., London.

Jenny, H. (1941), *Factors of Soil Formation,* McGraw Hill Co., New York.

Kellaway, G. P. (1957), *A Background to Physical Geography,* Macmillan and Co., London.

Kendrew, W. G. (1953), *The Climates of the Continents,* Oxford University Press, London.

King, C.A.M. (1962), *Oceanography for Geographers,* Edwin Arnold Publishers Ltd., London.

Koppe, C. E. and G. C. Delong (1958), *Weather and Climate,* McGraw Hill Book Co., New York.

Koppen Geiger (1954), *Klima der Erde (map)* Justs Perthes, Darmstadt, Germany. American Distributor, A. J. Nystrom and Co., Chicago.

Lake, P. (1961), *Physical Geography,* Macmillan and Co., Calcutta.

Lee, L. Don and Judson, S. (1954), *Physical Geology,* Prentice Hall, London.

Lehmann, I. (1968), *Interior of the Earth as revealed by the Earthquakes, in Study of the Earth, Reading in Geological Science* ed. White, Prentice Hall of India, New Delhi.

Longely, R. W. (1970), *Elements of Meteorology,* John Wiley & Sons Inc., London.

Longwell, C. R. Knopf and Flint, R. F. (1946), *A Text Book of Geology,* John Wiley and Sons.

Monkhouse, F. J. (1958), *The Principles of Physical Geography,* University of London.

Mukerjee, P. K. (1976), *A Text Book of Geology,* The World Press Private Ltd., Calcutta.

Navarra, G. J. (1979), *Atmosphere, Weather and Climate : An Introduction to Meteorology,* W. B. Sounders Company, London.

Oxburgh, E. R. (1973), *Plate Tectonics in Understanding the Earth* ed. Gass Smith and Wilson, Open University Press, Sussex.

Petter Harris (1973), *The Composition of the Earth in Understanding the Earth,* Ed. Gass Smith and Wilson, Open University Press, Sussex.

Petterson, S. (1958), *Introduction to Meteorology,* McGraw Hill Book Co., New York.

Preece, D. M. & Wood, H. R. B. (1938), *Foundation of Geography Modern*

Geography Book I, University Tutorial Press Ltd., London.

Robinson, H. (1978), *Physical Geography,* McDonald & Evans, Plymouth.

Sharma, R. C. and Vatal, M. (1962), *Oceanography for Geographers,* Chaitanya Publishing House, Allahabad.

Singh, S. (1993), *Physical Geography,* Prayag Pustak Bhawan, Allahabad.

Small, R. J. (1982), *The Study of Landforms,* Cambridge University Press.

Smart, W. M. (1951), *The Origin of the Earth,* Cambridge University Press.

Stamp, L. D. (1956), *The Earth Crust,* George, G. Harrap and Co. Ltd., London.

Steer, J. A. (1964), *The Unstable Earth : Some Recent Views in Geomorphology,* Kalyani Publishers, New Delhi.

Strahler, A. H. and Strahler, A. N. (1977), *Geography and Man's Environment,* John Wiley and Sons, New York.

Strahler, A. N. (1963), *The Earth Science,* Harper & Row, New York.

Strahler, A. N. (1965), *Introduction to Physical Geography,* John Wiley & Sons Inc., New York, London, Sydney.

Tarbuck, E. J. and Lutgen, F. K. (1979), *The Atmosphere,* Prentice Hall, New Jersey.

Tarr, B. S. and O. D. Von, Englen (1948), *New Physical Geography,* The MacMillan Co., New York.

Taylor, F. G. (1965), *Elementary Meteorology,* Prentice Hall Inc., New Jeresy.

Thompson, C. J. & Miller, A. (1970), *Elements of Meteorology,* Charles, E. Merril Publishing Company, OHIO.

Thornsbury, D. W. (1969), *Principles of Geomorphology,* John Wiley & Sons Inc., New York.

Trewartha, G. T. (1953), *An Introduction to Climate,* McGraw Hill Book Co., New York.

Turner, F. J. & Verhoogen, J. (1960), *Igneous & Metamorphic Rocks,* McGraw Hill, London.

Wooldridge, S. W. and Morgan, R. S. (1937), *The Physical Basis of Geography, An Outline of Geomorphology,* Longman Green & Co., London, New York, Toronto.

Wooldridge, S. W. and W. G. East (1951), *The Spirit and Purpose of Geography,* Hutchison's University Library.

INDEX

Abrasian, 122, 131, 154
Abyssal plain, 307
Advection fogs, 200
Advectisonal inversions, 224
Aeration zone, 140
Aeolian, 136
Agricultural practices, 14-15
Aircraft industries, 194
Air masses,
catgories, 248
characteristics of, 245-47
classification of, 248-49
modification of, 247-48
significance of, 245-46
types of, 248-49
Alaska, bering glacier of, 121-22
Al-Masudi, 10
Allissow, B., 296
Alluvial fan, 105
Alluvial terraces, 110
Alluvium soils, 185
Andes mountain, 95
Angara land, 39
Animal rearing, 14-15
Antarctica, 35, 39, 87, 121, 310, 311, 312
Anti-cyclonic region, 196-97
Appleton layer, 205
Aquaculture, 15
Aquiclude, 141
Aravali hills, 81
Arete, 127
Artesion wall, 141
Atlantic ocean, 310
cold ocean current, 333
ocean currents in, 332-40
warm ocean current, 332
Asthenosphere, 44
Atmosphere,
absolute stability of, 273-74
characteristics of, 203-204
composition of, 198-199
heating and cooling of, 213-14
instability of air, 274-78
significance of, 198-99
source of water vapours in, 255-56
stability effects of, 274
stratification of, 206
structure of, 202-203
transparency of, 210-11
Atmospheric disturbances, 279-94
classification of, 279-80
extra-tropical cyclone, 280
area of occurrence, 283-86
origin of, 281-82
size and structure of, 284-85
tropical disturbances, 287-92
major, 287-90
minor, 290-92
Atmospheric equilibrium, 271-78
absolute stability of air, 273-74
determinants of, 272
dry adiabatic rate, 272
normal laps rate, 272
significance of, 272
wet adiabetic rate, 273
Atmospheric humidity, 257-58
layers, 200-207
pollution, 272
Atmospheric pressure, 188, 225-40
determinants of, 227-28
distribution of, 228-32
measurement of, 226-27
significance of, 225-26
variation of, 227

Atoll, 352
Attrition, 131, 154
Automobiles industries, 200

Backwash current. 152-59
Barometer, 188
Batholith, 52
Beaufort, Admiral, 188
Beniof zone, 46, 74
Bergchurund, 126
Berring, 308
Black soils, 178
Block mountains. 68
Bloom, A.L., 8
Bog soils, 185
Bottom of Sea, feature of, 208
Braided river, 108
Burrow, H.H., 6
Buffon, Frenchman, 10
Burchan, 135

Californic cold current, 335
Canyons, 308
Carl Ritter, 2, 3
Catostrophism, 10
Caverns, 146
Central place theory, 3
Chemical percipitate, 349
Chemical weathering, 99
Chemically formed sedimentary rocks, 57-58
Chenozem soils, 183-84
Chestnut soils, 184
Chinook, 218
Christaller, Walter, 2
Cirques, 126-127
Climate, element of 186-89
- categories of, 296-97
- classification of, 295-302
- condition, 138
- Koppen's system of, 296

Clouds, 261-63
- families of, 261-62
- formation of, 261, 267
- kinds of, 262-64

Cloudiness, 189-90
Coastal rocks, 153-54
Coastlines, 159-62
- emergence of, 160
- groups of. 159-60
- submergence of, 159-60
- types of, 162

Cold air mass, 248
Cold front, 252
Cold Ocean current, 333-34
Commercial crops. 15
Condensation, 259-61
- categories, 259-62
- kinds of, 259-30
- latent heat of, 214
- types of, 201

Cone, 105, 106
- conservative margin, 48

Continental drift, 41
- glaciers, 121
- shelves, 303-306, 346
- slope, 306-307

Convection, heating pressure, 213-14
Convectional thermal currents, 89
Coral reef, 349-54
- ideal conditons for, 350-51
- kinds of, 351-52

Corbonation, 99
Crag and tail, 128-29
Crests, 32, 46, 149-52
Crystallization of salts, 98-99
Cycle of erosion,
- Davis Concept of, 164-65
- mature stage, 166
- old stage, 166
- Penck's model, 167-68
- youth stage, 165

Cyclones, 188, 203, 247, 280-93
Cyclogenesis, 285-87

Dalmation coastlines, 160
Daly, R. A., 93, 94, 354
- glacial control theory of, 354
- hypothesis, 95

Darwin, Charles, 352

subsidence theory of, 352
Davis, W. M., 11, 109, 163
cycle of erosion, evaluation of, 166-67
Deep sea plain, 307
Deflation, 130
Deforestation, 14
Deltas, formation of, 108-109
Denudation force, 163
groundwater as agent of, 144-45
types of, 100
Desert soils, 184
Determinism, 3
Diseases, 200
Doline, 146
Drainage basin, 111-12
pattern, 114-16
Drinking water, 196
Dry air, composition of, 198
Dukuchaiev, 177
Dulton, G.E., 11
Dust cloud hypothesis, 27-28
Dust particles, 201-202, 259-60
Dyke, 53
springs, 143

Earth, indirect sources, 29-30
pressure, 30
Earth crust,
faulting in, 68
elements, 49-60
materials of, 49-60
Earth movement, 61-80
causes of, 61-62
tectonic, 73-74
types of, 62
Earthquakes, 46, 48
areas of, 78
causes of, 76-77
effects of, 77-78
elements of, 77
waves, 30
zones, 15
Earth surface,
factors of, 208-212
features of, 35-36
insolation distribution of, 208-210
shortwave radiation, 215
Easterbies (trade winds), 188
Ellen Churchil Sample, 5
Enclosed water, 158
Entwickelund (development), 168
Environmental geomorphology, 12
Epirogenic earth movement, 62-63
Erastothenes, 2
Erosion, cycle of, 109-110, 163-70
process of, 130
works of, 121-22
Evaporation, 195-257
Exploitation, 98
Exosphere, 206
Extra-tropical cyclone, 280-82
area of occurrence, 285-86
movement of, 284-85
origin of, 281-82
size and structure, 283-84
weather conditons, 284-85
Extra-tropical anticyclone, 286-87
Extrusive igneous rocks, 50-51

Faults, kinds of, 66-67
Fault springs, 143
Faulted shorelines, 62
Fertile alluvium soils, 172
Fiord coastline, 160
Firn, 120
First World War, 2
Fissure spring, 142
Flohn, H., 296
Flood plain, 107
new channel of, 110
Flooding, 196
Fluvio-glacial deposition, 123-24
Foehn winds, 243-44
Fogs, 260-61
advection, 260
orographic, 261
radiational, 260
upslope, 261-62
Folds, groups, 63-64

Fold mountain, 82
Fossil fuels, 200
Fronts , kinds of, 246-47
Frontal fogs, 261
Frontogenesis, 249
Frost wedging, 99, 122, 126
Frozen rivers, 196

Gaddes, Sir Patrick, 3
Gases, kinds of, 199-200
Gaseous mass theory, criticism, 19-20
Geography, approaches of, 6
    classification, 4-5
Geographic information, 16
Geomorphology, 8-9, 15-16
    prospects of, 11-13
Geomorphological features, 12-14
Geosynclinal sediments, 95-96
Geyser springs, 144
Gilber, G. K., 11
Glacial drift or tills, 122
Glaciers, 120-22
    deposition, 122
    denudation works of, 121
    erosion works, 121-22
    landforms formaion of, 125-26
Gradation, process of, 97-100
Grassland, temperature of, 218-19
Guesses, 29
Gulf stream current, 218

Hanging valley, 125
Heat Budget, 215
Herodotus, 9
Hertshorn, 4
Hetrosphere, 206
Hettner, Alfred, 2, 4
Hipparchus, 2, 3
Holdridge, L., 296
Holman, Arthur, 88
Holmes, 88, 90, 92, 93
Homosphere, 206
Horris, Peter, 32
Horst, 68
Horticulture, 14

Hot springs, 144
Hudson, 308
Human geography, 5-10
Humboldt, Alexander van, 2, 3
Humidity, 189, 255
    measurement of, 256-257
Huntington, 3, 5
Hutton, James, 9, 10, 167
Hydration, 190
Hydrological cycle, 195
Hygroscopic nuclei, 259
Hypsographic curve, 303
Hypabyssal intrusive rocks, 51

Ibn-Sina, 9
Igneous bodies, classification of, 51-52
    distinguess, 54-55
Igneous rocks, characteristics of, 51-54
Incised meanders, 110
Indian ocean currents, 335-36
Individual ocean, relief features, 309-316
    salinity in, 323-24
Inorganic deposits, 348
Insolation, 186-87, 208-220
    distribution of, 214-15
Instability, 276-78
    conditions of, 278
Inter-Tropical Convergence Zone (ITCZ), 289, 290, 291, 293
Interior of Earth, structure and compositon of, 31-32
Intrusive igneous rocks, 51
Ionosphere, 205
Isotherms, 221

Jean Brunches, 5
Jeffrey, 85, 92, 93
Jet plane, 204
Jhemchung, 308
Jig sa fit, 42
Joly, 90, 91, 92, 93

Kant, Immanuel, 18, 25
Karst Geomorphology, 163

Karst topography, 14-15, 148
Kashmir, 116
Kennelly Heaviside layer, 205
Kitab-al-Hind, 9
Kober, 83-85, 192-93
    Geosynclinal orgon theory of, 82-83
    Hypothesis, 84
Koppen, Waldimir Peter, 296
    climate classification, evaluation of, 301-302
Kupfer, E., 296
Kuroshis warm current, 335

La Blach, Vidalde, 3
Lagoon, 158
Land forms, 8, 177
    analysis of, 10
    global distribution of, 83
    kinds of, 82
    types of, 12
Land Breezes, 242-43
Latitudinal heat balance, 215-16
Latur earthquakes, 78
Laurentia, 39
Leeward (rain shadow zone), 13, 134
Length of days, 210
Leonado da Vinci, 10
Limestone cave, 148
Lobeck, A.K., 9
Local winds, 241-42
Loess plain, 136
Loo, 195
Lopolith, 52
Lowthian Green, 36
Lowland Coast Line, 160

Magma, 52, 58, 60, 69, 70
    instrusion of, 95-96
    kinds of, 72-73
Major climate types, sub-groups of, 297-98
Mantle, 33
Mattern horn, 127
Mechanically formed sedimentary rocks, 56-57
Medial moraines, 123
Mediterranean, 270
Mesosphere, 204
Metamorphic rocks, 58-60
    characteristics of, 58-60
Misfit river, 117
Mississippi, 308
Monsoon of,
    African countries, 239-40
    Australia, 238-39
    East Asia, 238
    genesis of, 239-40
    South Asia, 238
Morains, 123
MOHO, 33
Mountains, classification of, 80-81
Mountain Building, 83-94
    continental drift theory of, 87-88
    continent sliding hypothesis of, 93-94
    development thesis of, 83-86
    thermal contraction theory, 85-89
    radioactive theory, 90-91
Mountain Breezes, 242-43
Mountain soils, 185
Mountainous topography, 12-13
Munsel colour chart, 175
Murray, John, 354
Mushroom rocks, 132

Nabular hypothesis, 18-19, 27-28
    criticism, 19-20
Neo determinism, 3, 5
Neutral shorelines, 162
Newton's Law, 18
Newsea, 89
Niagra water fall, 103
Noctilucent, 205
No-strain zone, 85
Notch, 154

Oasis, 134
Ocean basins, 35-40
Ocean currents, 218, 327-28

characteristics of, 330-31
determinants of, 328-29
distribution of, 331-32
effects of, 337
Ocean deeps, 307-308
Ocean deposits, 344-48
classification of, 344-45
Ocean tides, genesis of, 338-39
Ocean terrigenous deposits, 345-46
Ocean water, characteristics of, 315-326
circulation of, 327-40
distribution of, 317-18
physical characteristics of, 315-20
salinity, 320-36
temperature of, 315-17
vertical distribution of, 319-20
Oceanographer, 308
Occluded fronts, 253-54
Okhotosk cold current, 335
Old sea cliffs, 162
Organic deposits, 347
earth movement, 63-64
Orogeny, 80-96
Orographic fogs, 261
Oscillations, 31
Otto Schemidsts' Inter-Steller dust cloud hypothesis, 24-25
Oxbow lakes, 105-106
Oxidation, 99-100
Oyashio Cold Current, 335
Ozone gas, 200
Ozonesphere, 200

Pacific ocean, 310, 324-36
cold ocean current, 335
warm ocean current, 334-36
Pakistan, 68
Panthalusa, 87
Pedalfer soils, 180-81
Pedestal rocks, 132
Pedocols soils, 183-84
Pelagic deposits, 346-47
Penck, Albrecht, 9
Penck, Walther, 11, 163
cycle of erosion, 167-70
model, 170-71
Perched water table, 140
Percolating water, 140
Permanent water table, 140
Permeability, 137
Physical environment, 198
Physical geography, sub-division of, 5-6
Physical weathering, 98
Piedman glaciers, 121-22
Pirate river, 117
Planetesimal hypothesis, 21-22, 26-27
Plate tectonic movement, 74-77
theory, 44-48, 95-96
Plutonic rocks, 51
Pollen, 259
Polar zones, 214
Polje, 146
Pot holes, 105
Powell, J., 11
Prairie soils, 184
Precipitation, 138, 201, 264-65
distribution of, 268-70
types of, 266-67
Pressure, 188-89
Ptolemy, 1, 2, 3
Push waves, 31

Quantitative revolution, 12
Quarring, plucking of, 122
Queensland, 354

Radiational fogs, 260
Radiation process, 213
Ramsay, 11
Randkettens, 84
Rapids, formation of, 104-105
Ratzel, F., 2, 5
Reforestation, 16
Regional geography, 7-8
Residual monodnock (Peneplain), 168
Ria coastline, 160
Richothofen, 11
Riding landmasses, 93-94

Rift valley, 68
Rivers, 101-103, 109-10, 112-14, 117-20
  captule, 117-19
  development, 110-11
  erosional and depositional works by, 102-103
  landform formation fog, 110
  meanders, 105-106
  pattern, 114-16
  process of erosion, 101-103
  rejuvenation, 109-10
  systems, 112-14
  transportation by, 102
  types of, 117-19
Roche mountonnes, 12
Rock cycle, 60
  classification, 50-54
  disintegration of, 98-99
  dislocation and displacement, 73
  formation, 49-50
Rockies mountains, 95, 242-44, 281, 286
Running water, 100

Salinity, 321-26
  composition, 321
  determinants, 322-23
  distribution of, 322-24
  sources of, 321-22
Saltation, 131
Sand dune, 134
Sand plasting machine, 132
Sargasso sea, 332
Saturation, 189, 258-59
Scarp foot springs, 142
Sea arch, 156-57
Sea Breeze, 242
Sea cave, 156-57
Sea cliff, 155-57
Sea waves, 151-54
  depositional features formed by, 158-59
  erosional features formed by, 154-57
  hydroulic action, 153-54
  movement of water particles in 151-52
Second World War, 11, 249
Secular earth movement, 62
Sedimentary rocks, 54-55, 129
  characteristics of, 57-58
  classification of, 56-57
  process of formation, 55-56
Sedimentation, 163
Seif sand dune, 135
Seismic waves, 30
Shake waves, 31
Shinook winds, 243-44
Short wave radiation, 215-17
Sial, 49, 87
SIMA, 33, 87, 90, 91, 92
Sink holes, distribution of, 145
Siwalik range, 103
Slack water (flood water), 107
Soils, characteristics of, 173-74, 182-84
  biological factor, 178-79
  colour of, 175
  classification of, 180-85
  element of, 172-73
  formation, 179
  forming process and factors, 177-78, 180
  minirals, 178
  organic materials, 173-74
  profiles, 175-76
  texture classes, 174
Solar constant, 208
  energy, 208
  radiation, 209-12
  system, characteristics of, 17-18
Spits, 158
Springs, 142-43
  types of, 142-44
Stack, 156
Stalactite, 148
Stalagonite, 148
Stamp, L. D., 4
Stationary fronts, 254
Strabo, 1, 9
Stratosphere, 204

Subduction. 95
Submarine canyons, 308
ridges, 309
Substratum, 91
density of. 91
solidification. 92
Sudden earth movement, 62, 68
kinds of, 68-69
Suess' classification. 39
Sun rays. angle of. 209
Sun spots. 212-13
Sunda Islands, 83
Surf, 152
Surface transportation, 194
waves, 31
Swallow holes, 146
Swash current. 152-60
Systematic geography, 6-7

Tarn. 127
Taylor, 5
Temperature, 186-88, 208-220
adiabatic change in, 259-60
distribution of, 219-20
factors determine, 217-18
horizontal distribution of, 219-20
normal laps rate of. 223-24
radiational inversion of, 224
vertical distribution of, 223-24
Temporary water table, 140
Tennesse Valley Authority (TVA), 16
Terrigenous deposits, 345
Tetrahedral theory, 36-37
Thermal anomaly, 221
Thronthwaite, C., 296
Thunderstorm, 189-90, 247, 291-93
distribution of, 293-94
types of, 292
Thythes sea, formation of, 89, 95
Tidal currents, 342-43
Tidal hypothesis, 22-23, 27
criticism, 23
waves, 343
Tides, 337-340
classification of, 339-40
duration of. 347-48
types of. 339-40
Topography. 4, 138
Tornado. 189, 247, 294
Transform faults, 48
Transportation, 13-14, 122
process of. 131
system, 197
Tributaries, 113-15
TRIO of Davis. 164
Trambols. 188
Tropical cyclone, 285-86
Tropical distribution, 287-90
Troposphere, 202-203
Tropopause, 204
Trough, 199, 152
Typhoon, 287-89
Tythes sea, 39

Underground water, 138-48
denudation work of, 148
depositional feature of, 146-47
elements of, 137
hydrological cycle. 138-39
landform formation of, 145-46
sub-surface water zone, 139-40
Upslope fogs, 261
U-shaped valley, 125
Uvala, 146

Valley Breeze, 242-43
Valley glaciers. 120-21
Valley inversions, 224
Vaporisation, 257
Varnius, Bernhard, 2, 5, 7
Vegetations, 130
Velocities, 249
Venezuela cold current, 333
Vibration, 30
Victoria fall, 103
Videl de la Blache. 5, 6
Voilent turbulents, 247
Volcanic eruption, 61, 76
materials, 346
Volcanism, 15

Volcanoes, causes of, 73-74
classification of, 69-73
eljected materials, 70-72
geographical distribution of, 75-76
Von Thunen, 3
Vosges, 68
V-shaped valley, 103

Warm air mass, 248
Warm front, 250
Warm ocean current, 332-33
Warcester, P. G., 8
Water, distribution of, 195
Water fall, 103-105
percolaion, 138
table, 140
vapour, 201
Watery planet, 17
Wave action, 149-60
formation of, 149-50
kinds of, 151-52
Wave-cut plateform, 155-57
Weather and man, 190-91
and culture, 192
and health, 195
and human activities, 196-97
and hydrology, 195
and transportation, 193-94
phenomena, 203
Weathering, types, 97-99
Wegner, Alfred, 87
continental drift theory of, 38-44
criticism, 43-44
Wells, types of, 141-42
Winds, 130-32
erosional works by, 131-32
depositional works, 134-35
determinents of, 233-34
kinds of, 188, 234-35

Yardangs, 132
Youth stage, 165

Zaire, 308
Zeugen, 132
Zone of Illuviation, 175